The Geography of North America

The Geography of North America

Environment, Political Economy, and Culture

Susan Wiley Hardwick

Department of Geography, University of Oregon

Fred M. Shelley

Department of Geography, University of Oklahoma

Donald G. Holtgrieve

Department of Geography, University of Oregon

Upper Saddle River, NJ 07458

Library of Congress Cataloging-in-Publication Data

Hardwick, Susan Wiley.

The geography of North America: environment, political economy, and culture / Susan Wiley Hardwick, Fred M. Shelley, Donald G. Holtgrieve.

p. cm.

Includes bibliographical references and index.

ISBN-13: 978-0-13-009727-9

ISBN-10: 0-13-009727-6

1. North America—Geography—Textbooks. 2. Human geography—North America—Textbooks. 3. Environmental geography—North America—Textbooks. 4. North America—Economic conditions—Textbooks. I. Shelley, Fred M. II. Holtgrieve, Donald G. III. Title.

E40.5.H37 2008
917—dc22

2007022128

Publisher, Geosciences and Environment: Dan Kaveney
Editorial Project Manager: Tim Flem
Project Manager: Debra A. Wechsler
Senior Managing Editor: Kathleen Schiaparelli
Marketing Manager: Amy Porubsky
Senior Operations Specialist: Alan Fischer
Director of Design: Christy Mahon
Art Director: Maureen Eide
Interior/Cover Design: Suzanne Behnke
Senior Managing Editor, Art Production and Management: Patricia Burns
Manager, Production Technologies: Matthew Haas
Managing Editor, Art Management: Abigail Bass
AV Project Manager: Rhonda Aversa
Art Studio/Cartographer: Spatial Graphics
Director, Image Resource Center: Melinda Reo
Manager, Rights and Permissions: Zina Arabia
Interior Image Specialist: Beth Brenzel
Image Permission Coordinator: Craig A. Jones
Photo Researcher: Teri Stratford
Copy Editor: Marcia Youngman
Editorial Assistant: Jessica Neumann
Cover Images: Map—SeaWifs Project, NASA/Goddard Space Flight Center and ORBIMAGE; Alaskan pipeline—Len Rue, Jr./Animals/Animals/Earth Sciences; moon over Vancouver skyline—© Corbis Images; Carlos Santana—Frank Micelotta/Getty Images, Inc.

Pearson Prentice Hall
Pearson Education, Inc.
Upper Saddle River, NJ 07458

Printed in the United States of America
10 9 8 7 6 5 4 3 2 1

ISBN-10: 0-13-009727-6
ISBN-13: 978-0-13-009727-9

Pearson Education LTD., *London*
Pearson Education Australia PTY, Limited, *Sydney*
Pearson Education Singapore, Pte. Ltd.
Pearson Education North Asia Ltd., *Hong Kong*
Pearson Education Canada, Ltd., *Toronto*
Pearson Educatión de Mexico, S.A. de C.V.
Pearson Education—Japan, *Tokyo*
Pearson Education Malaysia, Pte. Ltd.

Brief Contents

Contents

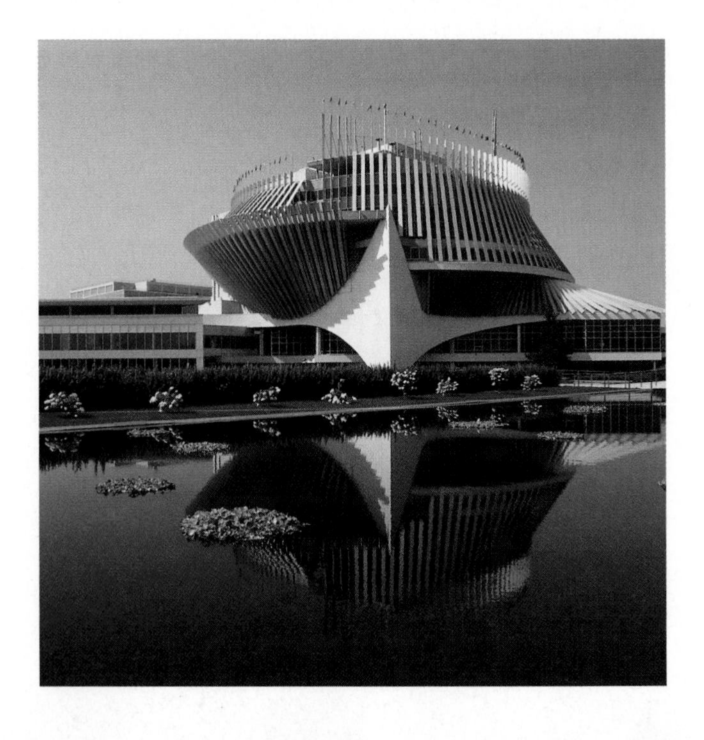

Megalopolis 118

8

The Great Lakes and Corn Belt 138

9

The Inland South 156

10

The Coastal South 174

11

The Great Plains 190

12

The Rocky Mountains 216

PUEBLO DON GASPAR
400

15

California 270

16

The Pacific Northwest 294

19

The Future of North America 350

Preface

Welcome to a study of the regional geography of North America. You may have traveled or lived in one or several parts of our continent, or you may be a resident or visitor in Canada, the United States, or Greenland from another part of the world. Whether you know a lot about the geography of North America or not—and whether or not you've taken other geography courses before enrolling in this one—the information discussed in each of the following chapters is designed with you in mind.

Your three authors have taught classes on the Geography of North America for many years at universities in Oregon, Oklahoma, California, Montana, Florida, and Texas. We've also gone on many enjoyable road trips to all of the regions discussed in this book. Thus, we're very enthusiastic about sharing what we've learned on the road and in our classrooms. You may want to add a copy of this book to your maps, GPS units, travel guides, and other materials on Canada, Greenland, and the United States the next time you have the chance to go on a road trip.

Each of the following chapters is designed to help provide you with new information about the geography of North America. We hope they also will help you understand and apply key geographic concepts, themes, skills, and perspectives in your other college or university courses and in life in general. Each of the maps, photographs, tables, and focus boxes also will be useful in expanding on what you learn from reading this text.

Acknowledgments

We appreciate the advice and support of reviewers of the book including Paul Adams, University of Texas–Austin; Thomas Bell, University of Tennessee; Mark Drayse, California State University, Fullerton; Alison Feeney, Shippensburg University of Pennsylvania; James Fonseca, Ohio University, Zanesville; Jay R. Harman, Michigan State University; Tom Martinson, Auburn University; Chris Mayda, Eastern Michigan University; Cynthia Miller, Minnesota State University–Mankato; Daniel Montello, University of California, Santa Barbara; Erik Prout, Texas A & M University; Alexander C. Vias, University of Connecticut; Gerald R. Webster, University of Alabama; and William Wyckoff, Montana State University. Their "geographer's eye" and attention to detail helped improve the content of the book significantly.

In addition to the reviewers who made the book more informational and readable, we owe a debt of gratitude to the graduate students and other colleagues who made important contributions to each chapter. They include our close colleague and friend, Andrew Marcus, for his help structuring the book and making it as student-friendly as possible, and Pat McDowell who provided support along the way; Lisa DeChano for her able and speedy assistance with Chapter 2; Ryan Daley for his insider writing about the Rocky Mountains in Chapter 12; Charity Book for her invaluable help with our popular culture inserts and overall editing of the book manuscript; Gordon Holtgrieve for firsthand information about Hawai'i and Alaska; Adrienne Proffer for her editorial assistance and work on the box on the television program *Lost* in the Hawai'i chapter; Grace Gardner for her expert cartographic assistance; Ginger Mansfield for her research assistance and web site support; Mary Crooks for her artistic advice related to illustrating each chapter of the book; James Book for his assistance with popular culture information and the surfer expertise needed to complete the Hawai'i chapter; Kimberly Zerr for her editorial assistance; Amanda Coleman for her help with the Inland South chapter (a place she knows very well); and Brittany Jones, Joanne Stanley, and Maureen Kelly for their creative, and at times quite tedious, work locating map and table data and photographs for the art manuscript. We are also indebted to Prentice Hall author, Robert Christopherson, for allowing us to use some of his powerful graphics in Chapter 2.

Our editors at Prentice Hall have been inspiring, patient, and helpful in all regards. We are especially grateful to Dan Kaveney for inviting us to write the book in the first place, and for guiding us patiently through the prepublication process (despite several author moves across country, promotion and tenure slowdowns, and changes in our responsibilities on the book and in our professional and personal lives). Along with Dan's efforts has been the efficient support of Debra Wechsler, Project Manager (production); Tim Flem, Project Manager (editorial); Marcia Youngman, copy editor; and Jessica Neumann, Editorial Assistant, whose efforts consistently made things work more smoothly at every level. To each of these helpful supporters and "textbook publication mentors," we owe a debt of gratitude for their patience, inspiration, and expertise.

Finally, this book couldn't have been written without the assistance of the literally thousands of students who have taken our Geography of North America courses over the past 25-plus years. Their feedback on earlier versions of these chapters helped us realize that things have indeed changed. Students need and deserve different kinds of learning materials today than they did even a decade ago. Due to their many suggestions and critiques, we have tried hard to provide a more innovative and slightly edgy approach to studying the geography of North America. After using this book, we really hope you find our conceptual and, at times, quirky approach helpful in understanding more about the continent.

The support of all of these editors, reviewers, cartographers, and other colleagues proved invaluable during the final year of work on the manuscript. Any and all weaknesses or errors in this first edition, however, are the result of our own shortcomings, not the fault of any of these helpful supporters.

So welcome to the world of regional geography! Please let us know if you have ideas for improving the text after you finish reading each of the following chapters. Your ideas, concerns, and inspirations for making this book more student friendly and content rich will improve it for geography students and faculty for years to come.

Susan Wiley Hardwick
Eugene, Oregon

Fred M. Shelley
Norman, Oklahoma

Donald G. Holtgrieve
Eugene, Oregon

To our fathers—

who learned by doing

and taught by example:

Asa G. Wiley, Fred Shelley,

and Edwin C. Holtgrieve

About the Authors

Susan Wiley Hardwick teaches at the University of Oregon where she specializes in the geography of the United States and Canada, urban and cultural geography, and the geography of immigration. Professor Hardwick is the author of six other books and a long list of articles on these related topics and is also the Past President of the National Council for Geographic Education. She is perhaps best known as the co-host of "The Power of Place," an Annenberg geography series produced for public television. Her latest book is a co-edited volume on immigration and integration in U.S. cities published by the Brookings Institution. She was awarded the statewide *California Outstanding Professor Award* out of more than 23,000 faculty when she taught at California State University, Chico before moving to Oregon. She is the parent of four grown sons who all live on the West Coast (and a 100-pound-plus Newfoundland dog).

Fred M. Shelley received his Bachelor of Arts degree from Clark University, his Master of Arts degree from the University of Illinois at Urbana–Champaign, and his Ph.D. from the University of Iowa. Dr. Shelley became Chair and Professor of Geography at the University of Oklahoma in 2004. His research and teaching interests include political geography, world systems, cultural geography, and North America. Fred has published over seventy scholarly articles and book chapters as well as ten books.

Donald G. Holtgrieve has used his geography training in a variety of applied areas as well as in the classroom as a professor of geography and environmental studies at California State University's East Bay and Chico campuses. He founded and headed an environmental research firm, coordinated projects for citizen environmental groups, and was a consulting urban planner for several state and local public agencies. In addition, some of his more satisfying professional experiences were as an inner-city high school teacher, reserve state fish and game warden, Aikido teacher, and land planner for sustainable development projects (including four wildlife preserves). He now enjoys teaching and applied research at the University of Oregon in Eugene.

1 Introduction

This book provides information on some of the key environmental, cultural, economic, and political issues that face North America. To provide a larger context for understanding the patterns and processes shaping people and places in North America today, we also include a discussion of the historical processes that have helped to shape each region in North America in the chapters that follow. Because of the physical, economic, and cultural linkages of the United States and Canada to the Mexican borderlands, along with their environmental and physical connections to the island of Greenland, we include information on the United States–Mexico borderlands region and Greenland. In today's rapidly globalizing world, with the ever-increasing economic and cultural linkages among places in North America, we believe that it is essential to extend our discussion where possible to include every corner of this vast continent as shown on the map in Figure 1.1.

Why Study North America?

North America is huge! The two largest countries, Canada and the United States, cover more than 7.5 million square miles (20 million square kilometers). North America is the world's third largest continent. Canada now has jurisdiction over almost 6.7 percent of the world's land area and the United States controls nearly 6.4 percent. The territorial and political control of so much of the inhabited (and potentially inhabited) portion of Earth by only two governments makes the continent of particular importance. Added to the size dimension of these two nations is the critical importance of the very large island of Greenland. Not only is it the world's largest island (at $2\frac{1}{2}$ times the size of the second largest island, New Guinea, and 52 times the size of Denmark), it is also the most northern country on Earth. Similarly, at the southern edges of the United States border, the Mexican borderlands are also critical to our story of North America. Both of these parts of the continent are important not only because of their large size, close relationship, and proximity to Canada and the United States but also because their peoples, places, and global economic connections make them distinct and fascinating places for geographic study in their own right.

Politically, all North Americans share a colonial past with European countries and a related aboriginal heritage. After thousands of years of settlement by native peoples who first came to the continent between 50,000 and 25,000 years ago, the political structure and cultural and economic foundation of post-colonial America was primarily dominated by Great Britain; Canada by the British and French; Greenland by Denmark; and the southwestern United States and Mexico by Spain. Each of these colonial foundations was imposed upon the cultures and economies of groups of indigenous peoples already in place.

Central to understanding the current politics and socioeconomic and cultural geographies of each of these nation-states are the different ways that each emerged from its colonial past. In the United States, for example, a bloody Revolutionary War launched the nation's independence from England. Canada, in contrast, remained loyal to Britain for a longer period, until the Dominion of Canada was ratified peacefully in 1867 and Canada began self-rule with a British parliamentary-style democracy. The last of the continent's political regions to break with its colonial past was Greenland. In a nonviolent agreement, this island nation severed its colonial relationship with Denmark in 1979 when it was granted home rule (although the queen of Denmark remains Greenland's head of state). The head of government is a prime minister, who is usually the majority leader of parliament.

Each of these political regions also has unique but interconnected demographic and cultural characteristics. Although an increasing number of Canadians, Americans, Europeans, and people from other parts of the world have immigrated to Greenland during the past several decades, the vast majority of residents are a mixture of Inuit and Danish peoples. Their impacts are evident everywhere—in the shops of Greenland's picturesque coastal towns, in the island's distinctive language patterns, and in its mixed music, native costumes, and other cultural expressions (see Figure 1.2). Greenland is also connected to the geography of the rest of the continent of North America by its physical location and by current political issues. In 2006, it continues to be embroiled in a sovereignty dispute with Canada over the tiny Hans Island (located between Canada and Greenland in the Arctic region).

Mexico also still carries the vestiges of a past dominated by Spain blended with its many indigenous peoples and more recent migrants from Europe, Asia,

◀ Whistler Village and Blackcomb Mountain in British Columbia, site of the 2010 Winter Olympic Games, demonstrates a successful mix of sustainable land uses

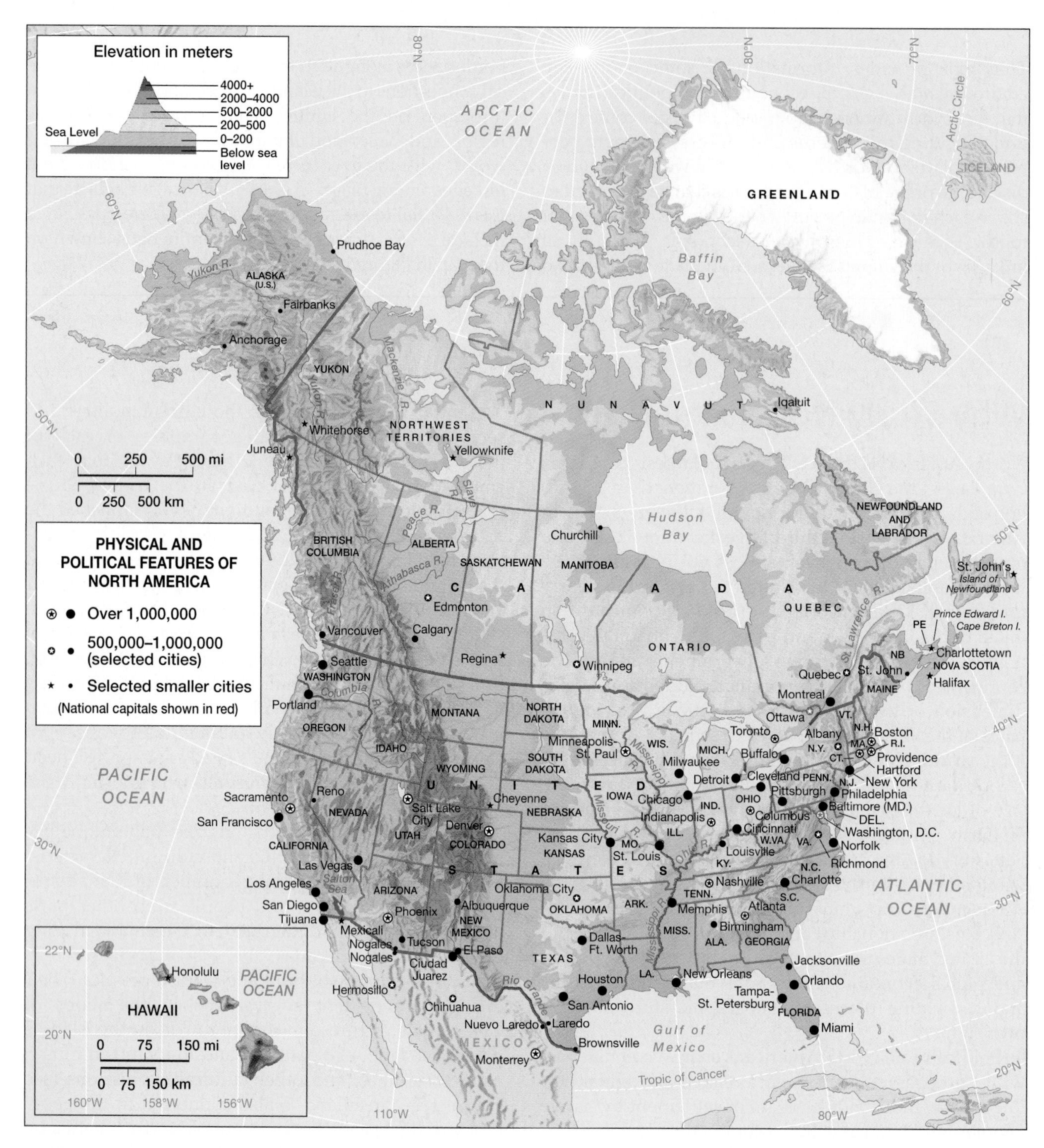

▲ **Figure 1.1** Physical and political features of North America

and Africa who have helped shaped MexAmerica's distinctive landscapes and cultures. The people who live in Canada and the United States are the most multicultural, polyethnic, and polyracial of any part of North America. Canada's major political and cultural distinctions are among the nation's French-speaking and English-speaking population and its aboriginal First Nations, Aleut, and Inuit peoples. And since the 1960s, an increasingly large overlay of cultures and peoples have relocated to both Canada and the United States from Latin America, Asia, and Africa as well as a long list of European countries.

The largest (in area) of the countries that make up North America is Canada. It is also the closest trading partner of the United States and shares the longest peaceful political border in the world. Yet despite its

▲ **Figure 1.2** Greenlandic people perform traditional music wearing traditional island costumes on special occasions

importance in world affairs and close proximity to the United States geographically, culturally, politically, and economically, the majority of the students who have taken our geography classes here in the United States seem to know very little about their own country's nearest northern neighbor. Canada stretches 3730 miles (5500 kilometers) across the continent from the eastern edge of Newfoundland, west across the prairies to the Canadian Rockies, then across British Columbia to the Pacific coast. Seven percent of Canada is covered with lakes and rivers and it has three of the world's 20 longest rivers. Overall Canada controls about 25 percent of the world's freshwater resources. Politically, Canada is divided into ten provinces and three northern territories. Its more than 30 million people mostly reside in towns and cities located near the border with the United States (Figure 1.3).

Canada currently has the highest rate of immigration (as a percent of total population) among the world's industrialized nations. To address the needs

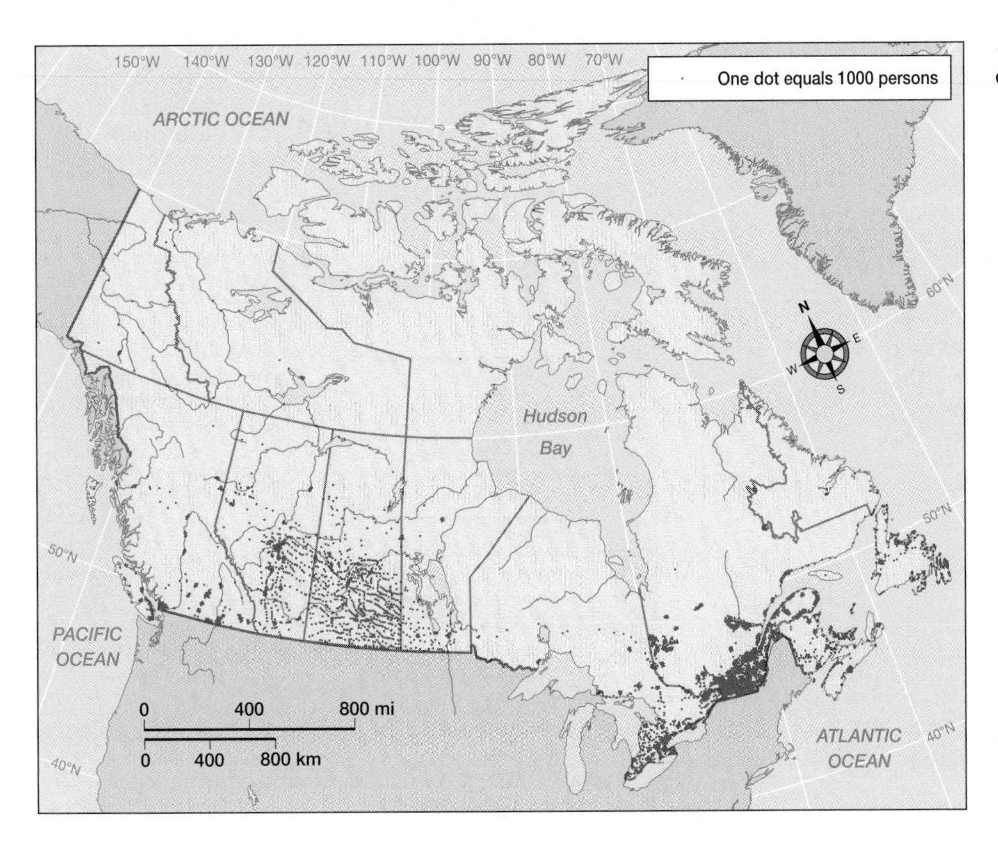

◄ **Figure 1.3** Population distribution in Canada

of foreign-born newcomers during the past 25 years or so, the Canadian government has allocated generous funding for immigrant and refugee resettlement efforts and also funded numerous studies of the settlement patterns, social processes, and adjustment experiences of Canada's foreign-born residents. Indeed, the word "multiculturalism" was first coined in Canada in the 1970s as a part of its many diversity policy initiatives.

As in all other parts of North America, Canada also has a long and colorful history of the settlement and survival of a diverse group of *First Nations* aboriginal people. In 1999, the new territory of Nunavut was approved by the Canadian government, a remote but quite large area of land in Canada's Far North. It is the only political region populated and governed by indigenous peoples on the North American continent (Figure 1.4).

As mentioned above, the United States is the second largest nation in North America. Its large land area is divided into 48 conterminous (or continental) states and two states, Hawai'i and Alaska, that are located far from the other states. This division is a result of the expansionist policies of the United States government and a desire to secure land at some distance from its originally bounded area. This expansion was dictated by a long held belief in **manifest destiny**, the quirks of history, the desire to exploit natural resources in Alaska, and the promising boon of tourism dollars in Hawai'i.

The United States has a system of government that is based more loosely on the British system than is the Canadian system. In both nations, the legal systems specifically provide for separation of powers between the federal governments on the one hand and state or provincial governments on the other. Regional and provincial governments are much more self-sustaining in Canada than are state or the few regional governments in the United States. As shown on the diagram in Figure 1.5, both Canada and the United States have a multi-tiered system of government with power flowing from the federal level at the top down to city governments at the bottom. One major difference in the political systems of these two North American countries is that the Dominion of Canada is more closely connected to Great Britain as one of its Commonwealth countries than the United States. As such, the queen of England is the primary figurehead over Canadian affairs of state with a Prime Minister serving as the elected,

▲ **Figure 1.4** The newest Canadian territory, Nunavut

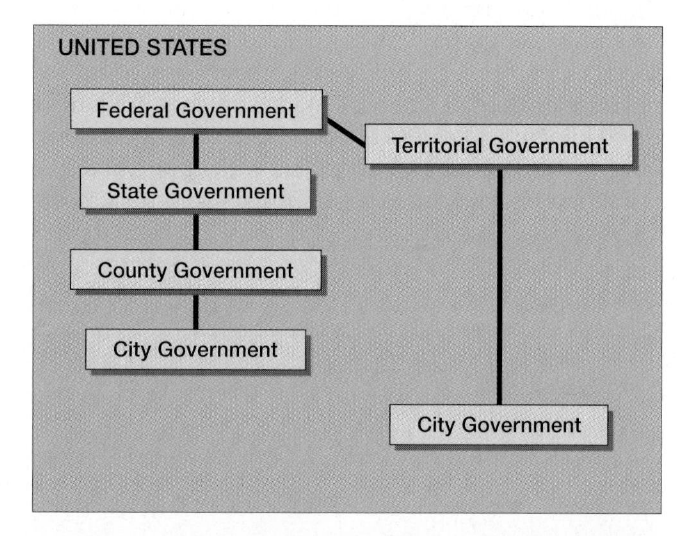

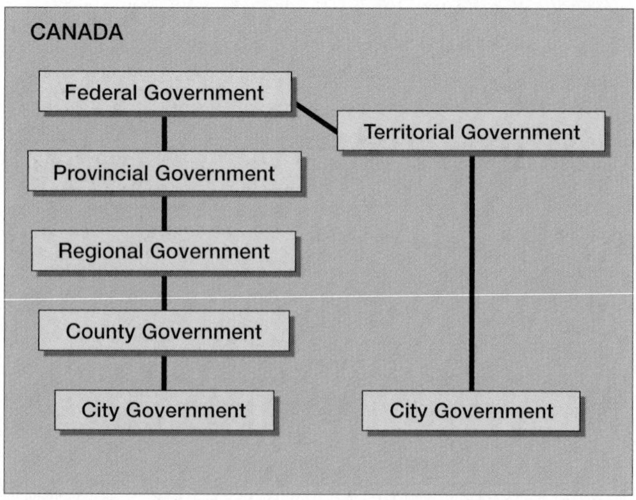

▲ **Figure 1.5** The differing governmental structures in the United States and Canada

in-residence head of the federal government. In contrast, the United States depends upon its federal government (and to a much lesser extent state and local governments) for its political decision making. The geographic intricacies of these political differences are discussed in more detail in Chapter 4 of this text.

Putting Geography Back on the Map

If you're an American student, there's a very good chance that you haven't taken a geography class in many years. In contrast, in Canada, Greenland, and Mexico, geography is taught to students more thoroughly during their precollege years. We applaud the political and educational decision makers in these and other countries in the world for realizing that it is critically important to enhance global awareness in students, as well as their understanding of local and regional people and places.

In this era of intensified interconnections and relationships on Earth, we believe that geography is one of the most important subjects in any school curriculum. Each of us is linked to other people and places in the world in ways we may take for granted as we shop for groceries imported from around the world, eat in "foreign" restaurants, and make cell phone calls to relatives and friends who formerly were out of reach. We use transportation connections that link the world's peoples as never before. In sum: Bring on more geography!

In the United States, geography has been a part of the much broader social studies curriculum for more than five decades. The result has been minimal attention to geographic learning for most American students for a long time period. Thus, your parents may have never learned much geography either. This problem was attacked head on by a group of K–12 teachers, administrators, and geography professors beginning in the mid-1980s and it continues to this day. The revolution in geography education was funded in large part by the National Geographic Society and was carried out by teacher-members of statewide Geographic Alliances that were organized in all the states of the United States, Puerto Rico, and Canada. It provides teachers with lesson plans, workshops, and institutes that offer support to learn (or re-learn) geographic concepts, themes, skills, and perspectives to teach to their own students. With help from other professional organizations such as the National Council for Geographic Education, the American Geographical Society, the Association of American Geographers, and the Geographic Education National Implementation Project, as well as the National Geographic Society, the impacts of this revolution continue to be felt today, but only in selected school districts and classrooms across the nation.

Working in tandem with the activist Geographic Alliance movement nationwide are other ongoing efforts to bring geography back to American schools at the political level. A few states such as Colorado require completion of a high school geography class for admission into state universities.

The **geographic education revolution** also made big inroads into spreading the word about the importance of geographic literacy and learning by producing a set of national standards, *Geography for Life: National Standards in Geography*. Students' knowledge and use of the standards are evaluated in classrooms with national examinations for 4th, 8th, and 12th graders as part of the National Assessment of Education Progress (NAEP). Geographers also provided content for the development and national dissemination of a College Board-approved *Advanced Placement Human Geography* course. In sum, some of the readers of this textbook who may have taken a geography class before enrolling in this one may have been the direct beneficiaries of this effort to bring geography back to the schools, and to the attention of the nation and world. If so, then the effort to bring geography back into school classrooms is making progress.

But much remains to be done. American students continue to rank near the bottom on Gallup Poll assessments and other international examinations in geography. In the latest test, many students were unable to answer questions about the location of Mexico as America's nearest southern neighbor. If you're one of these typical students who have never taken a geography class at the high school, college, or university level before, we're especially pleased to have you using this text. And for others of you who have benefited from the effects of the geographic education revolution by completing a course or courses before entering higher education—or even by declaring your interest in becoming a geography major or a professional geographer after graduation—your knowledge and understanding of core geographic concepts, themes, and skills will no doubt prove invaluable in mastering the contents of this textbook.

Why Study Regional Geography?

This text was written to help support geography courses that focus on the regional geography of North America. There are two broad types of geography—**physical geography** and **human geography**. Along with these overarching subfields of the discipline are courses that integrate information and skills from both human and physical geography with a focus on particular places in the world. This approach is called **regional geography**. Most colleges and universities offer human, physical, and regional courses as well as techniques courses such as **cartography** and **Geographic Information Systems (GIS)**.

This textbook is grounded in conceptually based regional geography—the study of selected parts of Earth that are defined and identified by certain unifying characteristics. North America as a continent is a region unto itself. And this large region also can be divided into political regions such as its four nation-states—Greenland, Canada, the United States, and Mexico. In turn, these four large political regions can be divided into other, smaller regions such as the Pacific Northwest, Mexamerica, and French Canada. This regionalizing approach will help make it easier for you to learn and apply the content of this book and make studying the geographies of each place more manageable.

Likewise, within many of these varying levels of regions and subregions are parts of other well-bounded and popularly known places such as the **Sun Belt** or the San Francisco Bay Area. These places may also be called regions because they have certain characteristics in common that help identify their distinctive geographies. The government also uses a regional approach to divide up territories such as its census regions, national parks and wilderness regions, and watershed regions. One example of these "regions within regions" at the urban level is shown on the image in Figure 1.6.

Geographers often divide regions into **formal**, **functional**, and **vernacular regions**. A formal region usually has an institutional or political identity and distinct boundaries (e.g., the New England region) whereas a functional region is described according to its interconnections or usefulness (e.g., the Salt Lake City Metropolitan Area). Vernacular regions such as "Dixie" are unified and distinctive areas defined by insiders who understand clearly their regional boundaries. These three types of regions can be **homogeneous** where their coalescent criteria are generally uniform (as in agricultural, religious, or linguistic regions), or **nodal regions** (where a core or central zone is most important, such as in an urban region).

▲ **Figure 1.6 The San Francisco Bay region viewed from space.** Urban areas appear in gray. The mysterious orange places are salt evaporation ponds in San Francisco and San Pablo Bays.

Approaches Used in This Book

This textbook, like many other regional geography books, is divided into a set of preliminary chapters called thematic chapters. Each of these deals with a particular theme such as **political economy** or **historical geography**. These thematic chapters are followed by a series of regional chapters that focus on one particular part or region of North America. Chapters 1 through 4 and our concluding chapter (Chapter 19) are thematic chapters. Chapters 5 through 18 are regional chapters that focus on the regions that we have defined for this book.

We begin with geographic regions that are located along or near the north Atlantic coast and then move across the continent from east to west using a broad brush approach (with a few "up close and personal" focal points along the way). The regions that we have defined to be discussed in this text and shown in Figure 1.8 include:

- The Atlantic Periphery
- Quebec
- Megalopolis
- Great Lakes/Corn Belt
- Inland South
- Coastal South
- Great Plains
- Rocky Mountains
- Intermontane West
- MexAmerica
- California
- The Pacific Northwest
- Hawai'i
- Far North

It is critically important to keep in mind that we could have defined the boundaries of the regions in this text differently by using the criteria that form the best and most cohesive units of space to determine the regional boundaries for Chapters 5–18. That is one of the exciting things about being a geographer! North America is a large and complex place with many different internal physical and human characteristics. Therefore, other geographers, historians, and popular writers have divided it into regions based on different criteria—and ended up with very different results. One of the best known examples of this process is the cultural regions of North America discussed in journalist Joel Garreau's book, *The Nine Nations of North America* (1981), shown in Figure 1.7.

Regional boundaries we have selected for these chapters are based on a combination of physical, political,

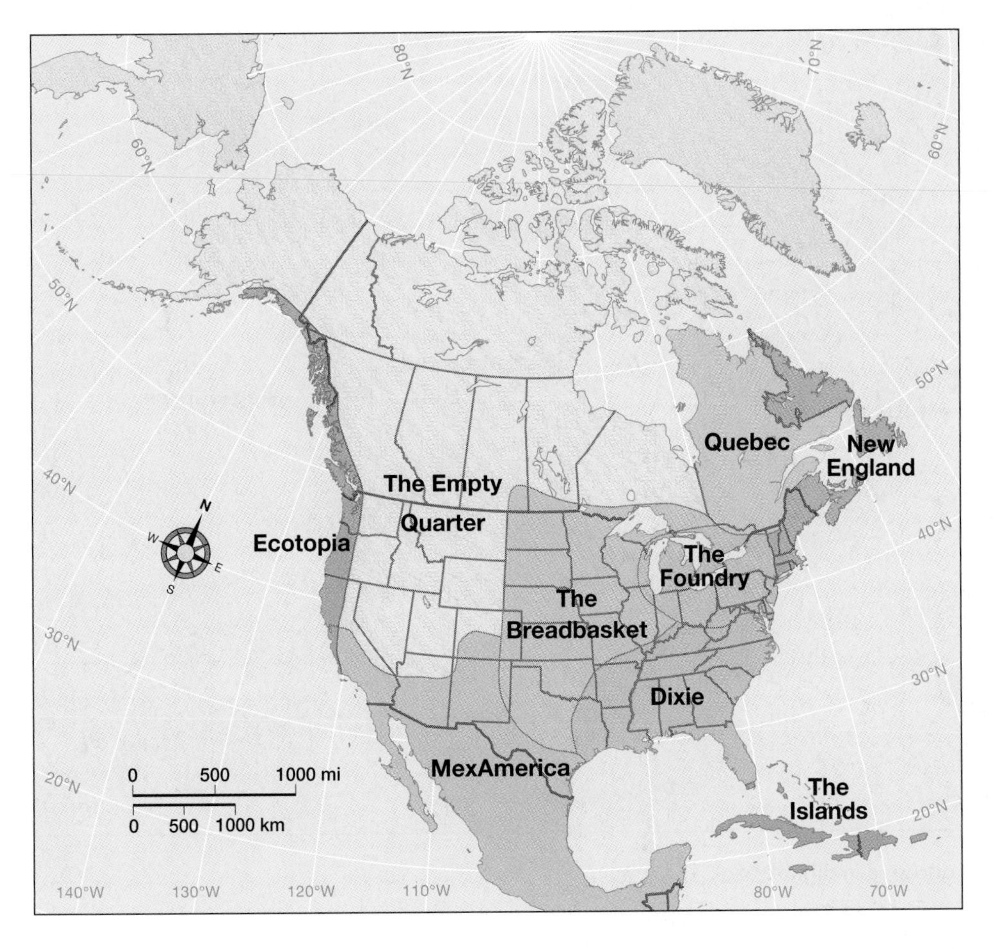

◄ **Figure 1.7** Garreau's North America regions

cultural, and economic boundaries as shown in Figure 1.8 below. The unifying criteria that help create boundaries for these regions are political, cultural, physical, or economic—or some combination of these.

Along with a discussion of a set of primary topics included in each of the following chapters, we also embed a detailed coverage of key geography concepts and themes throughout. To help you discern which of these are focused upon in each of the chapters, we identify them by using **bolded** text to set them off from the rest of the narrative. We also list all of the concepts discussed in each chapter with definitions at the end of the book in a Glossary. These conceptual links and lists should help you understand and be able to apply all of the key concepts embedded within the *National Geography Standards*—and prepare you to enroll in another geography course or succeed as a geography major or minor at your university.

We begin by presenting you with some thematic background information on the entire North American continent in the following chapter on physical geography. Chapter 3 then provides a broad overview of the historical settlement of North America, beginning with a discussion of the cultures, economies, social systems, and settlement patterns of its diverse aboriginal peoples, and then tracing the arrival and impact of later Euro-American colonizers and settlers. Chapter 4 presents an overview of the political economy of North

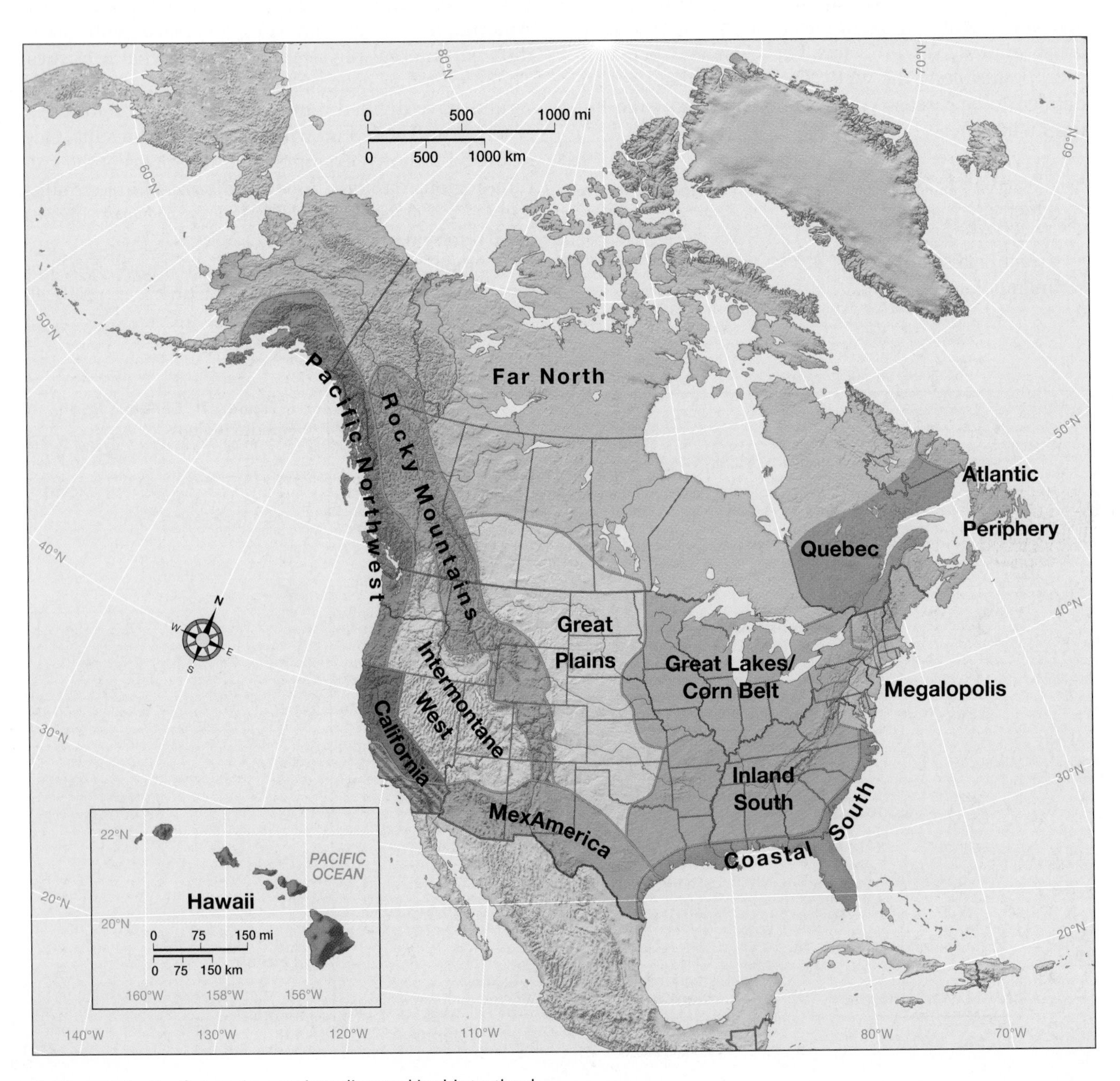

▲ **Figure 1.8** North American regions discussed in this textbook

America. Here we discuss the impacts of globalization and other theories and concepts related to the politics and economies of the continent, along with a discussion of the four levels of economic production—primary, secondary, tertiary, and quaternary.

As mentioned earlier, one of this book's innovations is including the Mexican borderlands and Greenland in our coverage of the geography of North America. We hope you find the inclusion of these often lesser known areas educational—as well as inducements to future travel to places you might not have considered within your reach before reading this text.

As a conclusion to this first chapter, we invite you to examine and reflect upon some of the other ways we might have divided up North America in this book. As shown on the maps in Figures 1.9 and 1.10, there are many different ways this could have been accomplished. The first map divides up the continent according to its landform regions, while the second map divides the same land area into agricultural regions. As you can see, regions are a shifting concept and the regionalizing of North America is not only ever changing but also can be accomplished in all kinds of different ways.

We invite you to begin your study of the geography of North America by taking another look at the map shown in Figure 1.8 earlier in this chapter. What regions might you change if you were asked to update this book? How and why are your regions different or the same as the regions shown on the maps in this chapter, and in some of the other regions mapped in later chapters? These and other related questions will guide our discussion and analyses of some of the peoples and places in North America in the chapters that follow.

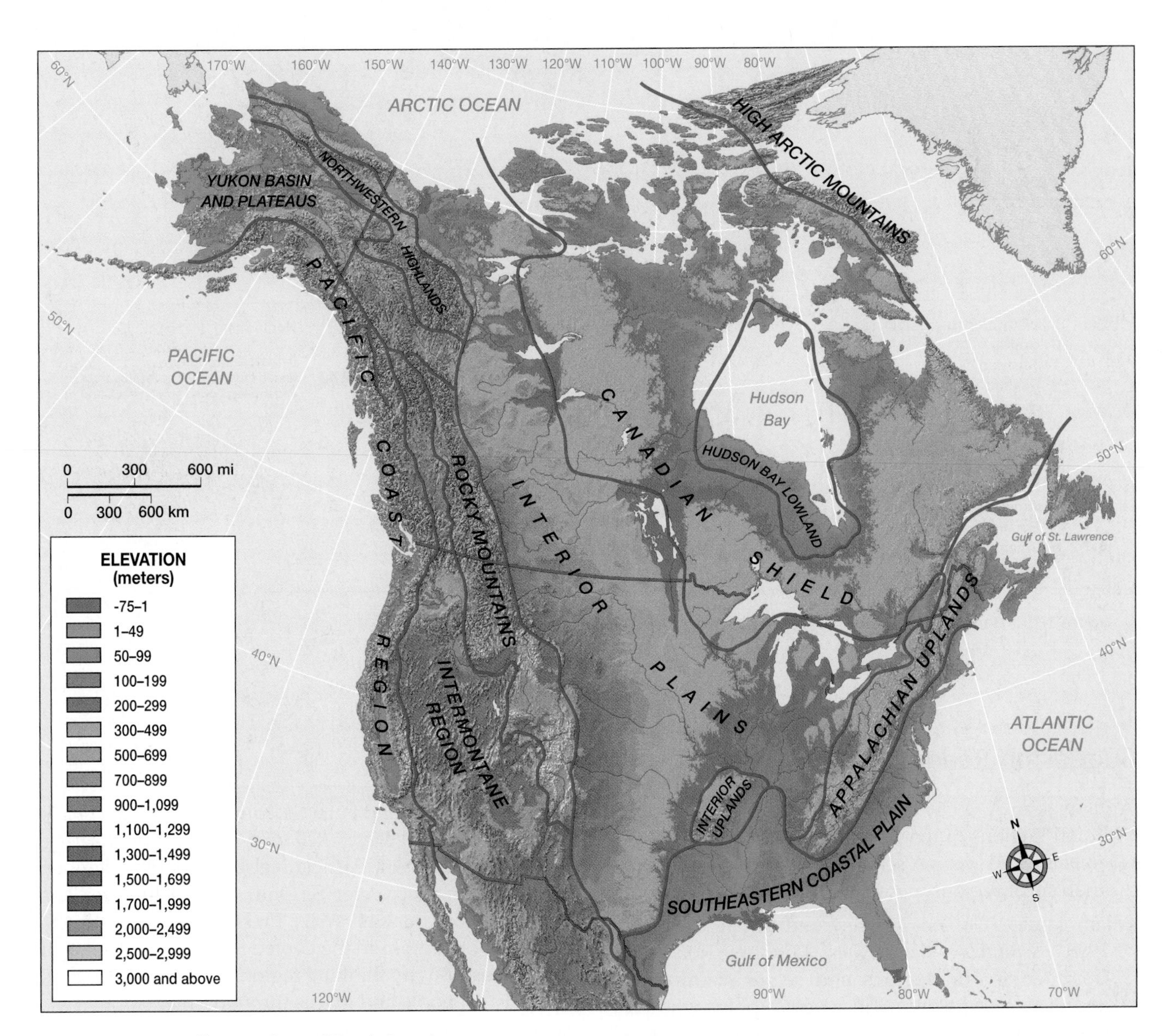

▲ **Figure 1.9** Landform regions of North America

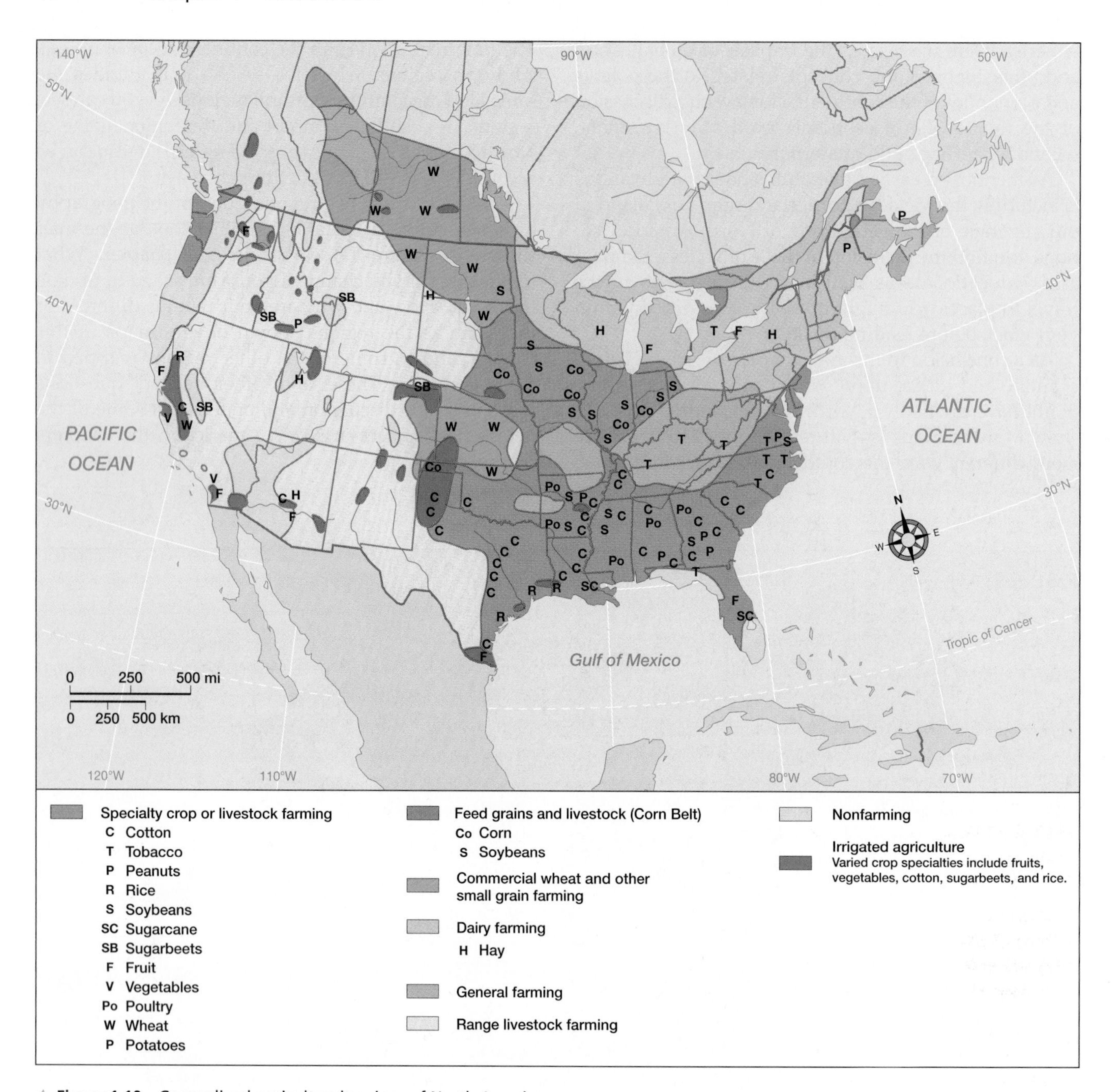

▲ **Figure 1.10** Generalized agricultural regions of North America

Suggested Readings

Agnew, John A. 1999. "Regions of the Mind Does Not Equal Regions of the Mind," *Progress in Human Geography* 23: 91–96. A scholarly and quite fascinating analysis of regions, perception, and mental maps.

Allen, James, Doreen Massey, and Allan Cochrane. 1998. *Rethinking the Region*. London: Routledge. Commentary on regions and regionalizing space and place by a team of British geographers who very effectively integrate theory and empirical data in their research.

Ayers, Edward L. and Peter S. Onuf. 1996. *Introduction to All Over the Map: Rethinking American Regions*, Edward L. Ayers, Patricia Nelson Limerick, Stephen Nissbaum, and Peter S. Onuf, eds. Baltimore: The Johns Hopkins University Press, 1–10. A historical "take" on regions in the United States . . . and some of the reasons why thinking regionally is a useful way to analyze people and places through time.

Buttimer, Anne. 1993. *Geography and the Human Spirit*. Baltimore, MD: The Johns Hopkins University Press.

Haggett, Peter. 1995. *The Geographer's Art*. London: Blackwell. A classic study of the field of geography with an emphasis on the geographic perspective and ways of thinking geographically.

Hardwick, Susan W. and Donald G. Holtgrieve. 1996. *Geography for Educators: Standards, Themes, and Concepts*. Englewood Cliffs, NJ: Prentice Hall. An introductory level textbook for geography teachers and future geography teachers structured around the National Geography Standards and five fundamental themes of geography.

Garreau, Joel. 1981. *The Nine Nations of North America*. Boston: Houghton Mifflin. A journalist's popular interpretation and defense of a whole new way of seeing regions in North America based on their cultures, economic and environmental characteristics and senses of place.

Geography for Life: National Geography Standards. Washington, D.C.: National Geographic Society.

Granatstein, J. L. and Norman Hillmer. 1991. *For Better or for Worse: Canada and the United States in the 1990s*. Toronto: Copp Clark Pitman, Ltd. A readable analysis of the relationship between the United States and Canada in the final decade of the 20th Century.

Kerr, Donald and Deryk W. Holdsworth, eds. 1990. *Historical Atlas of Canada: Addressing the Twentieth Century*. Vol. 3., Toronto: University of Toronto Press. A useful and comprehensive atlas of the economic, environmental, and cultural features of Canada at different time periods during the past century.

Martin, Geoffrey J. and Preston E. James. 1993. *All Possible Worlds: A History of Geographical Ideas*. New York: John Wiley and Sons. Classic overview of the discipline of geography that provides both introductory students and more advanced geographers with background in the key ideas and scholars important to the discipline's evolution.

Statistics Canada. Ottawa, Canada: Queens Printer, annual. The Canadian counterpart to the U.S. Census Bureau that publishes invaluable statistical analyses based on the Canadian census at different time periods.

United States Bureau of the Census. *Statistical Abstracts of the United States*. Washington, D.C.: U.S. Government Printing Office, annual. A useful statistical compilation of facts and figures on the population and other characteristics of people and places in the United States based on decadal census tabulations.

Warkentin, John. 1997. *Canada: A Regional Geography*. Scarborough, ON: Prentice Hall Canada. A textbook focusing on the geography of Canada that expands upon many of the key themes discussed in this book.

Van Loon, Hendrik Willem. 1932. *Van Loon's Geography*. New York: Simon and Schuster, Inc. Don't miss this fascinating "read" for a whole new way of visualizing and understanding more about our home on planet Earth.

Web Sites

The American Geographical Society
http://www.amergeog.org/

Association of American Geographers
http://www.aag.org

The Canadian Geographical Society
http://www.cag-acg.ca/ed

The National Council for Geographic Education
http://www.ncge.org

The National Geographic Society
http://www.nationalgeographic.com

The Royal Canadian Geographical Society (Société géographique royale du Canada)
http://www.rcgs.org/rcgs/

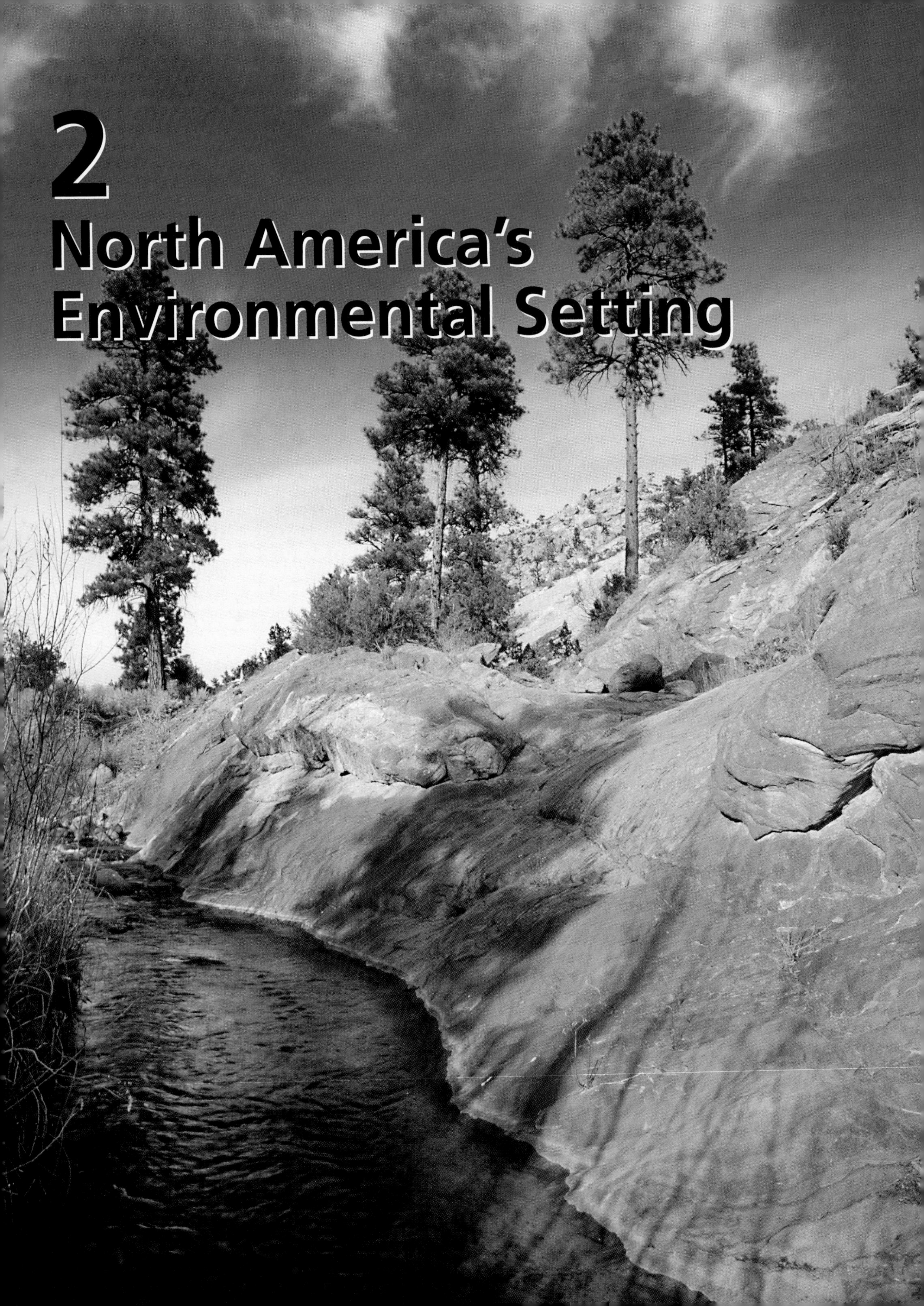

2
North America's Environmental Setting

Special thanks to Lisa DeChano for her contributions to this chapter.

PREVIEW

- Understanding geographic patterns and historical events requires knowledge of how Earth systems work and how they are presented on maps for comparison and analysis.
- Weather and climate can be explained with attention to four basic elements: temperature, atmospheric pressure, wind, and precipitation.
- North America's climate is greatly influenced by subtropical high pressure systems that form the westerly winds that move weather systems from west to east across the continent.
- Local area weather and climate are influenced by local environmental factors.
- Areas with interrelated patterns of geology, topography, and hydrology are known as **physiographic regions**.
- Another way to understand the physical geography of North America is to consider **watersheds** or drainage basins as a foundation for understanding land use and environmental planning.
- Because vegetation is a mirror of climate, hydrology, and soil types, maps of natural or native vegetation can help you speculate on what prehistoric landscapes may have been like.
- As people arrived in North America during the past centuries, they either adapted to the physical setting or began to significantly change it. In the 21st century, few populated areas on the continent look anything like the 16th century North American natural landscape.

Introduction

All events occur in a particular place and time. Before the history or geography of a continent can be understood, it is important to consider its environmental setting. **Physical geography** is the study of the physical (not human-related) characteristics of Earth and **human geography** focuses on human activity on the planet. Obviously the separation between the two is only conceptual because almost every place on Earth has been affected by human activity. In this chapter, we introduce the North American continent as it may have been before human contact. We provide answers to this big overarching question in this chapter: *What are the broad patterns of landforms, climate, hydrology, natural vegetation, and ecosystems of North America and how have these patterns helped shape human settlement?*

The terrestrial portion of the North American continent encompasses nearly 6.7 million square miles (17 million square kilometers). Its physical geography and human use of the land is very diverse because it is located mostly in the midlatitudes, where its climate, vegetation, and soils are quite varied. And because the United States, Canada, and Greenland also include tropical Hawai'i and polar regions, this text includes examples of most of Earth's climatic patterns and most of the major landform regimes covering Earth's surface. These various landforms, climates, and water bodies affected how soils were formed and what kinds of vegetation grow in each area. The combined matrix of landforms, hydrology, climate, soils, and vegetation, in turn, sets the stage for the support of animal life and the more than 100 identified **ecosystems** in North America.

Understanding the physical geography of North America provides a great deal of insight into where human settlements were located historically and what environmental constraints influenced human decision making over the years. Therefore, each of these environmental variables is discussed here with reference to how they affected historic and present day human landscapes.

Setting the stage for understanding geographic patterns and historical events requires some understanding of how Earth systems work. This chapter begins with a brief explanation of North America's climate systems, followed by a discussion of its landforms, hydrology, soils, vegetation, and ecological factors. It is a good idea to compare the continental maps of these physical elements in each section of this chapter. Notice the similarities of the physiographic provinces and climate zones. Then compare those patterns with the map of vegetation biomes. Note the differences, too. How does the pattern of natural vegetation change when viewed from different elevations or at different latitudes? What are the factors that forested areas have in common? What are the similarities between the patterns of soil and natural vegetation on the Great Plains? Do the maps accurately reflect your own observations about the physical geography of your home town?

◄ Pine Creek cuts through rock embankments in Escalante, Utah, and exhibits climatic, geomorphic, and ecological processes that shape the physical landscape

We can also compare the regional maps in this chapter with the map of regions discussed in this text as shown in Chapter 1 and with the "up close and personal" regional scale maps at the beginning of Chapters 5–18. This will make it possible to relate the topics discussed in this chapter to our focus on some of the ways humans have impacted upon the North American landscape over the last four centuries discussed in subsequent chapters of the book.

Landforms, Hydrology, and Soils

Climate, soils, and vegetation are all influenced by **geomorphology**, or the pattern of landforms on the surface of Earth. Geomorphic processes such as uplift, **glaciation**, **volcanism**, and **erosion** have helped shape the North American landscape into what it is today. Because geomorphic processes play such a significant role in shaping the North American landscapes, we discuss each of the major processes here briefly.

Topography

The United States Geological Survey is the government agency that does most of the topographic and hydrologic mapping in the United States. The Centre for Topographic Information does this in Canada. They produce the maps that are commonly used by hikers and for other outdoor activities. About one-third of the **topography** of North America is mountainous. The older, more worn features are in the East and the newer, more dramatic ones are in the West. Most tend to have a north–south orientation (except the Ozarks and Oachitas and Alaska's Brooks Range). In general the eastern side of the continent is down warping and the eastern coastline is slowly sinking while the West Coast is uplifting. The East Coast shore is almost flat, whereas most of the West Coast is backed by mountains. A large central plain covers the interior part of the continent. Plate tectonics and volcanism are also at work, especially in the West, creating the Cascade and Coast Range mountains among others. All the while the forces of erosion and deposition have sculpted the Rocky and Appalachian Mountains, as well as valleys and coastal plains, most visibly in the East. Ice in the form of glaciers has also been at work over geologic time creating and moving soil deposits and carving out the Great Lakes.

We will see in the next chapter that topographic features can be historic barriers to travel and settlement but can also provide resources for successful settlement. For example, if you look at a map of cities on the East Coast you will notice that Trenton, Philadelphia, Wilmington, Baltimore, and Richmond form a line between and paralleling the coastline and the Appalachian Mountains. This has been termed the **fall line** because each new settlement that later became a city was founded at the head of navigation, the point where sailing ships had to stop because the water from rivers entering harbors became shallow. These were also points where a change in geology created waterfalls (usually not very high) where waterwheel-powered mills could be built. With a landing and a mill site, a new community had advantages for growth.

An example of topography as a barrier is also seen in the historical perception of the Appalachian Mountains. They are, by western standards, not very high (about 2500 feet or 762 meters) but were a considerable barrier to travel and trade for wagon drivers (teamsters). This was enough to keep settlement on the eastern side for decades. During the colonial period the crest of the Blue Ridge was the formal demarcation line which was to assure Indian groups that colonists would keep to the east and everything west was Indian Territory. However, we know what happened to the "barrier" when Daniel Boone opened up the Kentucky bluegrass country. Other historic and sometimes continuing physical barriers were the Rocky Mountains and the Sierra Nevada. Resources taken from mountainous areas include coal, precious metals, timber, furs, and most important of all, water.

Topography has been used as a basis for political boundaries such as part of the boundary between the state of Maine and Canada. Another such division was the use of the Bitteroot Mountains to separate Idaho from Montana. While probably a good idea because mountain crests usually separate regional populations, these boundaries were extremely difficult to survey.

Tectonic Activity Earth's surface is composed of over 24 tectonic plates that move against one another, producing folding, faulting, earthquakes, volcanoes, and their corresponding landscapes (Figure 2.1). Most of the North American continent lies on the North America Plate except for parts of the West Coast, which lie on the Pacific Plate and are influenced by the Juan de Fuca Plate. This western margin is the most tectonically active region of the continent.

The plate boundary between the North American and Pacific plates is called a transform fault. These two plates slide past each other instead of one subducting under the other. The most famous fault along this transform fault boundary is the San Andreas Fault, which extends from the Salton Sea area of southern

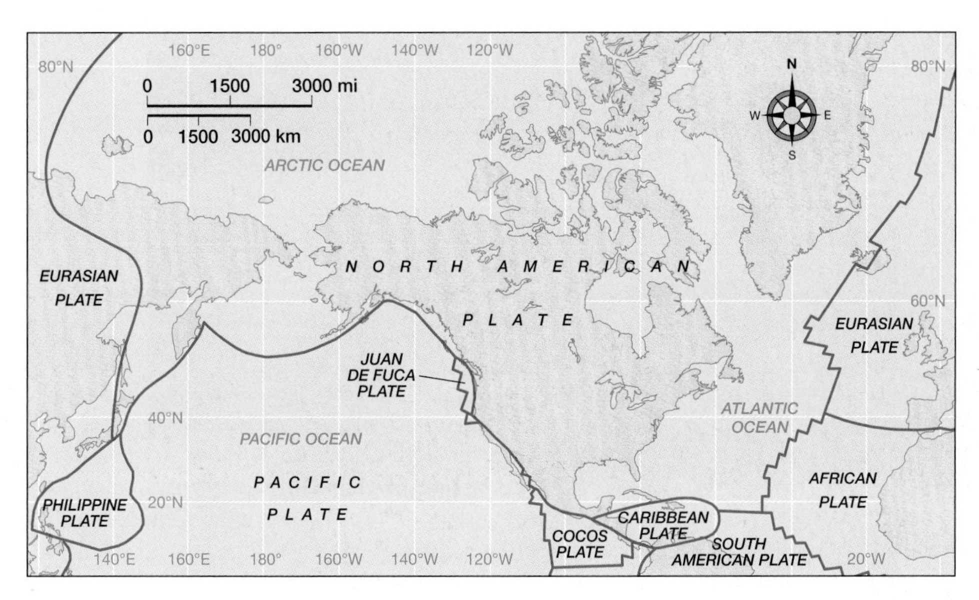

Figure 2.1 The tectonic base of the North American continent

California to north of San Francisco. The only place where this sliding does not occur is where the Juan de Fuca Plate and the North American Plate collide. From northern California to southern British Columbia, the Juan de Fuca Plate subducts under the North American Plate while it pulls away from the Pacific Plate. This **subduction** process results in **orogenesis**, or the creation of mountain ranges. A distinct landscape is found in this area, which includes the still active Cascade Ranges. Orogenesis is also responsible for creating the Rocky Mountains, the Sierra Nevada, and the Appalachian Mountains. The Appalachians are much lower than the other ranges listed because the orogenesis responsible for their creation occurred much earlier in geologic time, and so they were exposed to weathering and erosion over a longer period of time.

Crustal movement (either sliding or subducting) builds up tremendous amounts of friction because the movement is not a smoothly gliding motion. Strain builds up in the rocks until the frictional force is overcome. When this occurs suddenly, great energy is released and felt as an earthquake. The seismic release of energy is sometimes measurable at Earth's surface and physical infrastructures are damaged. In some cases, personal injury and loss of life result. Significant earthquakes have been recorded along the West Coast of North America such as the 1906 San Francisco earthquake and the 1994 Northridge earthquake. Earthquake risk is a fact of life on the West Coast. Preventing large earthquakes is impossible, but it *is* possible to help alleviate significant damage from earthquakes through monitoring and warning systems. These areas are continuously monitored by the U.S. Geological Survey, and, when possible, warnings of an impending large earthquake are broadcast. In many cities and counties, building codes require additional requirements for withstanding seismic movements and zoning codes prohibit construction of dwellings or schools in identified fault zones.

Volcanism Western North America boasts approximately 70 volcanoes but many of these volcanoes are inactive. Mount St. Helens, which experienced a major eruption in 1980 and more minor (but still quite dramatic) eruptions in 2005, is perhaps the most famous of these western volcanoes. It is currently the most active volcano in the chain of volcanoes along the Cascade Mountain Range (Figure 2.2). These volcanoes formed from magma moving toward the surface through a central vent from deep inside Earth. In volcanic areas along the Pacific Coast as well as in Yellowstone National Park, the tremendous heat from the rising magma boils groundwater, creating geothermal energy that can be seen as geysers and thermal springs. When conditions are right eruptions occur, spreading lava, ash, and other materials onto the landscape.

Volcanoes in the Cascade Range are known as **composite cones** and erupt explosively, meaning that they will explode and shoot matter, gasses, and heat into the atmosphere. If this material reaches high altitudes, long distances may be covered before this material is deposited. Mount St. Helens is one of more than a dozen potentially active volcanoes in the Pacific Northwest stretching from Mount Baker near the U.S.-Canadian border to Mount Lassen in northern California, including Mount Rainier near Seattle and Mount Hood near Portland, Oregon. These volcanoes are currently inactive, but it is known that most of them have erupted within the past few thousand years and it is probable that they could erupt in the future. Geologists estimate that, on average, about two major eruptions will occur in the Cascades every hundred years.

▲ **Figure 2.2** Mount St. Helens in the Cascade Range, 2004

Hawai'i is also made up of a series of volcanoes, but they are different from those of the Pacific Northwest. Hawai'i's volcanoes are **shield volcanoes**, which do not erupt explosively. They exhibit effusive (lava flow) eruptions and will sometimes shoot fountains of lava into the atmosphere. Shield volcano eruptions are not particularly violent, but they can produce enormous volumes of lava. These lava flows have a low viscosity that allows them to move easily over the landscape. As the lava travels into the ocean it cools and creates new land. Kileauea, on the big island of Hawai'i, has been erupting since 1997. Other famous volcanoes of Hawai'i are Mauna Loa and Mauna Kea on the Big Island, and Haleakala on Maui. Discussion and photographs of Hawai'i's volcanic landscapes are found in Chapter 17.

Glaciation The process of ice shaping the landscape is known as **glaciation**. This is the dominant geomorphic process in higher latitudes and higher elevations. Glacial ice forms when snowpack in high latitudes or high elevations does not completely melt during the summer; more snow is added in winter, and again the snowpack does not melt. This layering technique builds up snow layers that will eventually, under their own weight, transform into glacial ice. When enough ice has formed, the glacier will begin to move downslope.

Much of North America was covered in continental glaciers at periodic intervals that began approximately 1.5 million years ago and ended about 10,000 years ago. During this time a great expanse of ice, centering on what is now Hudson Bay, covered all of Canada and Alaska and extended as far south as the Ohio and Missouri rivers and the middle Columbia River (Figure 2.3).

Landscapes in the southward path of a glacier were sculpted by both erosional and depositional processes. For instance, continental glaciers carved out the Great Lakes, which are still rebounding from having the great amount of weight lifted when the glaciers retreated. **Alpine glaciers** (most of which are currently melting) are responsible for the spectacular landscapes of the Rocky Mountains, having carved steep slopes near the peaks. In some of the northern states such as New York and Michigan, parallel hills called **drumlins** and extensive systems of scoured and infilled glacial drift sediments cover the landscape. **Moraines** dot the landscape of much of the Midwest, where glaciers dumped their debris during periods of melting ice. Interestingly, there is one area in Wisconsin, known as the driftless area, which appears to have been spared all effects of glaciers.

Hydrology, Water in Action A careful look at a map of river discharge shows that the largest drainage in North America is the Ohio-Missouri-Mississippi **watershed**. These immense rivers were important for early transportation and cities were often located on their banks and at their confluence. These patterns are still visible on maps today. The downside of living next to rivers was the danger of flooding. Where there were many rivers in close proximity there could be canals between them for transportation and trade. Almost all of the canals built in the United States were located east of the Mississippi River. In several parts of the United States and Canada, rivers have been selected (usually by decision-makers hundreds of miles away) as boundaries between states or provinces. While such boundaries are easy to survey, they are often not very permanent. The lower Rio Grande international boundary was the subject of dispute for more than a century before an agreement for a permanent boundary between the United States and Mexico was established in 1964. Rivers in the West were also seen as a source of irrigation and for consumption in urban areas. The headwaters of major rivers are located in areas of high precipitation, while those same rivers may flow through desert conditions downstream. The founding of towns in the arid West was often based on the availability of water supply and the ease of transportation afforded by rivers.

Running water has the power to shape landscapes in powerful and significant ways. **Fluvial processes** have the power to erode the top layer of soil and deposit the sediment several hundred kilometers from where it originated. In areas of dry climate such as the western Great Plains and the Southwest, **riparian vegetation** occurs only along rivers and streams, although people frequently plant, water, and cultivate trees, shrubs, and grasses elsewhere. Water can also modify glacially formed landscapes; in fact, over time, it is often difficult to tell that glaciation was the original shaping process.

In North America east of the Continental Divide, two prominent drainage systems exist: the Great Lakes–Saint Lawrence system and the Mississippi-Missouri system (Figure 2.4). The Great Lakes–Saint Lawrence system is responsible for draining areas in southeastern Canada and the northeastern and north central United States. The Great Lakes, Saint Lawrence Seaway, and adjoining waterways have become the transportation arterial that supports the entire industrial region around them, linking the entire central lowlands to the Atlantic Ocean and the rest of the world. The more southerly central United States and parts of southern Canada are drained by the Mississippi-Missouri system. The main stem of the Mississippi River begins in Itasca State Park in central Minnesota and flows to the Gulf of Mexico near New Orleans. This is the other great waterway that affords bulk shipping access to the central part of the continent and historically has been the gateway to the western part of the continent via its tributary, the Missouri River, which joins the Mississippi River near St. Louis. Another major tributary, the Ohio River, flows into the Mississippi River from the east at Cairo, Illinois, approximately 140 miles (225 kilometers) south of St. Louis. The Ohio River drains the northern Appalachian area and the eastern portion of the Midwest. It was a prime

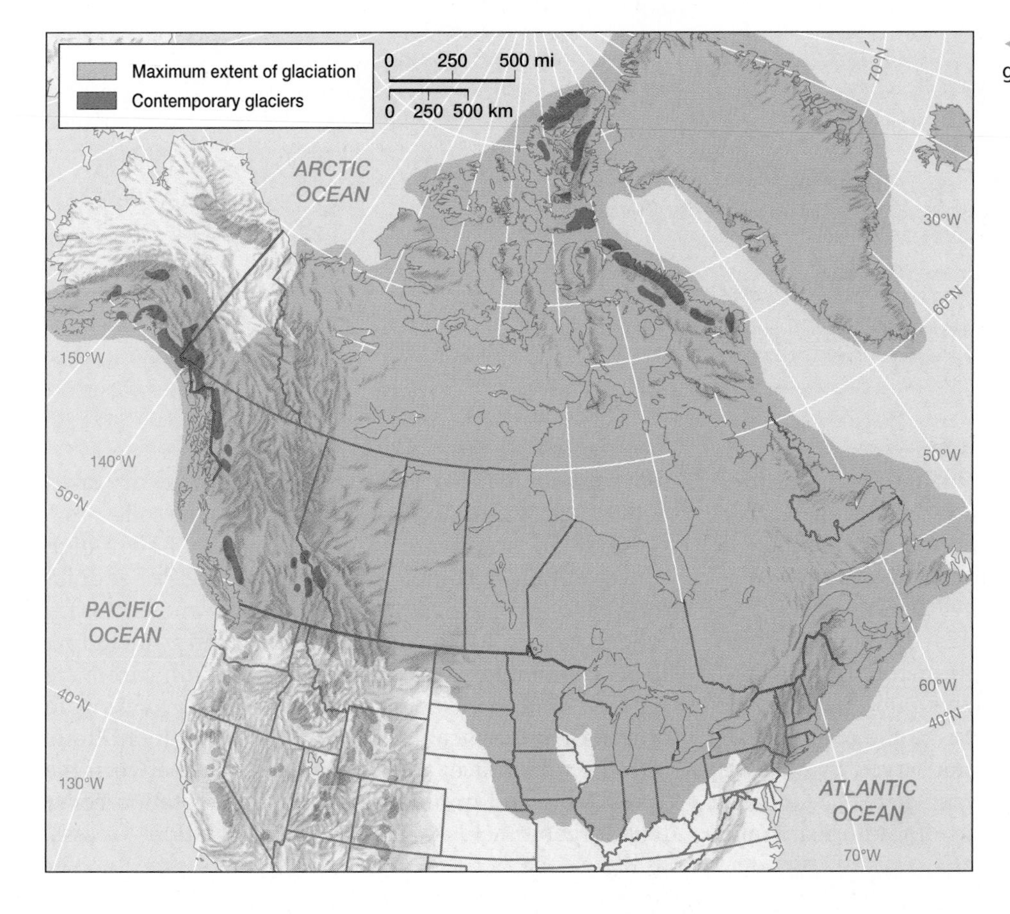

Figure 2.3 Prehistoric glaciation in North America

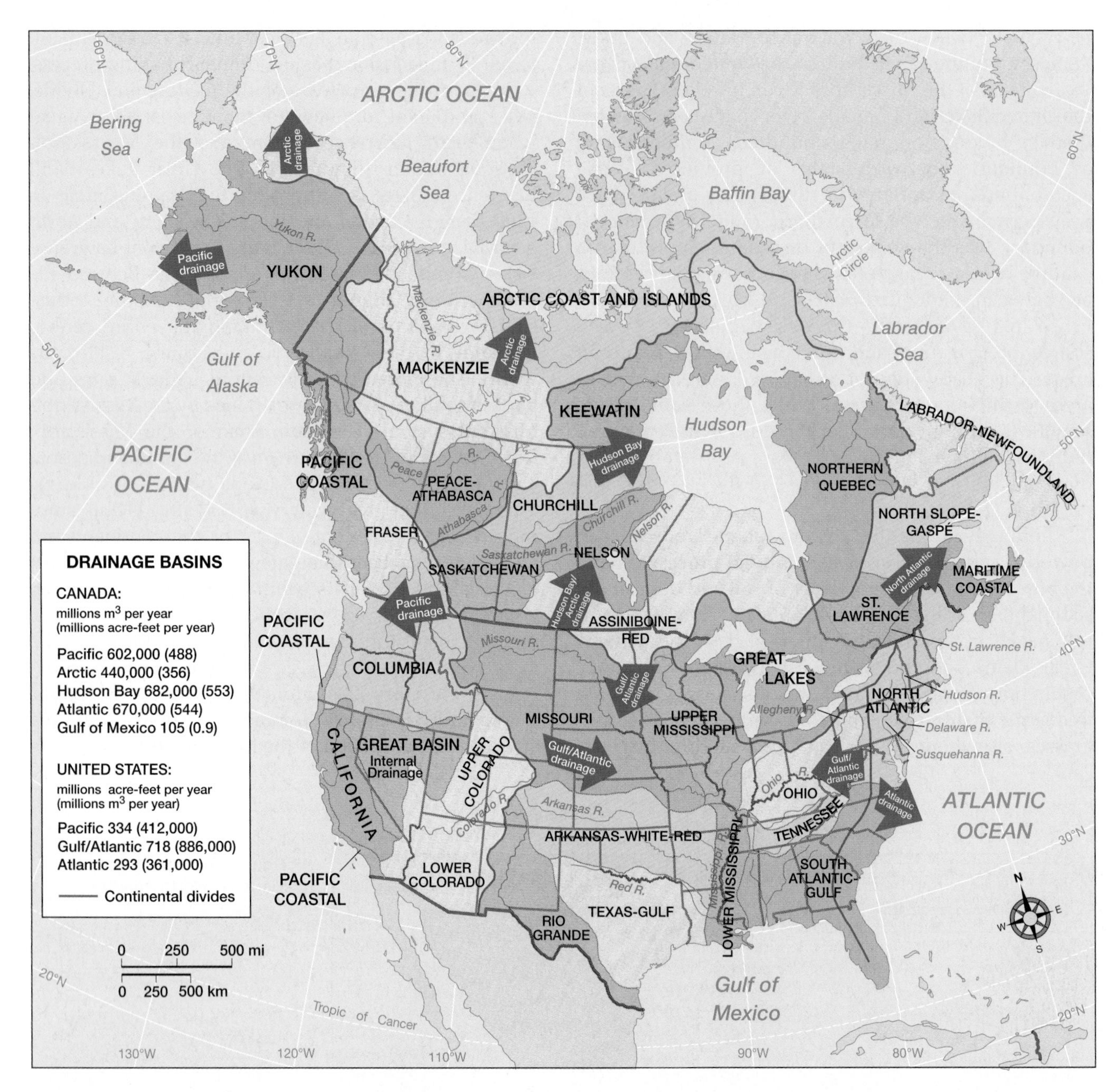

▲ **Figure 2.4** Major drainage basins of North America

location for steelmaking in the 19th and 20th centuries by receiving coal and iron from the Appalachians. The western United States drains to the Pacific Ocean via the Fraser, Colorado, and Columbia Rivers as well as other, smaller stream systems. The Pacific Ocean is also the outlet for rivers and streams draining western California, Oregon, Washington, British Columbia, and Alaska. The Mackenzie River flows into Hudson Bay, draining most of the Arctic regions of the continent. A few areas of North America are not served by external drainage. This type of drainage occurs in the Great Basin states of Nevada, Utah, and a part of California. Water that comes into these areas flows to shallow or dry lakes and then is evaporated or percolated into the soil.

Karst landscapes are found in areas that have a high concentration of soluble rock such as limestone, dolomite, and gypsum. Several conditions are necessary for processes to form karst topography, including rocks containing at least 80 percent calcium carbonate (in limestone areas); intricate patterns of jointing in otherwise impermeable rock so that water can infiltrate into the subsurface; a zone containing air between the water table and ground surface; and vegetation cover with organic acids to enhance the solution process.

Limestone solution occurs most often in humid regions (Kentucky and Appalachian valleys, Ozarks) but can also occur in arid locations such as southern New Mexico's Carlsbad Caverns. Stalactite and stalagmites formed from the solution process as shown in Figure 2.5 are common in Carlsbad's Temple of the Sun.

Karst processes do not only change subsurface landscapes. The solution process can weaken layers of soil above the water table and the surface material begins to sink forming a circular depression or a sinkhole. At times these sinkholes collapse and leave a deep depression in the landscape. When this occurs under a road surface, cars are in danger of falling in without prior warning. In areas where sinkholes are prevalent, such as Florida, Kentucky, and Indiania, autos, houses, and businesses have been destroyed and the hazard is a constant reminder that subsurface processes are ongoing.

Coastal fluvial processes are other types of erosional and depositional power that helps form distinctive landforms at the edges of continents. Waves are perhaps the most effective landscape-shaping agent along shorelines, as they can bring sediment ashore with the swash, straighten coastlines, and erode beaches with the backwash. Erosional coastlines in North America, particularly on the West Coast, have high relief and are rugged. Landforms created by erosion of the coast include sea cliffs; wave-cut platforms, which create a stair-step-looking landscape; sea caves; sea arches; and sea stacks. Depositional features of coastal processes include beaches; barrier spits—arms of deposited material attached to the shore; bay barriers—spits that have grown to completely cut off the ocean from the bay; lagoons—bodies of water formed behind barriers; and **tombolos**—deposited material that connects a sea stack to the main shoreline.

Other coastal processes form unique habitats for biological productivity, including wetlands, salt marshes, and mangrove swamps. Wetlands occur when the ground is saturated with water most of the time, typically found in areas of poorly drained soils. The 30th parallel of latitude is roughly the dividing line between salt marshes to the north (as in places such as Delaware and the Carolinas) and mangrove swamps to the south (for example, in the Everglades of Florida). This line of latitude marks the occurrence of freezing conditions, which controls mangrove seedling survival. Mangrove swamp areas accumulate enough sediment over time that trees and shrubs become established and form islands.

Wind Landforms shaped by wind are best observed in arid areas and along coasts where there are few obstructions. **Deflation** and **abrasion** are the two principal aeolian processes. Deflation is the process of lifting and removing loose material. Fine particles can be caught up in suspension and carried long distances before being deposited. Abrasion, or sandblasting, grinds away rock surfaces with particles

▲ **Figure 2.5** Temple of the Sun, Carlsbad Caverns, New Mexico

already in the air. Aeolian process can sculpt rocks into distinct angular landforms such as can be seen in Arizona and Utah. When particles are deposited on the landscape dunes can form and/or **loess** deposits can build up. Dunes can typically be found along the Atlantic and Pacific coasts as well as along the eastern shores of Lake Michigan. Loess deposits may originate great distances from where they are found and are typically high in clay content. These particles bind together, weather, and erode into steep bluffs. In North America extensive loess deposits are located in the central Plains, in the Palouse Region of Washington and Oregon, and in some lower reaches of the Mississippi River. Smaller pockets of loess deposits can be also be found in southern Alberta and Manitoba, and along the Missouri River in Iowa and South Dakota.

Soils North America contains some of the most productive agricultural soils on Earth. Soil is arguably the most important natural resource on the continent and the least understood. The characteristics of soils are very site specific but may be generalized into the classifications shown on the small scale (large area) map in Figure 2.6.

It is useful to stop reading here to compare the soils map shown in Figure 2.6 with the map of natural vegetation shown in Figure 2.15. Then compare these two related maps to the continental scale map of present-day agriculture shown in Figure 1.10 in Chapter 1. The success of agriculture in certain parts of the continent is the result of human perception and use of these fertile soils. The relationship between farming and the nature of soils is illustrated by the work of the U.S. Natural Resource Conservation Service (formerly the Soil Conservation Service), which has assisted farmers at the local level since the days of the Dust Bowl. Almost all soils are unique, having been formed by weathering of various surface rocks, the mixing of this parent material with various forms of organic material, and the presence of a range of moisture conditions. Climate is perhaps the most influential factor in determining the geographic distribution of soils in North America. Temperature and moisture determine how weathering takes place and how much moisture will support the development of biota (living organisms). Other variables in soil formation include topography, solar radiation, animal activity, and time.

There are 12 major types of soil in North America and the classification system for them can be easily researched with reference to a physical geography textbook or to the Internet. **Chernozems** and **Planosols** are some of the best soils on Earth for important crops such as corn and wheat. **Latosols** found in the Southeast are more limited due to more water content and less organic matter than the grasslands soils. Interestingly, the superior soils of the Great Plains were not used for farming until the technology existed for drawing well water, plowing deep grassland soils (sod), and fencing large areas from livestock. **Ultisols** are found in the southeastern United States, and due to iron and aluminum oxide residue these soils display a reddish color. They are low in fertility and have historically been misused through agricultural practices that did not allow for nutrient regeneration and planting crops such as cotton that expose soil to erosion. **Vertisols** have high clay content and shrink and swell as they become wet and dry, thereby causing cracks in the ground and in building foundations. In North America these soils can readily be found in Mississippi, Alabama, and Texas. It might be interesting for you to do some research on the soil types that are common in your home town.

At the local scale, soils are mapped and used by land use planners and land managers as well as farmers. Soils are important to farmers for housing nutrients and holding or draining water. Soil cover also supports vegetation that stabilizes slopes and supports building foundations. On steep slopes, soils tend to be thinner because water carries material in solution and by suspension over the surface of the slope instead of down through the **soil profile**. Soils at the bottom of the slope are thicker and wetter because they accumulate water and sediment from surface runoff. Because of the thinner soil on the slope there is usually less vegetation, leading to more soil erosion. On more gentle slopes, thicker soils develop, with greater organic matter content and more lush vegetation. Topography also affects how much solar energy is absorbed by the top layers of the soil in the landscape. In North America, south-facing slopes face the incoming solar rays of the sun, and so are usually drier and warmer than north-facing slopes. Soils on south-facing slopes often contain less organic matter due to more water loss through evaporation. In colder, wetter climates, south-facing slopes may be the more agriculturally productive; but in warmer, drier climates south-facing slopes are not as conducive to cultivation due to desiccation. This principle also explains why the first flowers in the spring in much of North America are usually found on the south side of houses, fences, or walls. They offer protection from the cold north wind and feature the welcome warmth of the sun.

Weather and Climate

Weather is what you experience when you walk out the door in the morning. **Climate** refers to long term patterns of weather and atmospheric conditions on Earth's surface. Climatologists deal primarily with four basic elements of climate: temperature, atmospheric pressure, wind, and precipitation. Of these, **precipitation** was one of the most important factors for historical

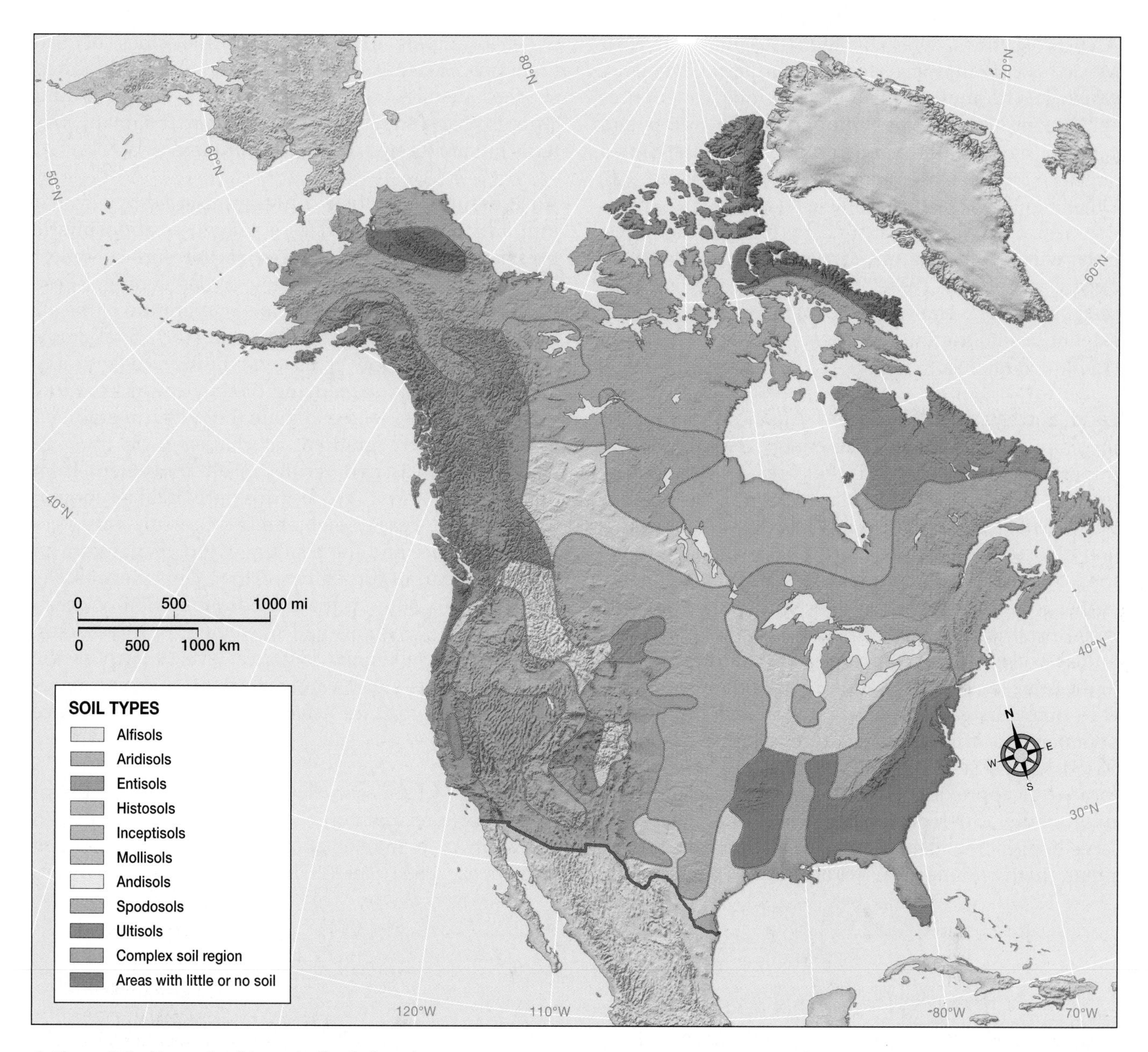

▲ **Figure 2.6** General soil types in North America

agricultural development in certain places. Temperature range between daily maximums and minimums was also important because that determines the **growing season** for crops.

The factors that influence weather and climatic atmospheric processes are numerous and interrelated. It is important to recognize that climatic variation is studied in varying scales. We may be discussing climatic events at a global scale or at a local scale much like your daily weather report.

Latitude **Latitude**, which measures north–south location, affects how much **solar radiation** is received at selected points and is a primary control on climate, especially temperature. Latitude also determines the length of daylight in each 24-hour period. That is why areas near the equator are generally warmer than places located far from the equator because they receive more direct rays of the sun through the atmosphere. Likewise, polar areas receive less direct solar energy, particularly in winter, when they experience days of 24-hour darkness. In general, the farther a location is poleward from the equator, either to the north (say in Anchorage, Alaska), or south, the less solar energy is received throughout the year. These areas are significantly cooler than the equatorial regions (e.g., American Samoa), where there is an energy surplus, and polar areas also have a larger annual range in temperature.

Because North America spans 50 degrees of latitude, the continent displays a large number of different climates. Parts of northern Canada and Alaska experience

Arctic conditions, while southern Arizona and New Mexico are semitropical deserts. Northern Greenland, which gets no sunlight at all in midwinter, experiences subfreezing temperatures throughout the winter and early spring, while the southernmost place in the United States in Hawai'i has a tropical climate (Figure 2.7). Other significant factors influencing climate patterns is the west–east variation caused by high altitude westerly winds, topographic barriers, and proximity to large water bodies. These "big picture" patterns can be seen on forecasts shown on the Weather Channel using satellite photography, radar imaging, and computer simulation models.

Land and Water Differences After considering the intensity and duration of solar energy, the second most important climatic control is the type of surface receiving the radiation. Land surfaces have a low specific heat, do not allow radiation to pass beneath the surface, and only slowly conduct energy. Land surfaces, therefore, both heat and cool rapidly. Continental climates, which cover a large part of the North American land mass, experience large temperature ranges both daily and seasonally when compared with maritime regions at the same latitude. The land–water proximity factor is easy to understand if you were to research the average summer and winter temperatures of places such as Fargo, North Dakota, and Seattle, Washington. Both are located at approximately the same latitude but have dramatically different climates. The further away from large bodies of water a place is located, the greater the range (high summer–low winter) of its temperatures. Other examples might compare the cities of San Francisco, St. Louis, and Washington, D.C. (all located near 38 degrees north latitude), or Vancouver and Toronto that also have radically different climatic patterns but are located at similar latitudes.

Air Pressure The force or weight exerted by air on a unit area on Earth's surface is known as **atmospheric pressure**. Pressure differences at the surface reflect whether the air is slowly rising or descending. These vertical motions often reflect the temperature of the air and are measured with a barometer. Horizontal gradients from high to low pressure at the surface (or at any altitude) result in air moving directly toward the low pressure as it seeks to eliminate the pressure gradient. Such a **pressure gradient** is characteristic of large geographic areas and results in air movement from "highs" into "lows." We feel this horizontal air motion as the wind. Areas of high and low pressure vary over time and space and the resulting wind changes are an important part of our daily weather. On a global scale, this airflow results in the subsidiary motion of ocean currents. The locations and movement of these ocean currents affected major historical events such as the "discovery" of America and the Atlantic Ocean connections necessary for the establishment of the later slave trade.

Elevation Changes in elevation also affect the distribution of precipitation. Temperatures decline with elevation at a rate of about 3.5 degrees Fahrenheit with every 1000 feet (2 degrees Celsius with every

▶ **Figure 2.7** The southernmost bar in the United States at 19° N latitude

300 meters) of elevation. This rate of decline is called the **normal lapse rate**. Thus, Denver at 5280 feet (1610 meters) elevation has an average temperature in January of 29.7°F (–1.3°C), whereas nearby Aspen at 7907 feet (2410 meters) has an average January temperature of only 20.7°F (–6.3°C).

Topography **Orographic precipitation**, caused by cooling air as it is uplifted, is especially common in parts of the world where moist air masses come in contact with high mountain barriers. One example of this is the impact of the Sierra Nevada and Cascade Mountains along the west coast of North America that block the passage of maritime air to the interior. This creates the immense, dry Great Basin desert on the downwind side of these mountains. Other topographic local winds also affect local areas such as the **chinook winds** in the Rocky Mountains and the **Santa Ana winds** in southern California. These are warm, dry regional winds that descend from high to low elevations over a mountain barrier or a high plateau. The chinooks have a reputation for creating domestic problems and even violent behavior, at times, among local residents. Descending warm air also is perceived as a negative environmental factor in southern California when the dry Santa Ana wind blows out of the desert. It has been documented that homicide and assault rates increase when this fierce wind blows. Numerous wildfires and losses of residential homes in the mountains rimming the Los Angeles basin also are associated with these often damaging winds.

Frontal storms are very common as well, and you have probably heard the term "front" during weather forecasts. A **front** is the point of contact between two differing air masses. When the warmer (and usually moister) air mass is forced to rise above cooler air, a discontinuity of surface temperature and air pressure is created. Precipitation is also common with this type of front. As warm air is pushed off the ground, cold temperatures and gusty winds mark the passage of the front. Various cloud formations mark the progress of frontal storms as well.

When climate is discussed in terms of global-scale air masses, which exhibit varying wind patterns, levels of moisture, and relative temperature, a rule to keep in mind is that warmer air can carry more moisture than the same amount of cooler air. Average summer and winter temperatures in North America are shown in Figure 2.8. Contour lines of equal temperature, as shown in the figure, are called **isotherms**.

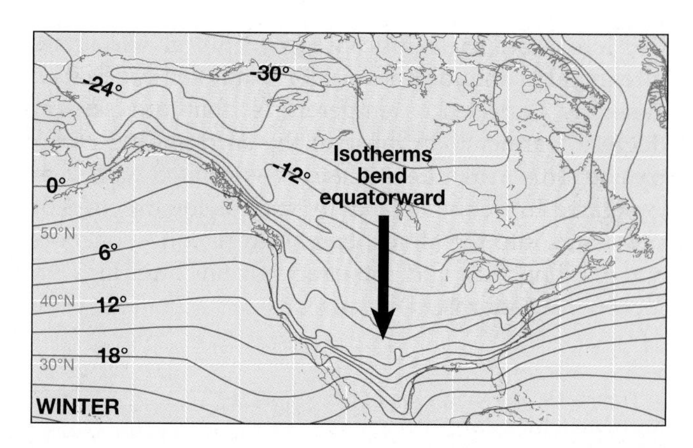

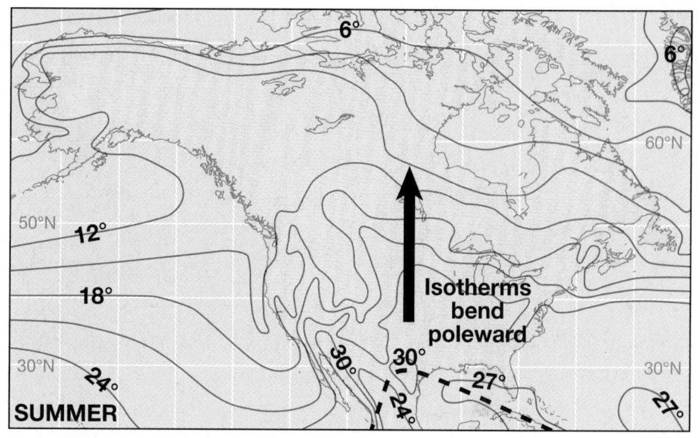

▲ **Figure 2.8** Mean annual temperatures (in centigrade) in January and July

Temperature Temperature is also influenced by cloud cover which inhibits direct solar rays from heating Earth's surface during the day, thereby keeping daily maximum temperatures lower than would be the case if clouds were absent. Conversely, at night clouds act as a blanket insulating Earth and producing higher nighttime minimum temperatures through more re-radiation than would occur off of the clouds than if no clouds were present. Close proximity to water will also modify land and air temperatures so that the difference between the daily minimum and maximum temperature is not as great as that for a more inland location. This can also be seen between inland and more coastal locations on an annual basis. For instance the maritime city of Vancouver, British Columbia, experiences a July mean temperature of 71°F (22°C) and a January mean temperature of 42°F (6°C), giving an annual range of 29°F (16°C). The July mean temperature in Winnipeg, Manitoba, more than a thousand miles (1600 kilometers) inland from the Pacific Coast, is 79°F (26°C) and the January mean temperature is 9°F (–13°C). The continental location of Winnipeg translates into a 70°F (21°C) temperature range between January and July. A combination of high latitude, high elevation, and distance from water has contributed to the coldest recorded temperature in continental North America at –81°F (–63°C) at a place called Snag, in the Yukon, and –87°F (–66°C) in Nerhice, Greenland! A dramatic combination of low altitude, low precipitation, and mountain barriers characterizes the place with the highest recorded temperature in North America, 134°F (73°C), at Death Valley, California. This is also the driest place on the continent, with an average rainfall recorded at less than one inch per year.

There are also significant geographical variations in climate at the local level. For example, large cities generally experience temperatures that are several degrees warmer than those in the surrounding countryside. This **urban heat island** is associated with heat generated from human activity such as the injection of pollutants into the atmosphere, heat retention of roofs and parking lots, and the effects of tall buildings on local wind patterns.

Air Masses As mentioned previously, air masses are very large bodies of relatively stable air that develop over a source region which provides them with specific moisture and temperature characteristics. Air masses are called **continental** or **maritime**. Their identities are based on moisture characteristics and are designated **equatorial**, tropical, and **polar**, based on temperature. Generally speaking, maritime air masses forming over the ocean or other large bodies of water are wet, while

▶ **Figure 2.9** Air mass regions of North America in (a) winter and (b) summer

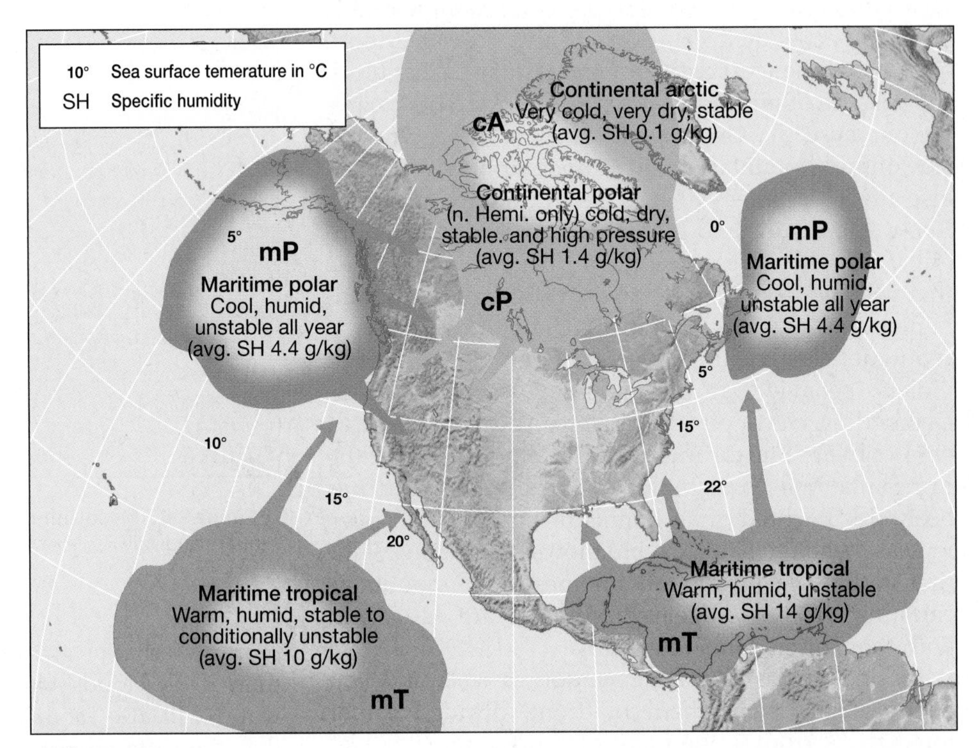

a) Winter pattern

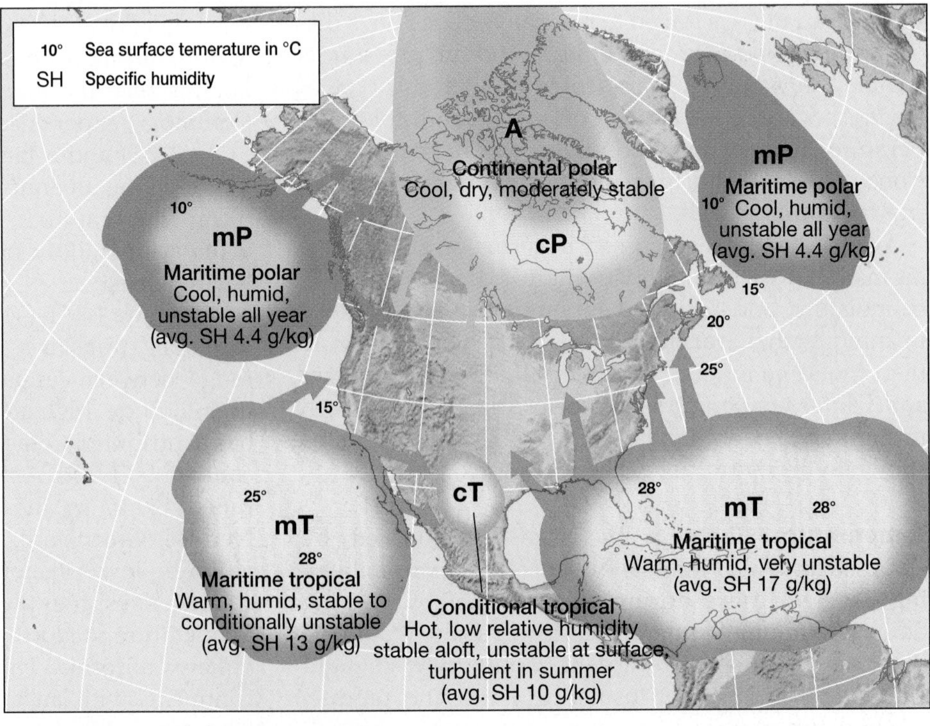

b) Summer pattern

continental masses are dry. The temperature of an air mass is usually determined by latitude. Warmer air masses can carry relatively more moisture than colder ones. Figure 2.9 shows the general patterns of summer and winter air masses over North America. Polar air masses form at high latitudes (centered at approximately 55° N) and tropical air masses form at low latitudes (centered at approximately 25° N). These characteristics combine to form different air masses that then dominate weather and climate in North America. Table 2.1 describes the characteristics of each of these air masses.

High latitude air masses are associated with cool summer weather and bitterly cold winters in the northern interior. They filter air into northern Canada bringing extremely frigid temperatures but not producing much, if any, precipitation. Locations such as Fairbanks, Alaska, may experience temperatures as low as –50°F (–46°C) when continental Arctic air masses arrive in winter. The **continental polar** air masses spill from their source areas into southern Canada and the northern United States. In winter, the weather in northern West and East coast locations is often dominated by **maritime polar** air masses. Locations that are influenced by maritime polar air masses, such as Juneau, Alaska, tend to experience wet winter weather and cool summers.

Maritime tropical air masses generate strong flows of warm, wet air into areas of the southern United States. These air masses are responsible for the extremely warm, humid conditions experienced in the American Southeast during the summer in places

TABLE 2.1 Air Masses of North America

Air Mass	Source Region	Temperature and Moisture Characteristics in Source Region	Stability in Source Region	Associated Weather
cA	Arctic basin and Greenland ice cap	Bitterly cold and very dry in winter	Stable	Cold waves in winter
cP	Interior Canada and Alaska	Very cold and dry in winter	Stable entire year	a. Cold waves in winter b. Modified to cPk in winter over Great Lakes bringing "lake-effect" snow to leeward shores
mP	North Pacific	Mild (cool) and humid entire year	Unstable in winter Stable in summer	a. Low clouds and showers in winter b. Heavy orographic precipitation on windward side of western mountains in winter c. Low stratus and fog along coast in summer; modified to cP inland
mP	Northwestern Atlantic	Cold and humid in winter Cool and humid in summer	Unstable in winter Stable in summer	a. Occasional "nor'easter" in winter b. Occasional periods of clear, cool weather in summer
cT	Northern interior Mexico and southwestern U.S. (summer only)	Hot and dry	Unstable	a. Hot, dry, and cloudless, rarely influencing areas outside source region b. Occasional drought to southern Great Plains
mT	Gulf of Mexico, Caribbean Sea, western Atlantic	Warm and humid entire year	Unstable entire year	a. In winter it usually becomes mTw moving northward and brings occasional widespread precipitation or advection fog b. In summer, hot and humid conditions, frequent cumulus development and showers or thunderstorms
mT	Subtropical Pacific	Warm and humid entire year	Stable entire year	a. In winter it brings fog, drizzle, and occasional moderate precipitation to N.W. Mexico and S.W. United States b. In summer this air mass occasionally reaches the western United States and is a source of moisture for infrequent convectional thunderstorms.

like Miami or St. Thomas in the U.S. Virgin Islands. **Continental tropical** air flows from central Mexico into the interior of the United States but is typically only a major influence on weather during the summer months. A hot July day in Brownsville, Texas, comes to mind when remembering the impact of this type of weather system.

Wind Patterns The high-altitude jet stream often talked about on weather broadcasts is responsible for moving air masses from their source regions. It can bend as far south as Texas while funneling cold air into the United States in the winter and influencing eastward moving storm systems. At other times, the jet stream moves as far north as the prairie provinces in Canada, bringing warm, dry weather to southeastern Alberta, Saskatchewan, and Manitoba. During the summer, the jet stream has a lesser influence on storms because it stays in the higher latitudes.

North America's climate is greatly influenced by subtropical high pressure systems that form the basis of the westerly winds that move weather systems in a west-to-east pattern across the continent. These pressure systems migrate seasonally so that the subtropical high pressure cell that lies just to the southwest of California during the winter shifts northward and westward as July approaches. More northerly locations face weather and winds associated with the subpolar low pressure systems. Because winds blow from areas of high pressure to areas of low pressure, interior Canada's variable weather is influenced by both the subtropical high and the subpolar low pressure systems.

The westerly wind system brings precipitation to the continent from the Pacific Ocean. The Pacific Northwest region in the United States and Canada receives the most annual precipitation from this flow. Henderson Lake in British Columbia holds the record for the greatest average annual precipitation of 262 inches (665 centimeters). North America's highest snowfall in one season (1027 inches or 2600 centimeters) was in the same general area in the Cascade Mountains of Washington State.

The southeastern states in the United States generally receive the second highest annual precipitation (Figure 2.10). The driest part of North America includes parts of eastern California at Death Valley, as mentioned previously, and Nevada, Utah, and Arizona, where continental air masses dominate weather patterns and mountains block the arrival of maritime air from the Pacific Ocean. In contrast, the windward side of these barrier mountains at Donner Pass, California, receives the highest annual snowfall in the world. Also for the record books, Mt. Waialiale in Hawai'i boasts the highest rainfall in the world, with an average of 460 inches (1170 centimeters) per year. Hawaiian climate and weather patterns are described in detail in Chapter 17.

All of the above processes interact to enable the creation of a map of climate regions as shown in Figure 2.11.

Climate Zones in North America

Based on the work of early climatologist Vladimir Koeppen, most physical geographers recognize six major climatic zones, each associated with a letter of the alphabet. These zones include (A) **tropical**, (B) **arid** and **semiarid**, (C) **mesothermal**, (D) **microthermal**, (E) **polar**, and (H) **highland** climates. All of these major climate zones are present in North America although some are larger in geographical extent than others. The locations of each of these zones as well as a few of the more localized anomalies that are present in some regions are discussed here. Because zones A, C, D, and E appear sequentially as our attention moves northward from the equator toward the North Pole, we discuss these four types first, and then consider arid, semiarid, and highland climates. Figure 2.11 shows the distribution of these climatic zones.

Tropical "A" Climates Tropical climates are those associated with tropical air masses throughout the year. In North America, tropical climates are limited to southern Florida and Hawai'i. Both of these areas are influenced by maritime tropical air masses bringing moisture-laden warm air into the region. However, because both areas are nearly surrounded by water, temperatures are not as hot in summer as is the case in the continental southeastern United States, and hence many A climate dwellers regard their weather as less oppressive.

Mesothermal "C" Climates Mesothermal climates are located primarily along the Atlantic and Pacific coasts of the continent. This climatic zone runs the entire length of the West Coast of North America from southeastern Alaska to southern California. East of the Rocky Mountains, mesothermal climates can be found in a region from the Atlantic to about the 98° W longitude (that is, as far west as Oklahoma City, Dallas, and San Antonio) and northward to about the 40th parallel of latitude (that is, as far north as Philadelphia, Indianapolis, and Kansas City).

There are several distinct types of mesothermal climates. The southeast region is classified as having a **humid subtropical climate**, while **mediterranean climate** and **marine west coast climates** dominate along the West Coast. The humid subtropics receive precipitation all year, with mild winters and hot summers. The moist unstable air mass brought in by the wind from the warm water source region of the southern Atlantic Ocean and Gulf of Mexico can produce

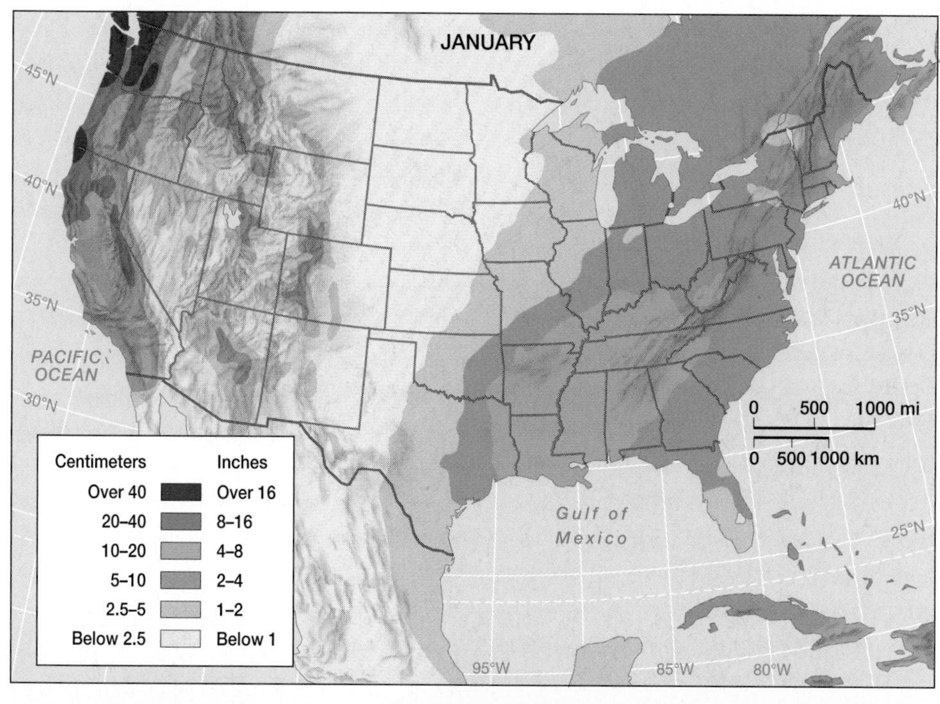

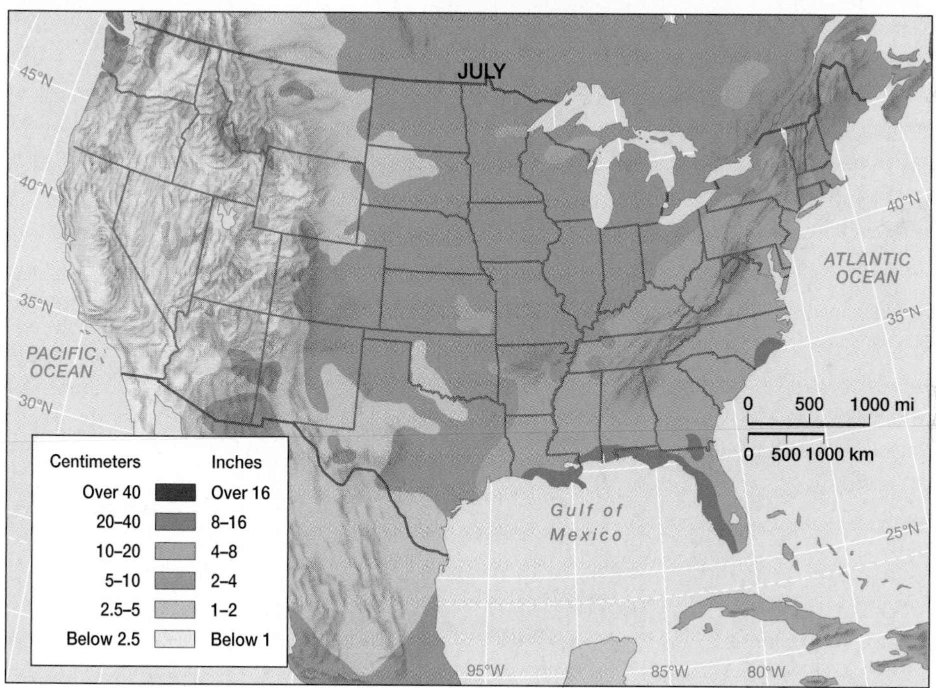

▲ **Figure 2.10** North American precipitation patterns in winter and summer

convectional rain showers over the area. Hurricanes can dump large quantities of rain in this area, many times causing severe flooding due to the already saturated soils as occurred in the devastating Katrina storm of fall 2005 on the Gulf Coast of the United States and in the earlier hurricane-flood of 1900 in Galveston, Texas. Other severe weather is generated from cyclonic storms or frontal activities produced from the clash of a continental polar air mass from the north and a maritime tropical air mass from the south. Most cities in the southeastern United States, including Houston, New Orleans, Memphis, Atlanta, and Charlotte, experience this humid subtropical climate. In portions of the Appalachian highlands, higher elevations result in lower summer temperatures. The Great Smoky Mountains National Park and nearby communities such as Asheville, North Carolina, and Gatlinburg, Tennessee, are popular with tourists wishing to escape

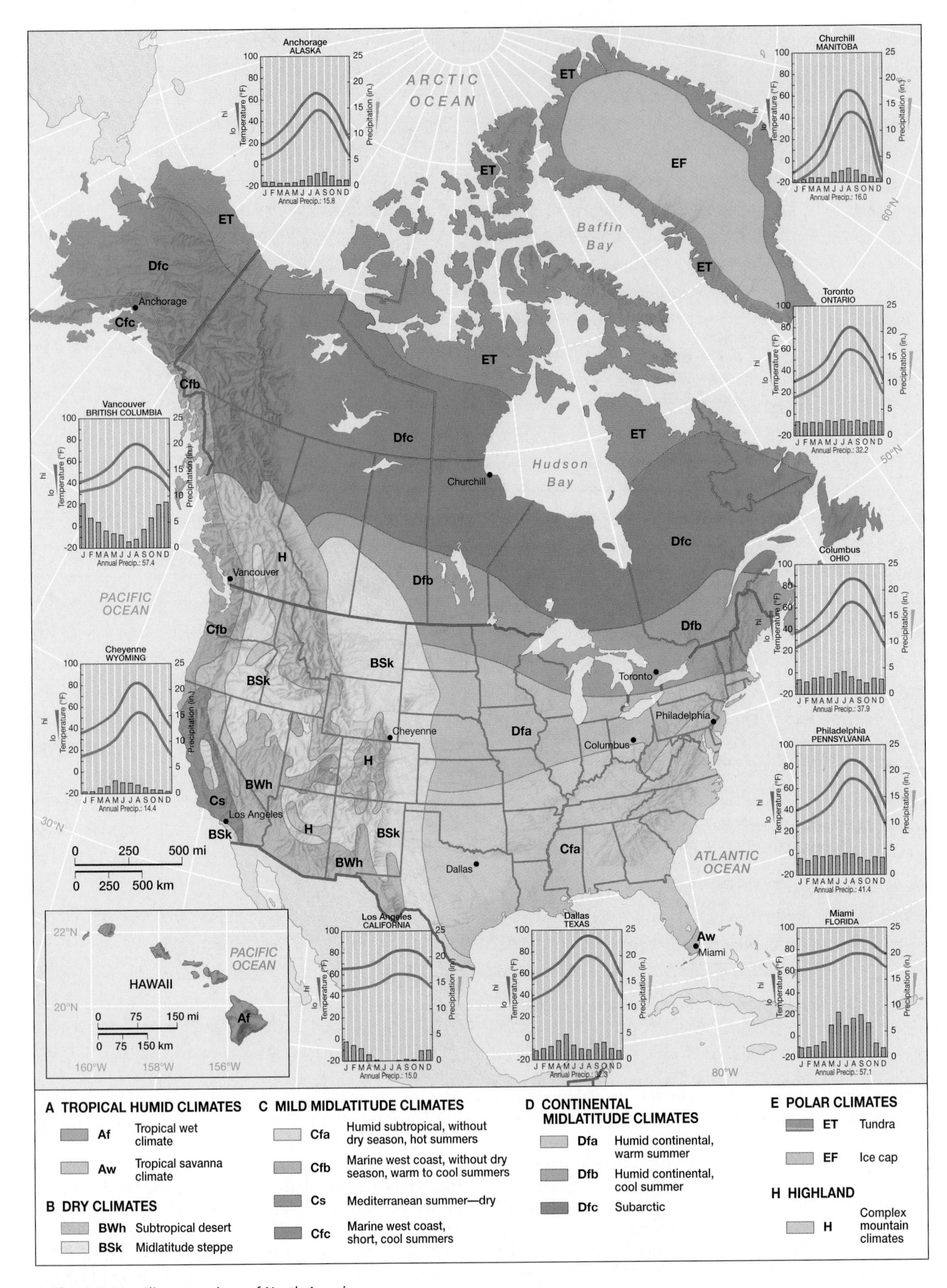

▲ **Figure 2.11** Climate regions of North America

from heat and humidity at lower elevations. Tornados, another form of midcontinent extreme weather event, are described in Chapter 11.

Most of coastal California, including the metropolitan areas of Los Angeles, San Francisco, Sacramento, and San Diego, has a mediterranean climate. Places with this type of mesothermal climate receive most of their annual precipitation during the winter months and experience a dry summer season, which is the opposite of most other areas of North America. This pattern of precipitation is due to subtropical high pressure blocking winds that would otherwise bring moisture to the area from the maritime polar air mass during the summer. In winter the subtropical high pressure shifts away from the coast and allows moisture-laden air to flow in from the Gulf of Alaska. Summer fogs often occur along the West Coast, in places with mediterranean climates. The term *mediterranean* refers to the fact that this climate is also characteristic of countries like Spain, Italy, and Greece, which border the Mediterranean Sea.

Farther north, in the Pacific Northwest, the maritime polar air masses dominate for longer periods of time over the course of the year, resulting in a marine west coast climate. This climate is associated with cooler summers, rainier winters, and more unpredictable weather patterns as compared to the humid subtropical climate and the mediterranean climate. Winter fogs occur frequently. They are created by the flow of relatively warmer moist air flowing over the moderating affect of very cold water. The marine west coast climate is associated with coastal areas of Oregon, Washington, British Columbia, and southeastern Alaska and includes cities such as Portland, Seattle, Vancouver, and Juneau.

Microthermal Climates Microthermal climates are found along the northern tier of the United States from the East Coast to the upper Midwest and throughout most of Canada. Microthermal climates are associated with longer, colder winters relative to mesothermal climates. Typically, average temperatures are below freezing for several months each winter. Microthermal climates include (1) **humid continental** (hot summer and cool summer climates), and (2) **subarctic climates**. The humid continental hot summer locations are influenced by the continental polar air mass but can also be affected by continental tropical and maritime air masses throughout the year. When the colder, drier air from the north clashes with the warmer, wetter air from the south violent storms can erupt, dumping large quantities of precipitation on the landscape, similar to what occurs in the Southeast. Cities in the central area of the United States such as Des Moines and Omaha have humid continental climates with hot summers and cold winters.

The higher latitude microthermal climates (humid continental mild summer) are characterized by a frost-free period of at least three months and less precipitation than the humid continental hot summer climates or the mesothermal climates. These microthermal climatic subzones receive a large quantity of precipitation annually, which is important for recharging soil moisture. In North America, the Great Lakes moderate winter cold and reduce summer heating in places like upstate New York and southern Ontario. Thus, while the "lake effect" snow in winter months may pose challenges for residents, moderate temperatures support vineyards for wine grapes in this part of North America. Much of New England and the Atlantic provinces of Canada also experience this type of climate.

The subarctic subregion of the microthermal climates lies poleward of the humid continental mild summer subregion. Subarctic climatic locations experience dry conditions compared to their lower latitude counterparts, cool summers with short growing seasons, and long, cold winters. This is also the **permafrost** zone, where soils are totally or partially frozen all year (Figure 2.12). The short growing season here often makes agriculture risky and unprofitable, so most areas with subarctic climates support very few people. Exceptions are communities whose economies are not related to agriculture, such as Schefferville, Quebec, and Thompson, Minnesota.

Polar Climates In the extreme northern portions of Alaska, Canada, and Greenland, as would be expected, **polar climates** are dominant. Locations that have polar climates are strongly influenced by the continental Arctic and continental polar air masses. They do not experience a true summer, and their monthly average temperatures do not rise above 50°F or 10°C. Because of the extreme northern location, the sun does not rise for the better portion of six months, producing continuous night during the winter. Snow covers the landscape for eight to ten months of the year, producing either permafrost or ground ice conditions. When the snow melts, tundra plants appear such as sedges, mosses, lichens, and some flowering plants. Few settlements dot the polar climate landscape. Polar regions are also low precipitation areas, less than 10 inches (25 centimeters) per year, and are technically deserts.

Arid and Semiarid Climates In the desert Southwest of the United States, **arid** and **semiarid** climates dominate. Subtropical high pressure systems bring subsiding air with low relative humidity into these areas. **Adiabatic** heating (the warming of air as it descends in elevation) also adds to the arid and semiarid conditions of this geographic region. As mentioned earlier in this chapter, **orographic** uplift (air that is forced up by blocking landforms) pushes moisture-laden air over the western mountains. As these parcels expand and cool, water vapor changes from a gas to a liquid, resulting in clouds and then precipitation. The air then descends on the leeward side of the

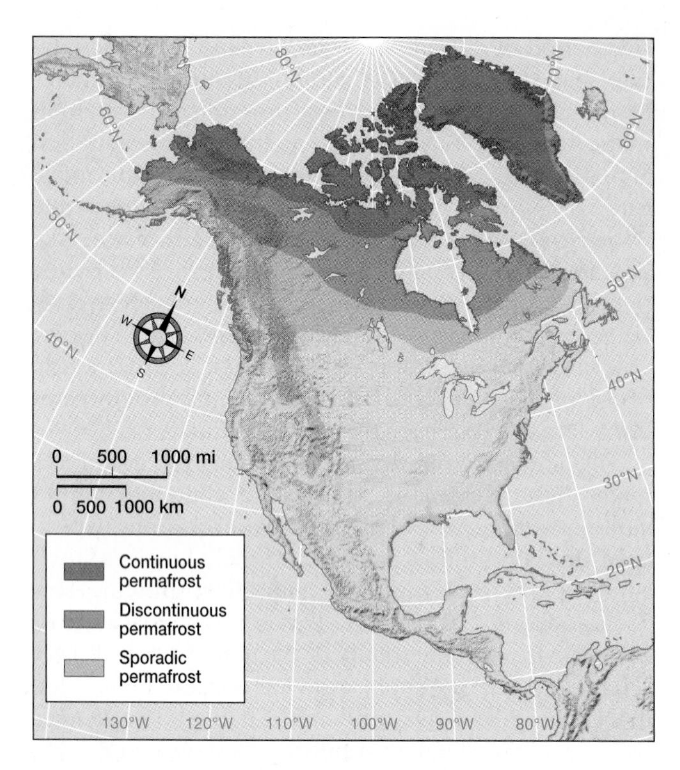

▲ **Figure 2.12** Permafrost zones in North America

mountains and is warmed in the process. The capacity of descending (warming) air for holding water vapor is increased so the land in the rain shadow is relatively dry. Such arid conditions are found in southeast California, along the southern and central portions of Arizona, in New Mexico, and along the southwest margin of Texas, spilling over into Mexico. The southern portions of Utah and Nevada also experience desertlike conditions. These arid climates are characteristic of cities such as Phoenix, Las Vegas, and El Paso.

At higher latitudes and elevations, the descending air is not as warm and dry, so semiarid conditions exist. Semiarid climates in North America include areas to the east of the Rocky Mountains through northern New Mexico and north and west Texas, as well as the western margin of the Great Plains. Cities such as Denver, Cheyenne, Great Falls, and Calgary have semiarid climates.

Highland Climates The **highland climate** realm is located in places where the presence of high mountains causes extensive local climatic variation. There are a few pockets of highland climates in North America, including a finger that extents from southeastern California and northwestern Arizona through western Nevada and eastern Oregon and Washington; a larger area in eastern British Columbia, mid- and southern Alberta, Idaho, western Montana, and northern Utah; and a pocket in Colorado. Resort communities such as Aspen, Vail, and Steamboat Springs, as well as many popular national parks in the United States and Canada, including Yellowstone, Glacier, Jasper, and Banff, are included in the highland climate region. These locations are unique because, although some are located at fairly low latitudes, their high elevations produce climates similar to those found at higher latitudes. Landscapes at these higher elevations mimic those of the polar climates, ranging from short grasses, sedges, and mosses to a constant cover of snow and/or ice.

North American Physiographic Regions

Areas with interrelated patterns of landforms, vegetation, soils, and hydrology are known as **physiographic regions**. Twelve major physiographic regions are recognized and shown as a result of these processes. As discussed in Chapter 1 (see Figure 1.9), the mapping of North America into specific regions is often based on this division of physiographic regions and provinces. You also can see in other chapters how mapped areas of indigenous peoples, economic activities, and land uses often are closely related to the physical geographic landscape.

The **Atlantic and Gulf Coastal plain** of the southeastern United States flanks the Atlantic Ocean from New York to Florida and the Gulf of Mexico from Florida to Texas. The coastal plain meets the Piedmont of the Appalachian Mountain system at the fall line. It is characterized by some of the flattest terrain on the continent, gently sloping toward the sea. In relatively recent geologic time, this area was submerged, and many wetland areas still provide habitat for waterfowl and other wildlife in **estuaries** (mouths of rivers mixing salt and fresh waters), swamps, marshes, and lagoons. These include estuaries such as Chesapeake and Delaware bays and well-known swamps such as the Great Dismal Swamp of Virginia and Okefenokee Swamp of Georgia. These landforms are typical of low-lying areas where large rivers and their corresponding tributaries meet the ocean. The coastal plains of the Southeast include the mouth of the Mississippi River, which deposits vast amounts of sediment into the Gulf of Mexico. As the velocity of the river slows due to the low slope of this area, a heavy sediment load is deposited in Louisiana and forms the Mississippi Delta. Some would say that the delta includes areas as far north as Arkansas. Recent floods in this area are reminders of the importance of land rejuvenation through sedimentation. While on the subject of flooding, it is significant that part of the coastal plain includes the offshore islands (fittingly called **barrier islands**) that protect coastal shipping and shoreline land development from Atlantic storms. The importance of this physiographic province to current human activities on the land is noted in Chapters 5, 7, and 10 of this text.

The **Appalachian Mountain System** is a geologically old system of parallel mountains and valleys trending southwest to northeast along the eastern portion of the North American continent. Its easternmost province is the Piedmont (literally, the "foot of the mountains"). The **piedmont** consists of low, rolling hills with moderate relief. As mentioned earlier, it is separated from the coastal plain by the **fall line**. This boundary more or less terminates at Columbus, Georgia, although some would argue that it reaches Montgomery, Alabama. The Piedmont was important to colonial cotton and tobacco farming, but places where these crops are grown have shifted through time as Piedmont soils were overused.

West of the Piedmont, the Appalachian region is the result of mountain-building processes that extend from Newfoundland to central Alabama. Rocks in the Appalachian Mountain System are crystalline in the north, sandstone in the middle, and sedimentary from Pennsylvania to the south. These rocks have weathered differently and offer very different landscapes from north to south. The highest region is the Blue Ridge (4000 to 6000 feet) and marks the drainage divide between easterly flowing streams that flow into the Atlantic and westerly flowing streams that eventually join the Mississippi River drainage system. It was labeled The Proclamation Line by colonial authorities and marked the beginning of Indian-owned lands to the west.

The sedimentary geology of the Ridge and Valley Province, including the Allegheny and Cumberland plateaus, defined an isolated and practically self-sufficient region until the 1930s. Coal mining was, and still is, a mainstay of the economy.

The Northern Appalachians extend from New York to Quebec and Newfoundland. Their glaciated metamorphic rocks, many lakes, and rugged topography have limited historical settlement and today provide popular outdoor recreational opportunities.

West of the southern Appalachians, and physically very similar to them, are the interior uplands. This region's Ozark section is located in southern Missouri and northern Arkansas and the Ouachita section in western Arkansas and eastern Oklahoma.

The **central lowlands** (sometimes referred to as the interior plains) are located due west of the Appalachian Mountain System and north of the coastal plains of the Southeast. Several different sections comprise this now greatly industrialized and urbanized physiographic province. The first is the Saint Lawrence Valley and transition zone between the Appalachian Plateau and the Mississippi River Valley. This area consists of gently folded and westward dipping rock strata of sandstone and limestone. It is subject to karst processes (formation of caves and sinkholes) because of the interactions of plentiful water with the limestone. It is also famous for its lime-rich soils, which create good pasture for raising livestock such as the famous racehorse farms of Kentucky Blue Grass country.

The second main subregion is the Mississippi–Great Lakes section. The flatness of this area originated from glaciation and is emphasized by a glacial drift soil blanket. This is also an area of extensive loess bluffs as sediment was deposited here by wind from places west. The Balcones Escarpment in Texas marks the southwestern boundary of this subregion and the southern boundary of the third main subregion, the Great Plains. The Great Plains is considered the Breadbasket of the continent, with rich grassland soils overlaying its geologic base. Wind and water are the main agents that created the gently rolling landscape that characterizes the region. Along the region's western border water and wind erosion is exemplified by the South Dakota Badlands and the Sand Hills of Nebraska. The Great Plains' geographic history in the United States is related to the use and misuse of the land, including cattle drives, railroad promotion schemes, and farming operations. The drought and dust storms of the 1930s and current overdependence on groundwater exemplify continuing human–environment issues. The final subregion of the Central Lowlands is the Interior Uplands or Ozark-Ouachita Section. This section is similar in appearance to the Appalachian Mountain System but the overall elevation and relief is less.

The **Rocky Mountains** or **Cordilleran Province** was formed by uplift, folding, and faulting of the rock strata. Two parallel mountain ranges formed, generally trending north to south. The Rocky Mountains, which extend from New Mexico to the Laird River in Canada, comprise most of this physiographic region. Mining, ranching, and recreation are significant economic activities here. In the far north of the province the Brooks Range extends the mountainous cordilleran terrain into Alaska. This province has also experienced volcanism as part of its geologic history. Lava flows have also been mapped within the Rocky Mountain region in New Mexico, as well as in Yellowstone National Park in Wyoming. The peaks of 13,000 to 14,000 feet (4000 to 4300 meters) in the Rockies (Figure 2.13) identify the continental divide where water falling on the east may flow into the Gulf of Mexico and water flowing off western slopes may find its way to the Great Basin, the Gulf of California via the Colorado River or, more likely, may be diverted to urban or agricultural uses in one of 11 western states. This area, much of it publicly owned, is more fully described in Chapter 12.

The **Intermontane West** is located between the Rocky Mountains and the Pacific Coast mountain ranges. This region has been shaped by fluvial and aeolian processes, thereby providing spectacular scenery but often difficult agricultural potential. Several subregions are important within this physiographic province. The Colorado Plateau has been uplifted while the Colorado River and its tributaries have

Figure 2.13 Rocky Mountains above the tree line have conditions similar to polar conditions

greatly incised the landscape to create steep canyons and various mesas and buttes (Figure 2.14). Wind erosion and volcanism have also helped shape this southwestern landscape. The material once covering these volcanic landforms eroded away leaving the more weather-resistant material exposed.

Adjacent to the Colorado Plateau is the basin-and-range area, which also includes the Great Basin and Death Valley. This distinct area is characterized by short, rugged mountain ranges intermingled with flat valleys and without drainage to the sea. Much of the annual rainfall in this region evaporates quickly, thereby creating the extremely dry conditions that exemplify the desert Southwest. In many places the landscape is barren of most vegetation and is subjected to extensive erosion by wind. The northernmost section of the Intermontane West in the United States centers on the plateaus and basins of the Snake River in southern Idaho and the Columbia River of eastern Washington. This landscape also shows evidence of the giant glacial Lake Missoula flood that occurred 15,000 years ago as it carved canyons and scoured soils away. These rivers have greatly incised today's volcanic landscape, creating deep canyons

Figure 2.14 Mesas and buttes in southwestern Utah

and dramatic waterfalls. The northernmost region of the Intermontane West lies across the Alaska–Canada border, extending throughout the central portions of Alaska and southward through the Yukon and British Columbia. Portions of this subregion are flat and others portions consist of dissected plateaus similar to its more southerly counterparts. This region also boasts mountainous terrain similar to the Rocky Mountains and the Pacific Coast mountain ranges; however, the Canadian mountains are less extreme. The role of physical geography in influencing the settlement and use of land in the Intermountane West is discussed in Chapter 12.

The westernmost **Pacific Coastland** physiographic province extends the length of the west coast of North America from southern California to western Alaska. Although it appears as one long province on a map, the landscape is varied due to the various geomorphic processes that have shaped it. All along the shoreline coastal processes are at work. Steep cliffs have been cut by wave action. Beaches are reduced or recharged by the swash of waves. In more northerly coastal locations bays, fjords, islands, and peninsulas have been created by these coastal processes.

Faulting is characteristic of much of southern California. The San Andreas Fault, along with the numerous smaller faults in this area, is a constant hazard to the residents of central and southern California. The Cascade Range, the Coast Ranges, the Sierra Nevada, and the Olympic Mountains provide evidence of volcanism and widespread uplift. This area is also marked with several structural depressions, including the Willamette Valley in Oregon, Puget Sound in Washington, and the Central Valley of California. Throughout history these depressions have experienced fluvial processes, mudflow, and deposits of alluvia. Some of these areas are wet and even submerged such as Puget Sound, while others are seasonally dry, fertile, and used for very successful agricultural production. The relationships between physical geography and the human occupance of the Pacific Coast region is discussed in Chapters 15 and 16.

The **Yukon Basin** and **Northwestern Highlands** occupy most of central Alaska and the southern part of the Yukon Territory. Land here is hilly and mountainous with many rivers.

The **Canadian Shield** covers more than half of Canada, all of Greenland, and includes the far northern parts of Minnesota and far upstate New York. This area is a wide expanse of ancient rock, older than a billion years, that has been greatly compressed by ice and then contorted creating a rugged landscape. It has valuable minerals such as iron ore, silver, nickel, and gold. Mining of these minerals is a major economic activity in this region. There are numerous lakes, which are connected by a network of streams sometimes called deranged drainage. The southern portion of the Canadian Shield is covered with slow-growing boreal forest vegetation, which gives way to bogs and tundra vegetation in more northern locations. Soils are mostly organic, such as **Histosols**. Land use is based on extractive or primary industries such as fishing, forestry, and mining. Settlements are small and distant from one another. North and surrounded by the Canadian Shield are the **Hudson Bay Lowland** and the **High Arctic Mountains**. Both are largely unpopulated. More about these physiographic provinces, as well as **Greenland**, is found in Chapter 18 on the Far North.

The Hawaiian Islands have a unique physical geographic history and are discussed in Chapter 17.

Biogeography and Ecology

Because vegetation is a mirror of climate, hydrology and soil types, a map of natural or native vegetation (Figure 2.15) can be compared and analyzed with the others in this chapter. Plant cover, or lack of it, was important to pioneer settlers. Forests were both a blessing and a burden in that they could be used as a source of building materials but had to be cleared for farming. Other forms of land cover affected land use such as grasses for livestock grazing and for wildlife habitat. Natural vegetation was often a key indicator of what kinds of soils were beneath and what kinds of climatic cycles to expect. The natural vegetation map in this chapter (Figure 2.15) may be compared to the soils and climate maps (Figures 2.6 and 2.11) to visualize some of these relationships. Annual rainfall totals are especially significant. Rainfall availability determines whether an area is desert, grassland, or forest. Generally an area of 10 inches (25 centimeters) or less annual rainfall will support only the most specialized drought-resistant plants. The relationship between precipitation patterns and natural vegetation distributions can be seen by comparing the precipitation map (Figure 2.10) with the vegetation map (Figure 2.15). This is true regardless of temperature, soil characteristics, or topography. Grasslands are found in regions that experience 10–30 inches average precipitation. In these areas rainfall is irregular, fires may be frequent, and grazing animals may inhibit tree seedlings. Temperate grasslands tend to have limited and irregular rainfall and a large seasonal temperature range (warm summers and cold winters). Most are located in the centers of land masses away from the moderating influence of oceans. They usually are very windy places; the American Midwest and Canadian Prairie Provinces are typical. Many temperate grasslands historically supported large herds of grazing animals such as the bison in North America. Soils in the temperate grasslands are

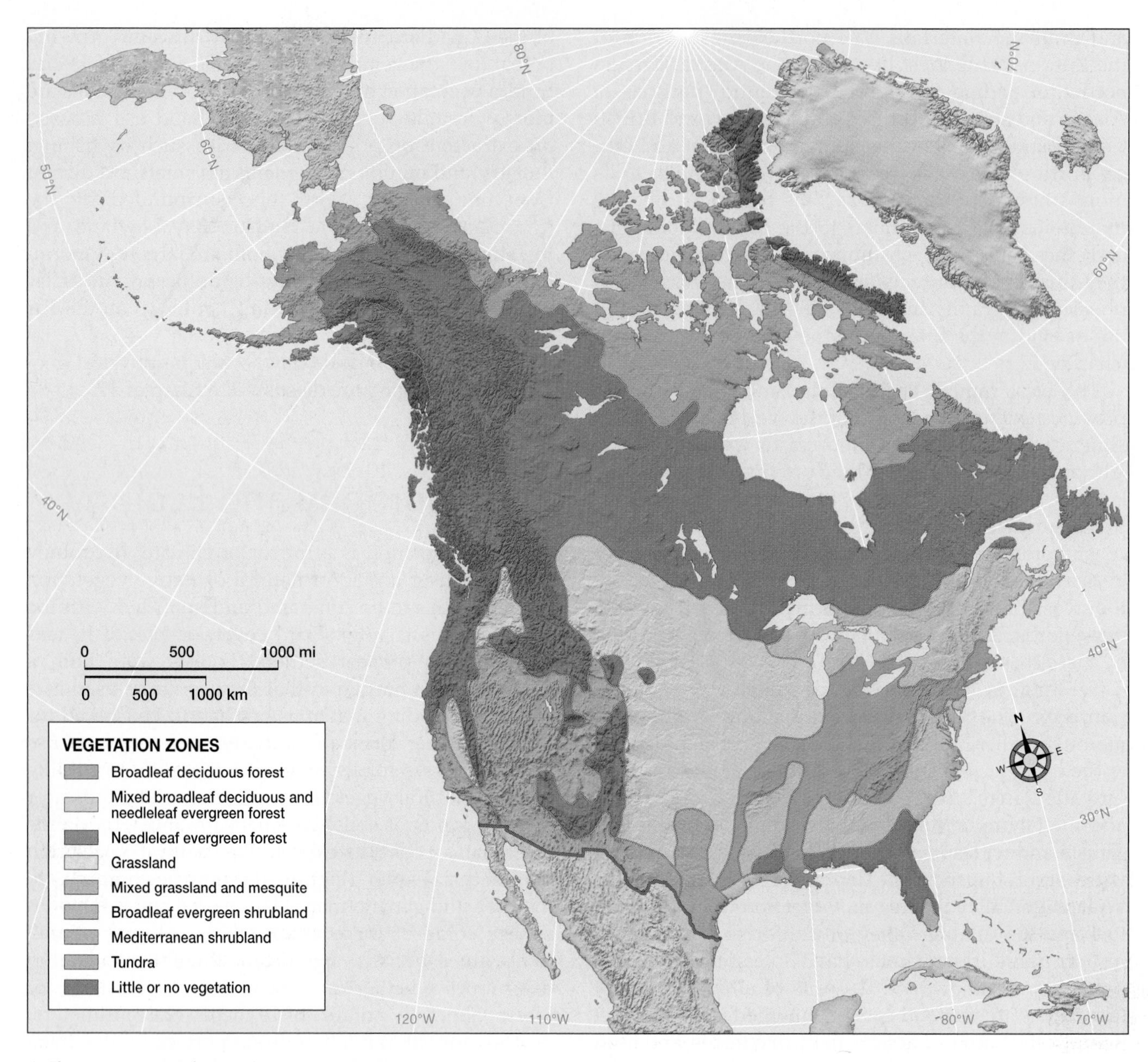

▲ **Figure 2.15** North American vegetation zones

the best on Earth for field agriculture and produce a major portion of the world's wheat, maize (corn), livestock, and vegetables. The deep topsoils are very fertile due to their large humus content. Nutrients are stored in the soil rather than in the living biomass (as in forest ecosystems).

Today, less than 1 percent of the original midcontinent grasslands in North America remains in its original state, having been heavily used because of high soil fertility, value for grazing livestock, and improved irrigation technology. Forests occur in undisturbed areas where rainfall patterns are regular and average over 30 inches (75 centimeters) per year. The primary requisite for tree cover is adequate year-round rainfall. As in the cases of other biomes, the number, range, and specific characteristics of species are influenced by soil, topography, temperature, and human influences. There are tropical rain forests in Hawai'i and boreal forests in Alaska. You may see from examining the vegetation map that deserts, grasslands, and forests occur naturally at most latitudes. Midlatitude temperate deserts such as are found in North America are generally hot in the summer and have cold winters. They support a variety of succulents, shrubs, and seasonal wildflowers. American deserts are some distance from water bodies. Desert soils almost always are nutrient deficient because the growth rates of plants are slow and there is little biomass to decompose into humus. Neither desert nor forest, the mediterranean scrub biome, found particularly in California and southern Oregon, is a greater reflection of climatic influence on vegetation than the other biomes. Its deep-rooted, small-leaved, perennial shrubs mixed with scattered, savannalike woodland is particularly adapted to summer drought and mild winters. With application of summer water this environment becomes highly productive for agriculture because many of its soils are deep and fertile. Warm, rainier summers have resulted in these regions becoming the focus of tourism and resort industries.

As a summary for this chapter, six broad **biomes** are presented here. They include **forest**, **tundra**, **grassland**, **scrubland**, **desert**, and **subtropical wetland**. As with the other large area maps in this chapter, local conditions may be very different than the patterns shown at macro scale. Many biomes blend across climate and soil type boundaries but their general shapes and sizes greatly overlap each other. The major biomes are shown in Figure 2.16.

Forests (Regions 1 through 5 and 19 through 21) The various forest types of North America include tropical and temperate rain forests, broadleaf deciduous forest, mixed broadleaf deciduous and evergreen needleleaf forests, and coniferous forests. The only location in our definition of North America that possesses a tropical rain forest is Hawai'i (see Chapter 17). Its warm, moist climate is conducive to the formation of Oxisols and the growth of tropical vegetation. At higher elevations, broadleaf evergreen forests are also present.

Temperate rain forests correspond with the marine west coast climate and the various soil orders (Oxisols, Ultisols, and Spodosols) associated with this climate. They are therefore found in the western portions of the Pacific Northwest of the United States and along the western margin of Canada. These forests support a lush mix of broadleaf and needleleaf trees. However, fewer tree species are found here relative to their tropical counterparts. These rain forests are unique not only because they are found at higher latitudes but also because they receive a large amount of moisture from summer fog and the maritime polar air mass. North America's temperate rain forests house the tallest trees found on Earth, coastal redwoods (*Sequoia sempervirens*). They also contain commercially valuable species such as Douglas fir, spruce, hemlock, and cedar. Few areas of this type of forest are native "old" growth; most are secondary growth forests, meaning that in the past these forests have been cut down and are either replanted or left to regenerate naturally. Poor timber management plans have plagued these areas in the past, but better management practices in recent years have been put in place in an attempt to achieve sustainable levels of production.

Vegetation in humid continental locations is typically a mix of broadleaf deciduous (oak, hickory, beech, maple) and needleleaf evergreen (fir, pine, spruce) forests. These mixed broadleaf deciduous and needleleaf forests are supported by areas that experience warm to hot summers and cool to cold winters. They are underlain by Ultisols, Alfisols, Inceptisols, and Spodosols. Along the Gulf of Mexico, such forests are comprised mainly of broadleaf tree species. In the southeast, pine species such as loblolly, longleaf, and pitch pine are predominant. White and red pines and hemlock abound in the forests of New England and through the Great Lakes region.

Continuing northward into cooler and drier regions, deciduous and mixed forests are replaced by needleleaf evergreen forest species. These forests are also termed boreal forests or taigas. This is perhaps the largest biome in North America area in terms of aerial extent. It extends over most of Canada, all of the Upper Peninsula of Michigan, and northern Wisconsin and Minnesota. Permafrost conditions exist in many boreal locations, which indicates that they are underlain by poorly drained soils such as Histosols and Gellisols. Spodosols can also be found under the boreal forest areas in the southern extents of this biome. Many boreal forests are beginning to show signs of global warming influence. The warmer temperatures are thawing more of the active layer of soil, causing waterlogging of soils, which these tree species cannot tolerate. The ultimate result is a dying off of these trees in response to the excess of water in many parts of the Far North.

Tundra (Regions 12 through 17) The tundra biome is found in the highest latitudes of North America that can sustain vegetation (see Chapter 18). In these areas the soils are poorly drained and exhibit a thinner permafrost layer than the soils of needleleaf forests. This biome consists of vegetation that can endure cold winters, low amounts of heating, and little sunlight with a short growing season. This biome includes grasses, sedges, lichens, and some low shrubs. Tundra locations are also excellent breeding habitats for waterfowl such as geese and swans, and grazing areas for mountain goats and bighorn sheep. Areas along the Arctic coast of Alaska and Canada contain this tundra vegetation.

Grasslands (Region 18) Another North American biome is the grassland, located in the Great Plains region of the continent (see Chapter 11). The 98th to the 100th meridian delineates the short-grass from the tall-grass prairie. This demarcation is caused by a difference in precipitation, with areas to the east experiencing more rainfall than evaporation. To the west, potential evaporation exceeds precipitation. In Chapter 11, we discuss major consequences in historical settlement of the Great Plains that are related to these phenomena. Naturally occurring trees found in the grassland biome are generally restricted to stream and river corridors. This biome is the most modified by humans because it is the area of greatest crop and livestock agricultural production. The Mollisol soils provide a rich layer of topsoil and is extremely fertile and conducive to large farms. People have also planted and cultivated large numbers of trees in cities and

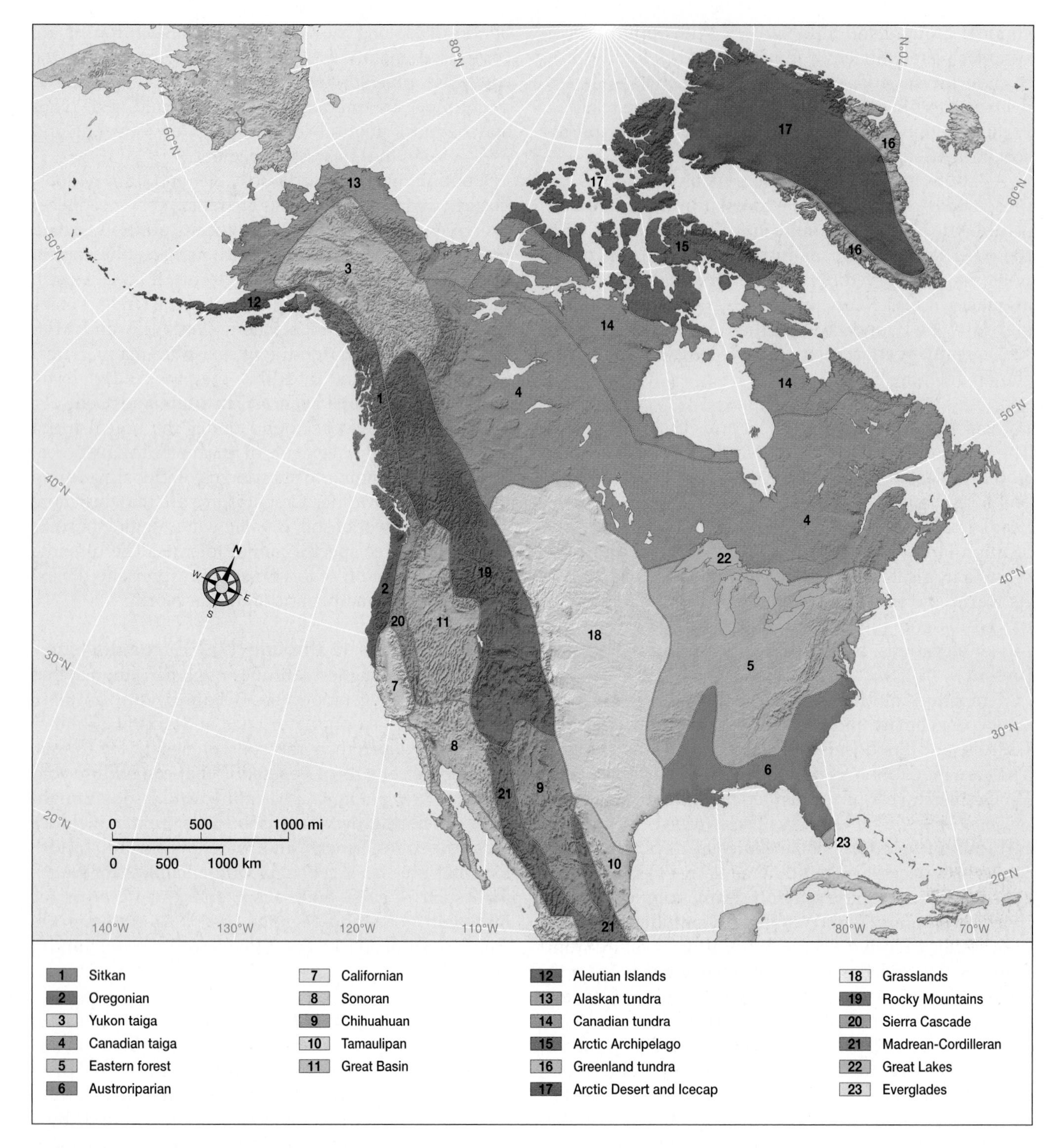

▲ **Figure 2.16** Bioregions of North America

towns and near farmsteads in order to provide shade, windbreaks, and for aesthetic reasons.

Deserts (Regions 8, 9, and 11) West of the grassland biome is the driest, least vegetated biome of the continent. In this region, annual rainfall totals are less than 10 inches (25 centimeters). This low amount of precipitation coupled with great potential **evapotranspiration** defines these dry areas as a desert biome. Vegetation that survives in this landscape must not only tolerate very little water on average but also must be rooted well enough to withstand flash flooding. This biome is increasing in areal extent as desertification takes place, due mainly to native vegetation removal, intensive agricultural practices, and poor soil moisture management, which can lead to increased erosion and **salinization**.

Mediterranean Scrub (Region 7) Moving westward, the desert biome gives way to the mediterranean scrubland of western California and southern Oregon (see

Chapter 15). This area corresponds with the mediterranean climate and lies north of the shifting subtropical high pressure cells. Because of the dry conditions that exist, fire is a constant possibility and, in fact, was historically part of the natural ecosystem. Vegetation in these fire-prone areas, called chaparral in California, is well adjusted to the hazard because it has deep root systems and the ability to re-sprout roots after a fire episode. Chaparral typically includes blue and live oak, toyon, redbud, manzanita, and many other species. The climate and soils of this region allow for subtropical fruits, vegetables, and nuts to be grown. European wine grapes are particularly profitable. Some other crops include citrus fruits, olives, avocados, artichokes, and almonds, most of which are grown only in this biome in North America as well as other parts of the world.

Subtropical Wetland (Region 12) This comparatively tiny region is critically important to the survival of many plant and animal species and consists mostly of protected marshes and mangrove zones.

Ecosystems and Watersheds

One difficulty with mapping physical geographic information lies in finding a way to combine all the components of a natural landscape into biotic communities or ecoregions that can be compared with maps of human occupancy and activity. Figure 2.17 shows 76 different **ecoregions** that have been identified by "overlapping" the distributions of precipitation, temperature, elevation, hydrology, geology, soils, vegetation, and human impact into one map. Another difficulty with mapping biogeographic information lies in selecting an appropriate scale that will be compatible with other topics of analysis, such as mapping of cultural or economic regions. An intermediate level between the very generalized descriptions of global biomes and the mapping of local, site-specific biotic communities is the ecoregion. The U.S. Environmental Protection Agency and other federal agencies use four levels of detail depending upon their needs. Figure 2.17 shows North America at level II. Level I is more generalized and level III is more regionally specific.

Estimates of the degree of correspondence or overlap of the regions provides some insight into what regional factors may have influenced human decisions regarding settlement patterns, land use, or cultural imprints. **Watersheds** are physical attributes of Earth's surface defined by common drainage systems. However, they can also be used by resource planners and managers because they are usually inhabited by people with common interests. There are almost an infinite number of watersheds in North America, but many of the major watershed systems have been identified for flood control, habitat restoration, and other conservation-related projects.

Conclusions

The physical geography of North America described here at a general scale were what indigenous peoples adapted to, based on their culture and available technology. As people from other places arrived in North America, particularly through the last four centuries, they either adapted to the physical setting or they began to significantly change it. In the 21st century, few populated areas on the continent looked anything like the 16th century landscape. This historical process of landscape modification sets the stage for the story about how this continent was settled by a remarkable mix of different people in its post-indigenous history. The rest of this book tells that story and describes their settlement patterns, cultures, and political–economic systems.

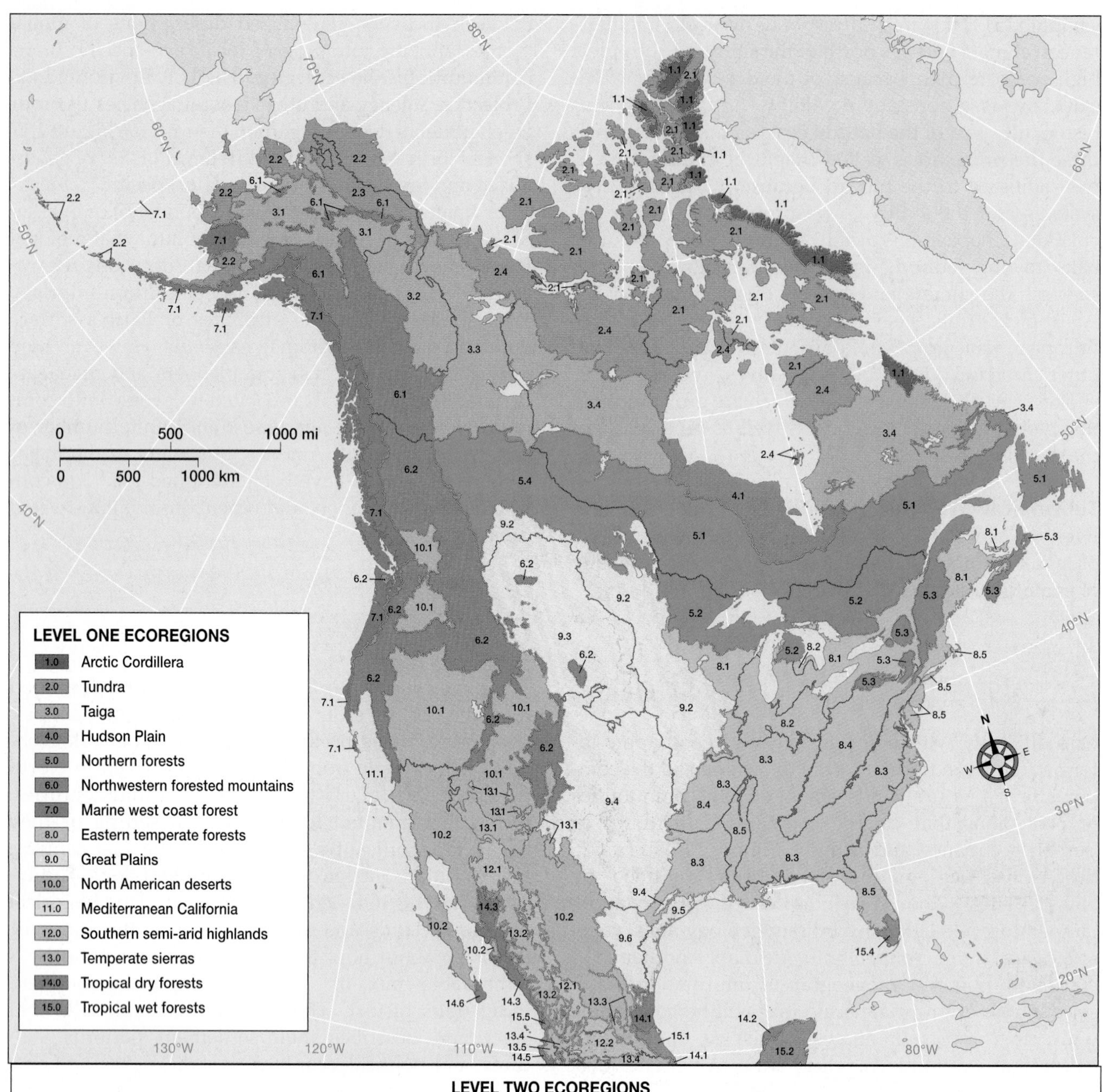

LEVEL TWO ECOREGIONS

1.1 Arctic Cordillera
2.1 Northern Arctic
2.3 Alaskan tundra
2.4 Brooks Range tundra
3.1 Alaskan boreal interior
3.2 Taiga cordillera
3.3 Taiga plain
3.4 Taiga shield
4.1 Hudson Plain
5.1 Softwood shield
5.2 Mixed wood shield
5.3 Atlantic Highlands
5.4 Boreal plain
6.1 Boreal cordillera
6.2 Western cordillera
7.1 Marine west coast forest
8.1 Mixed wood plains
8.2 Central USA plains
8.3 Southeastern USA plains
8.4 Ozark, Ouachita-Appalachian forests
8.5 Mississippi alluvial and southeast USA coastal plains
9.2 Temperate prairies
9.3 West-central semi-arid prairies
9.4 South-central semi-arid prairies
9.5 Texas-Louisiana coastal plain
9.6 Tamaulipas-Texas semi-arid plain
10.1 Cold deserts
10.2 Warm deserts
11.1 Mediterranean California
12.1 Western Sierra Madre piedmont
12.2 Mexican high plateau
13.1 Upper Gila mountains
13.2 Western Sierra Madre
13.3 Eastern Sierra Madre
13.4 Transversal neo-volcanic system
13.5 Southern Sierra Madre
13.6 Central American Sierra Madre and Chiapas highlands
14.1 Dry Gulf of Mexico coastal plains and hills
14.2 Northwestern plain of the Yucatan Peninsula
14.3 Western pacific coastal plain, hills and canyons
14.4 Interior depressions
14.5 Southern pacific coastal plain and hills
14.6 Sierra and plains of El Cabo
15.1 Humid Gulf of Mexico coastal plains and hills
15.2 Plain and hills of the Yucatan Peninsula
15.3 Sierra Los Tuxtlas
15.4 Everglades
15.5 Western pacific plain and hills
15.6 Coastal plain and hills of Soconusco

▲ **Figure 2.17** Ecoregions of North America

Suggested Readings and Web Sites

Books and Articles

Bailey, Robert G. 1995. *Descriptions of the EcoRegions of the United States,* 2nd ed. Washington, D.C.: USDA, Forest Service. As the title suggests, this small but fact-filled book includes a large reference map and details how all of the ecosystems in the country are classified.

Christopherson Robert W. 2003. *Geosystems: an Introduction to Physical Geography.* Englewood Cliffs, NJ: Prentice Hall. The best selling and most attractive textbook about physical geography based on a systems approach.

Diamond, Jared. 1997. *Guns, Germs, and Steel: The Fates of Human Society.* New York: W.W. Norton and Company. A Pulitzer Prize winning book that integrates the human and environmental history of Earth in an engrossing style.

Marsh, William M. 2005. *Landscape Planning: Environmental Implications,* 4th ed. New York: John Wiley & Sons. A "how to" manual on dealing with various environmental conditions in land use planning and development in North America.

Orme, Anthony R. ed. 2002. *The Physical Geography of North America.* New York: Oxford University Press. This is a set of 25 articles by well-respected authors that presents a detailed and current statement of knowledge about the systematic framework and regional environments of North America.

Ricketts, Taylor H. et. al. 1999. *Terrestrial Ecoregions of North America: A Conservation Assessment.* Covelo, CA: Island Press. The authors believe that by conducting an ecoregion-based assessment of biodiversity, the process of conservation planning and action will speed up and be more effective.

Vale, Thomas R. 2005. *The American Wilderness: Reflections on Nature Protection in the United States.* Charlottesville and London: University of Virginia Press. This book examines the various meanings we attribute to nature as expressed through protected landscapes in American at scales ranging from the wooded corners of city parks to vast wildernesses like Yosemite, the Everglades, and Okeefenokee.

Web Sites

Physical Geography Introduction/Overview
http://www.physicalgeography.net/home.html

National Oceanic & Atmospheric Administration (U.S. Department of Commerce)
http://www.noaa.gov/

Hurricane Information
http://hurricanes.noaa.gov/
http://cimss.ssec.wisc.edu/tropic/

U.S. Global Climate Change Research Program
http://www.usgcrp.gov/

Ecoregions of the United States
http://www.nationalgeographic.com/wildworld/terrestrial.html

3
Historical Settlement of North America

PREVIEW

- Today's North American landscape is the result of past cultural, economic and political decisions.
- Prehistoric people occupied most of North and South America and had considerable impact on the physical landscape.
- Explorers from outside North America brought their perceptions and beliefs from home to North America with varying degrees of success.
- Contact between native and incoming peoples was the beginning of the globalization process that continues today.
- Massive immigration to North America and internal migration have shaped modern society, cultural diversity, and political systems nationally, regionally, and locally.
- The immigration process has been shown to be a mixture of personal and group decision making involving cultural, economic, and political factors.
- Various groups who occupy a place over time usually leave a physical imprint and a cultural and institutional legacy.
- Cultural imprints on the land can be observed in survey systems, place names, buildings, other built features, and in the lives of the current occupants of a place.
- Modern metropolitan living is based on population movements, technological changes in transportation, resource use, and labor utilization.
- Differences in North America's 14 regions are attributed to various circumstances of its physical, cultural, economic and political geography.

Introduction

To some, it was a wild and forbidding land. Dense impenetrable forests and wilderness landscapes must have seemed formidable and sometimes quite scary to some of the outsider Europeans and others when they saw North America for the first time. Others reported that it was a paradise overflowing with wealth and promise. These contrary images helped create myths about pre-Columbian North America that many still hold today. The perception that it was a trackless and unsettled wilderness ripe for the taking by Euro-American frontiersmen remains. However, in reality, the continent was inhabited for at least 25,000 years prior to European contact with a diverse and sometimes quite dense population of aboriginal people already here.

Considering these mixed messages about its livability, and the large numbers of what would later be called Native Americans in the United States and First Nations peoples in Canada, why would thousands of immigrants from far away decide to relocate to such an uncertain place? And after they arrived, how were the original North American native peoples affected by the emerging aboriginal-Euro-American contact zone? How were today's local and regional economies and cultures shaped and changed by the 400-year influx of outsiders from Europe and elsewhere in today's Canada and United States? This chapter provides an introduction to these and other questions to give a historical geographic context for the information on each North American region presented in later chapters.

Early Perceptions

Geographer Richard Hakluyt (1552–1616) and statesmen Francis Bacon (1561–1626) wrote and spoke passionately about the importance of "planting" a new society in North America. Their views and others who followed them helped shape the common perception that this new continent was free and available for settlement by outsiders—a place with resources and open space just waiting to be conquered and "civilized" by Euro-Americans. In this view, planting takes place naturally as new groups arrive and expand the extent of land under their control. This deeply ingrained belief in **manifest destiny**, the right of the European (and later the American and Canadian governments) to control newly occupied land to the Pacific Ocean and beyond, was a manifestation of this perception as well.

Today it is essential that we rethink our understanding of some of the deeper reasons why the settlement of the continent happened the way it did. It is essential

◀ Founded in 1565, St. Augustine is the oldest continuously occupied settlement of European origin in the United States. Only the venerable Castillo de San Marcos, completed in 1695, survived the destruction of the city by invading British forces in 1702.

to include the experiences of Native Americans, First Nations, and other ethnic and racial groups, as well as women's settlement stories, in our understanding of the evolution of North American places and cultures. This discussion of national culture (if indeed there *is* a national culture) in Canada and the United States is ongoing. Some suggest that it was the rapid expansion of early English cultural systems in the United States and English and French culture in Canada that most influenced subsequent landscape tastes and the evolution of particular cultural landscapes. Whoever settled a place first usually had the greatest impact on shaping the landscape. Geographer Wilbur Zelinsky called this the theory of **first effective settlement**.

It may also be suggested that it was the strong individualistic attitudes of early immigrants that most affected the creation and evolution of new cultural landscapes in North America. In this view, settlers (no matter what their place of origin), were motivated by values centered on owning one's own land, occupying new places, and controlling their society through strong governmental structures. Most were also firm believers in individual and family rights (except, many thought, these rights did not apply to non-whites, women, and children).

Recognizable local cultures began to evolve all across North America as new immigrants survived their first critical years of contact with a new physical environment and with the cultures of aboriginal peoples. This happened through a combination of selection and maintenance of certain cultural traits from overseas; newcomers interacting with some aspects of their new physical environments; and the spatial juxtaposition of ethnic and racial groups who had been widely separated in Europe, Africa, and elsewhere. These processes led to cultural interchange and forging of unique national and local identities.

Despite distance and technology challenges at least up to the 20th century, North America was never really isolated from or uninfluenced by the cultural evolution of the rest of the world. The connections between immigrants and their homelands were never completely severed and some return migration took place. It is important to understand that the exploratory, colonial, and independent phases of nation building in North America were part of a global picture.

Exploration, Discovery, Settlement, Exploitation

Native American Settlement Patterns and Cultural Imprints

Small groups of hunters and explorers from Asia were the first to inhabit the North American continent. They either crossed the Bering Land Bridge during the interglacial period following the most recent Ice Age (when land connecting Asia and North America was exposed) or, as more recent scholars suggest, they traveled in small boats from Eurasia to both the Atlantic and Pacific coasts of North America. These groups then continued their migration along the coasts and into the interior, moving south in search of game and other resources (Figure 3.1). This migration happened, by some estimates, more than 25,000 years ago.

It has been customary in too many textbooks to treat aboriginal peoples as one homogeneous cultural group. The reality is that beyond a general shared outlook, the first residents of North America were more diverse than many of the immigrants who arrived centuries later. These early people are believed to be a large factor in the extinction of many Pleistocene animals such as the mastodon, dire wolves, saber tooth cats, and armored rhinos. With fire, they were able to change vegetation patterns and commence a long process of environmental change that continues today.

By the time European explorers arrived on the scene in the 15th through the 19th century, explorers' and trappers' journals and travel accounts reported that indigenous peoples lived almost everywhere—in coastal tidewater areas and swamps, near rivers and lakes, in dense forests, in dry desert areas, and on mountain slopes as well as on extensive prairies and grasslands. North American cities of several thousand people existed before the period of European exploration and settlement. Fundamental to the indigenous worldview was a nonlinear melding of environmental and social aspects of life in the landscape. This belief had important implications for the lack of interest of these groups in owning land as property (a strongly held European belief). Even so, by at least 10,000 years ago, the impacts of vegetation burning, planting, hunting, and building by native peoples would no doubt have been observable to newcomers. However, due to their worldview and their lower population numbers and density, the environmental impact of aboriginal people on the continent was minimal in comparison to that of later arrivals who utilized a comparatively higher level of technology.

Settlement patterns of various native peoples (referred to as **First Nations** in Canada and **Native Americans** in the United States) shifted through time. Many of the earliest groups preferred open environments with dense populations of animals such as the Great Plains and prairies. Others chose coastline and river fishing sites. Along the northwest coast, Native residents occupied a region with abundant resources for sustainability. Here, groups like the Kwakiutl invented complex rituals and belief systems. Along the California coast, native people such as the Chumash organized elaborate trading networks with each other and with interior tribes. In desert areas, population density was lower since survival depended upon

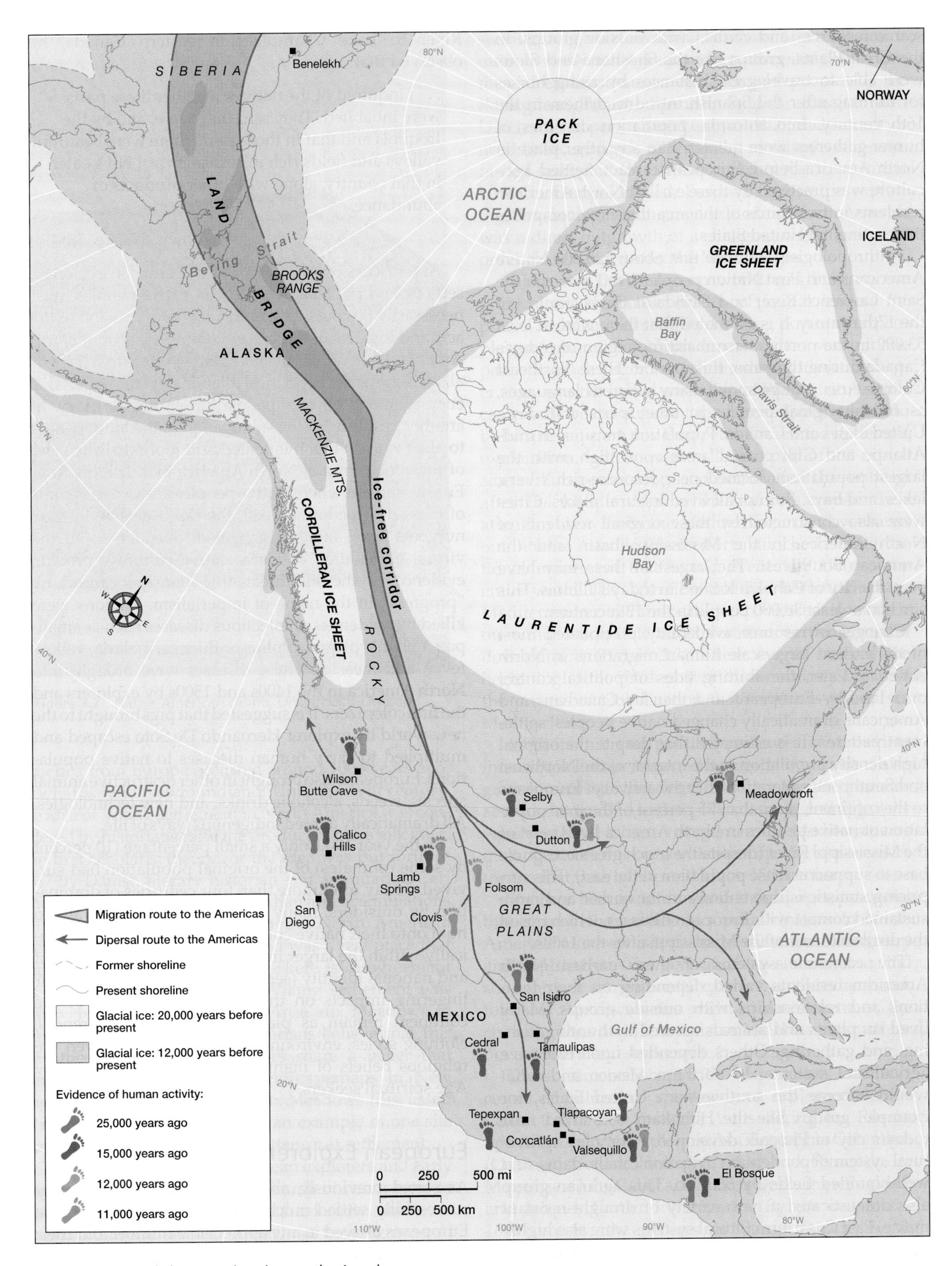

▲ **Figure 3.1** Early human migrations to the Americas

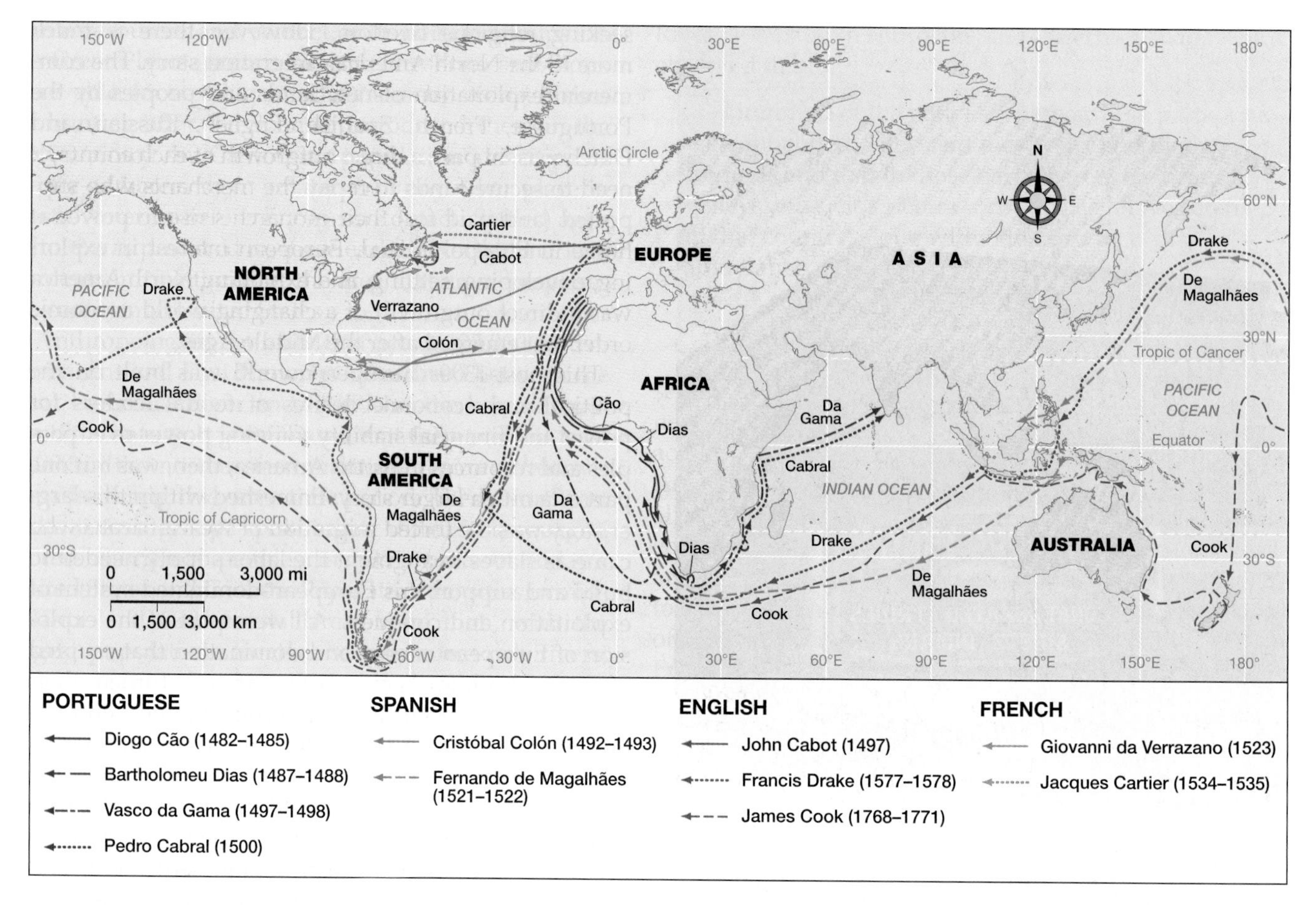

▲ **Figure 3.3** The Atlantic Ocean used as a European sea

called the **Columbian Exchange**. The term is now used to describe the ecological exchanges and impacts that Columbus and his successors initiated. These processes had lethal consequences for aboriginal people as well as natural landscapes in North America. And at least three more centuries of European, Native American, and African conflict lay ahead.

Colonial Settlement: New Land Uses, New Cultures

The evolving North American cultural landscapes in the centuries after the arrival of European settlers may be visualized as an ancient parchment called a *palimpsest* which contained layer upon layer of settlement histories. Aboriginal landscape features were removed as on one of these erasable parchments, as new landscapes emerged. In turn, Spanish, French, Portuguese, English, Dutch, Russian, and African newcomers stamped their own "landscape tastes" onto the North American scene (see Figure 3.4). Beginning in the late 1500s, a number of small-scale settlement schemes were attempted in North America such as Raleigh's efforts to "civilize" the woods of North Carolina and Gilbert's efforts to construct a settlement in Newfoundland. All failed except the Spanish town of St. Augustine, founded in 1565, the oldest continually occupied European settlement on the continent.

Portuguese and African Settlement Well before Columbus began his search for a new passage to the "Indies," the Portuguese had established a market economy that linked the Americas with Europe and Africa. By 1415, this emerging sea power had launched a colonial empire, first by conquering the Madeiras, the Azores, Cape Verde Islands off the coast of Africa, and land along much of the West African coast. They established trade with East Asia by circumnavigating the African continent and crossing the Indian Ocean. The Portuguese also were the earliest fishermen on the Newfoundland coast, predating the arrival of the British and French. The Portuguese-inspired system of using Africans as slaves to provide labor for development of their sugar plantations in South America and the Caribbean had a powerful impact on the politics and cultures of the entire North American continent for centuries. While Spain was focusing its energies on the search for gold in Mexico, Portuguese sea-faring predators carried slaves from West Africa and the islands just offshore to North and South America and returned in a regular cycle carrying sugar back to Europe. This

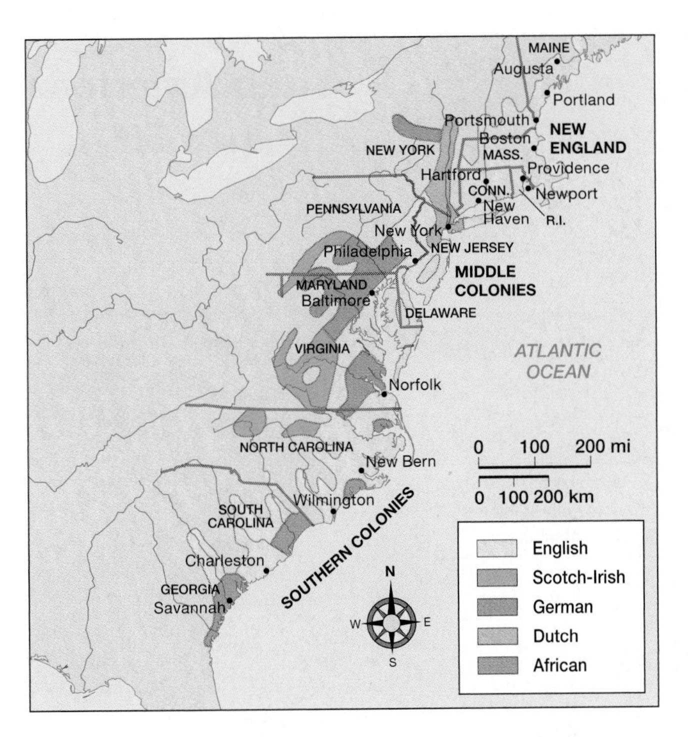

▲ **Figure 3.4** Source areas of early immigrants into the United States

system of forced migration brought tens of thousands of Africans to a place where they were viewed as the consummate *other*. As such, their cultures, beliefs, values, and desires were considered unworthy, unimportant, and in some cases, illegal. The following unique features of the American slave trade were a direct result of this Portuguese European-African-American system (see Meinig 1986, p. 20):

1. The extensive use of mass labor in specialized agricultural or mineral production for export;
2. The exploitation of such laborers as a commodity to be used up and replaced by purchase;
3. An extensive system to supply large numbers for sale annually at a reasonable price;
4. The heavy male bias in such use and trade, inhibiting the comprehensive formation of families;
5. The formalized debasement in custom, and in part in law, of such people as unworthy of acculturation and incorporation into the general community; and
6. The linkage of status with color, by which blacks but not whites were subject to slavery.

The integration of these attitudes, laws, and beliefs about slavery and racial divides were to affect American society and politics for centuries to come. Competition between Portugal and Spain for political control of new lands started with the Treaty of Tordesillas in 1494 (see Figure 3.5). By 1760 the Treaty of Madrid divided the known Western Hemisphere by giving Spain control over the western part of this territory and awarding Portugal the easternmost part of the continent. The impacts of this imprecise geographical decision are still visible in today's Latin America, with Brazilians (in eastern South America) speaking Portuguese and Spanish being the official language of most of the rest of Latin America.

French Settlement Meanwhile far to the north, the French were engaged in exploring northeastern Canada from Grand Banks, Newfoundland, to the inner territory along the Saint Lawrence River and beyond. Their search for resources resulted in the discovery of enough fish and furs to build a colonial economy. New-Found-Land was the term used generally for all of this part of the continent. French colonies located along the Saint Lawrence River which developed agricultural systems depended upon French-born cattle imported from the West Indies and the cultivation of hardy cereals, tobacco, and corn. Even more than the English, the French were dependent on the agricultural skills of native people already living in the area. Their most dense settlement area initially was located on the 200 miles of river country between present-day Quebec and Montreal.

Despite a rich potential for settlement in this part of Canada, the French voyagers were relatively slow to actually colonize North America. Individual French hunters, fishermen, and traders outnumbered colonists until the early 1700s. The first French farms in Canada were located near Quebec in the 1630s. By 1692, approximately ten thousand people resided on French farms in the area living on free land granted by the Crown. Thereafter, French settlements extended to trading towns on the outer edge of the Great Lakes and agricultural villages along the Mississippi and lower Ohio Rivers that extended as far south as New Orleans. Figure 3.6 shows the extent of French settlement in North America by 1700.

Throughout the French-dominated parts of North America, crops were planted and land ownership patterns were determined by the **long lot system** shown in Figure 6.5 in Chapter 6. This land division schema was especially useful along rivers because it allowed for as many people as possible to have access to river transportation. Thus, from Quebec to the Gulf of Mexico long, ribbonlike pieces of land were laid out along the water. Behind them, the French created open spaces of common land for the community and other open land reserved for the community's grazing purposes. The long lot system was implemented in other parts of New France as well. They still may be seen on survey maps in Vincennes, Indiana, and in the Mississippi delta country of Louisiana as well as the Saint Lawrence Valley.

Understanding the rise of New France is pivotal to understanding why most of Canada speaks English today, while French is the official language of Quebec. The end of French domination over all of eastern Canada as well as the Saint Lawrence and Mississippi drainage areas began in 1755 in Acadia. Here British officials who had long been concerned about the area's

▲ **Figure 3.5** Spanish and Portuguese influence in the Western Hemisphere

French-speaking population decided to deport them. Most went to Canada or down the Saint Lawrence River, with some traveling further south to New Orleans (where they came to be called **Cajuns**). Others were sent to British colonies in the United States or to England or France. Through the late 1750s, Britain continued to attack French-dominated territory in North America. In 1759, after a British victory partly due to the half-hearted effort by French defenders, the Treaty of Paris marked the end of the Seven Years War. France

▲ **Figure 3.6** France in North America, drawn in England by Edward Wells in 1700.

retained control of fishing rights in Newfoundland and two small islands in the mouth of the Saint Lawrence River. Spain was given control of Louisiana in a secret treaty, and the British ended up with all of Canada, including Acadia and southern Newfoundland in addition to all the French holdings in what was to become the United States. More about this drama is related in Chapter 6.

Russian Settlement Along the outer edges of the North American Pacific Rim, another story of European expansionism was unfolding at about this same time as Russians discovered and lightly settled the Aleutian Islands and today's Alaska as far south as the coast of northern California. Native Americans living on the Pacific Coast when the Russians arrived, led by Danish Captain Vitus Bering, had never seen Europeans. Ultimately a transnational fur trade was established that linked Alaska and the coast of northern California with Siberian towns such as Irkutsk and with St. Petersburg in European Russia. However, this Pacific connection from North America to Europe did not last long. The invasions of American and British trappers in the 19th century left the Russian effort overextended and underpopulated.

Spanish Settlement It was the Spanish who had the greatest and most indelible impact on cultural landscapes during this first century of European settlement in North America. Following Columbus's well-known first voyage in the name of Spain in the late 1400s, the passage of the *Laws of the Indies* in 1573 dictated the development and design guidelines for construction of presidios, missions, civil communities, and land grants in a huge area stretching from north of the Rio Grande to Tierra del Fuego. In support of the doctrine of first effective settlement, Spanish landscapes ultimately spread from far north of the present United States border to the southernmost tip of South America. Over the course of the next three centuries, these distinctive landscapes would evolve into a dramatic hybrid of Spanish and Native American design. Following the plan laid out by the Romans when they conquered the Mediterranean centuries earlier, Spanish domination of New Spain resulted from their efforts to impose a uniform ordering of all lands and peoples despite local and regional native diversity. After Cortes conquered the Aztecs in 1521, Spanish "influence by force" rapidly spread inland and along both Mexican coasts. The centerpoint of their influence was Mexico City, built on the ruins of the capital of the Aztec Empire. Subsequent exploratory trips by Cabeza de Vaca, Coronado, and others expanded the geographical knowledge of today's North American borderlands area. These expeditions nurtured false hopes in Spanish leaders who continued to believe in discovering the imaginary but supposedly fabulously rich Seven Cities of Cibola and other such treasures. These expeditions as well as Spanish slave raiding of Indian pueblos resulted in their military control of southern and western North America. The Spanish Empire continued to expand in size and influence into the early years of the 19th century.

New Spain was an empire of missions and towns as well as rural estates and ranches. Today's familiar grid street patterns and central plazas with important buildings reflect the sustained occupation of Spanish America. The continued presence of the Spanish language, Spanish-style architecture and Catholic religion remain as evidence of this long period of Spanish and, later, Mexican rule in the hemisphere. Chapter 14 expands this discussion of the borderlands landscape. Taking the lead in frontier settlement were missions for religious conquest and presidios for military control of local populations. In New Mexico and Florida, missions were built immediately adjacent to Native American settlements. Elsewhere, they were constructed as separate rectangular compounds used to convert local Indians. Los Angeles (actually *El Pueblo de la Reina de Los Angeles*) is the largest city to have been founded as a Spanish Pueblo.

Three types of land grants were also authorized to individuals by the Spanish government to be used as civic communities and agricultural development. They were known variously as *pueblos* (places), *villas* (villages), or *ciudades* (cities) depending on their size. In addition, larger land grants for cattle-raising were awarded in Mexico, Texas, New Mexico, and California. After Mexico became independent from Spain in 1821, two other types of land grants were laid out—*ranchos* for stock raising in California and large *colonias* destined for settlement by families and other colonists in Texas. These various types of land grants created a rectangular system of land ownership on the land. As compared to the French long lot system, Spanish-dominated territory in North America shows up on aerial photographs and road patterns as large blocky pieces of land with squared-off edges and distinctive boundaries (Figure 3.7). In other places in New Spain, especially in the creation of ranchos, the metes and bounds system was used.

Along with these distinctive land surveying patterns and architectural styles, the Spanish also initiated a two-way transfer of plants, animals, and diseases from Europe to North America—with maize (corn), cacao (chocolate), potatoes, and squash going to Europe and Africa—and wheat, oats, barley, horses, cattle, malaria, measles, and smallpox coming into North America. The lasting impacts and the citizenship of many of today's North American residents resulted from this early Spanish settlement in North America. As a result of three geopolitical decisions—the U.S. annexation of Texas in 1845, the Mexican cession in 1848, and the Gadsden Purchase in 1853—more than 100,000 Mexicans became U.S. citizens in the 19th century. American Indians, whose people had lived there for centuries, did not.

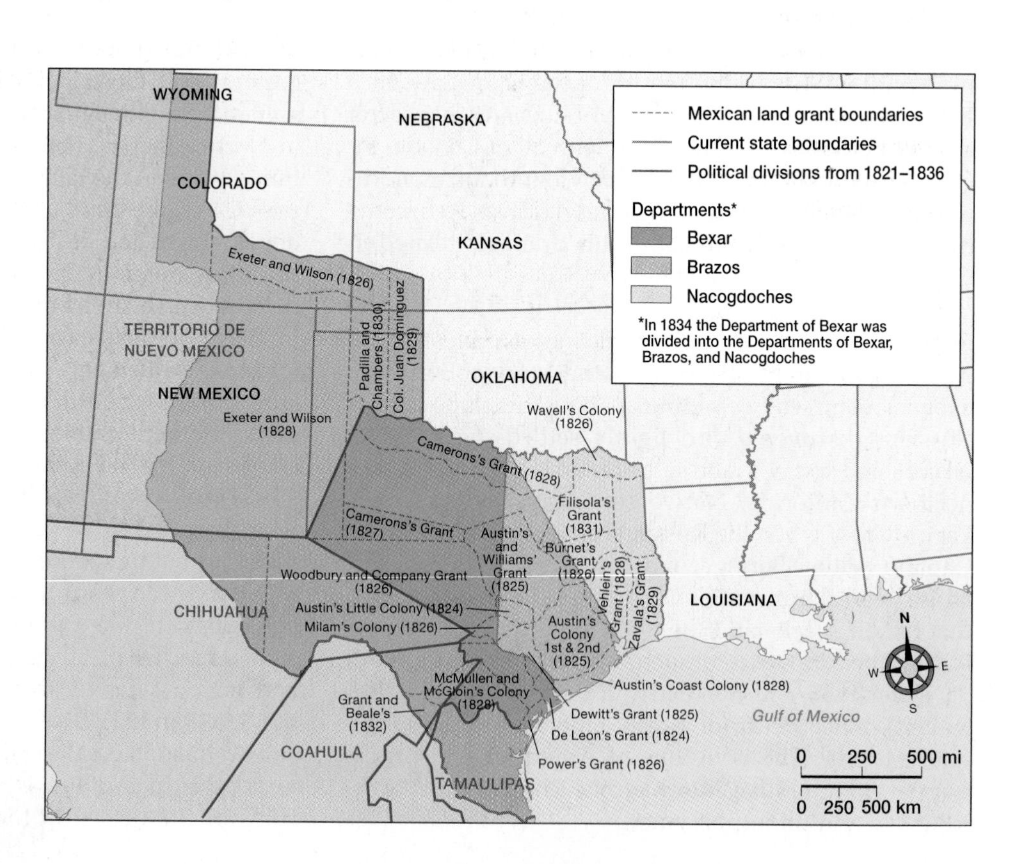

Figure 3.7 Historic land grant boundaries in Texas

Dutch Settlement Spanish, French, and British colonizers faced an unexpected contender for power in North America in the late 1500s when Dutch merchants set up trade networks with ports in Africa, South America, and Asia. By 1602, the Dutch East India Company was formed with state backing. Seven years later, it began to search for new trading opportunities in the Americas. Following the advice and experience of their British consultant, Henry Hudson, this Dutch company decided to focus its expansion plans on the Delaware and Hudson river valleys of New Jersey and Long Island. Control of this land provided access to all the land between Virginia and New England. At the same time, plans were made to colonize places in Guiana, Brazil, and the Caribbean West Indies.

Land for a new colony was purchased from local Indians and then the Dutch laid out their new city of New Amsterdam on today's Manhattan Island. A group of Flemish and Walloon Protestant colonists was recruited to relocate to this new land. With the administrative center of their colony at New Amsterdam, New Netherlands eventually stretched all the way from the lower Delaware River to Albany in upstate New York. Despite this well organized system, things didn't work out for the Dutch. While their fur trading and fishing colonies continued to expand, labor shortages in urban settlements, along with the Dutch company's feudalistic land grant system, resulted in the failure to create and support successful settlements in North America. Many of the new immigrants from Holland were attracted to the relative freedoms of British colonies in New Jersey and New York and moved nearby. Others blame the diversity of New Netherlands' cultures as the primary reason it did not succeed.

In 1664, the British captured all of New Netherlands and the 8000 plus people who lived there. Their new possession included such Dutch legacies on lower Manhattan Island as a series of well designed streets, storybook-looking houses, taverns, a large state building, and a canal and windmill all the way north to the site of today's Wall Street. Of particular note were the 200 red and black tile, two-or three-story, gabled row houses. Over the years, Anglo landscape tastes were slow to replace this very Dutch-looking city. As a result, Manhattan and Long Island emerged as a unique and diverse place in North America by 1700. These two places also housed the majority of the 19,000 residents of what had once been all of New Netherlands, but was now a part of Britain (Figure 3.8).

British Settlement and British Landscape Impacts Despite the dominance of British governmental systems, culture, and landscapes in much of Canada and the United States today, England entered the colonizing race relatively late. Yet, less than a century after its initial colonies were planted in the 1600s, there were at least 250,000 British settlers residing in small settlements and on farms near the Atlantic coast between South Carolina and southern Maine. Questions remain about the reasons *why* this one particular nation-state had such a powerful and lasting impact on shaping the emergence of Canadian and American cultures. Was it the comparatively large number of people from England, or was it their distinctive settlement system of encouraging more permanent families (instead of individuals) to relocate here that most influenced future cultural evolution? Did they modify their heritage to adapt to the peculiar characteristics of the new land itself, forging a new culture that was based on this adapted and fused British–North American identity? As historian Frederick Jackson Turner suggested, was it the frontier experience that most shaped emerging cultures? Did the mobility and freedom of the westward movement craft the cultural landscapes and values that characterize and define American and Canadian cultures?

We can be certain that the values and perceptions of provincial rural England came to North America with the first Puritan settlers and later English immigrants. They brought with them a firm belief in the rightness and righteousness of their Christian faiths, the right of individuals to own and manage their own land, and a sense that being truly "civilized" meant encompassing these English values above all others.

Whatever the exact reasons for the emergence and ongoing dominance of this British-inspired system in the two largest nations of North America, we can be confident that the British knew what they were doing from their first efforts to colonize the eastern seaboard. Applying what they had learned as longtime colonizers in other places, they developed carefully constructed plans for new land in North America. Thus, formal procedures were developed by the two companies responsible for Britain's launching of their first two settlements at Jamestown in Virginia and at Plymouth in New England.

Jamestown was established first in 1607. It remained a backwater location in relative isolation from all but the local Native American residents until the Dutch began to invest in creating settlements in the Hudson Valley. Less than two decades later in 1620, a group of strict Puritan Separatists established their colony at Plymouth just north of Cape Cod. Soon thereafter followed the establishment of the Maryland Colony and a large in-migration of new British settlers to Massachusetts. By the end of the 18th century, there were more than 250,000 Europeans living in an area that stretched from the tidewaters of North Carolina to the rocky coasts of Maine. Plymouth was the most English of all these earliest settlements due to its religious and socially cohesive population of Separatist Puritans and its relative isolation from other colonists. Here, new immigrants came in as families and even entire congregations to practice their faith freely and build a new society. Later, when the Massachusetts Bay Colony was founded, a series of distinctive and more dynamic settlements were formed by settlers who originated from

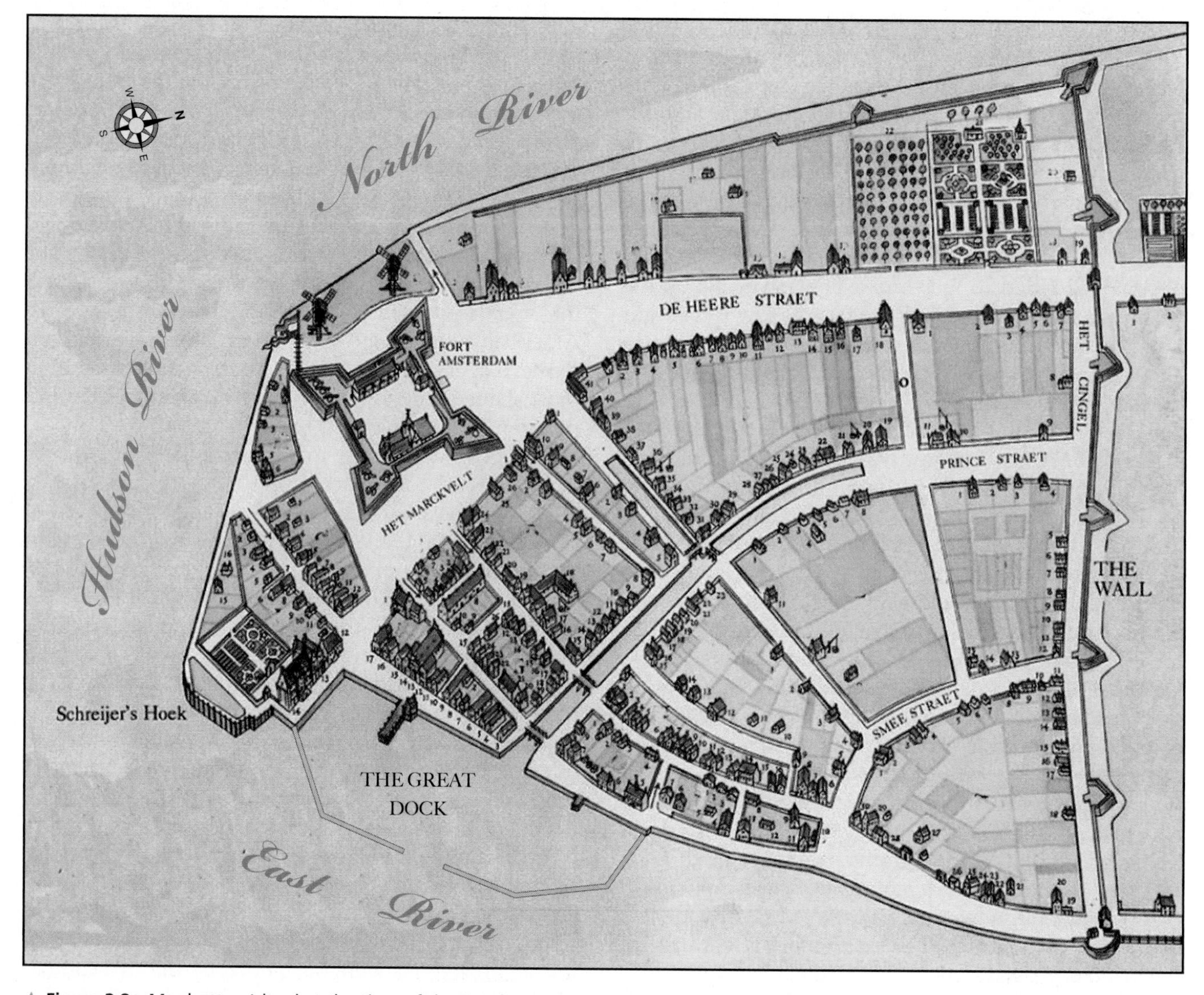

▲ **Figure 3.8** Manhattan Island at the time of the Dutch

many different parts of England, as happened in most other places. Between 1700 and the beginning of the American Revolution, the total population of the original 13 British colonies reached almost 2.5 million. Despite their common culture and economic ties with Britain, most had begun to define themselves as Americans first. By 1820, the population of America had exceeded England's.

Meanwhile, in other parts of northern and eastern North America, English settlements emerged as well. Canada was the agricultural heartland of all of British North America. After the Treaty of Paris gave ownership of land to Britain northward from Massachusetts to Acadia and Quebec to Britain, exact boundaries of this land remained elusive. Following the American Revolution, however, it became apparent that this boundary needed clarification to identify what the limits of new U.S. territory really were. Thus, in 1817 the boundary was determined along the Atlantic Coast and in 1842 a line was drawn inland dividing the United States from British- and French-held territory.

By 1826, there was much contention between the United States and Britain over claims to the Columbia River system on the Northwest coast. United States Secretary of the Treasury Albert Gallatin suggested a claim in favor of American ownership in terms of what he called the "principle of contiguity" in his statement called "The Land West of the Rockies." It argues that lands adjacent to already settled territory can reasonably be claimed by the settled territory. This argument was an early version of the doctrine of manifest destiny and became the legal premise by which the United States claimed lands in the west to the Pacific Ocean.

British-held fishing towns along the coast of Canada's northeast Atlantic periphery were multicultural mixing grounds. Here, fishing communities were home to not only the British but also French, Irish, German Protestants, Loyalists fleeing the American Revolution, free Blacks from Halifax, deported Acadians returning home, New Englanders, and thousands of Catholic Highland Scots. At the same time, lines of demarcation separating Quebec from the rest of British-controlled

Canada were defined in 1763. Shortly thereafter, British officials advertised free land to interested colonists. It was hoped that pioneer farmers and commercial and industrial newcomers would help build an empire for the British even in isolated places.

Even after passage of the Quebec Act in 1774 that laid out the boundary of British territory through the middle of Lake Ontario and Lake Erie as far west as Erie, Pennsylvania, questions about international boundaries remained. Left out of this division was an unclaimed portion of land located between the Appalachians and Lake Ontario. Thus, the Erie Canal region and some of the western parts of the states of Pennsylvania and New York would have remained in the hands of the British after the American Revolution if enough British colonizers had made it their home.

Despite their common ethnic backgrounds, British North America evolved into a religiously and culturally rich place. In Pennsylvania, for example, William Penn's unique style of leadership and his group of English Quakers soon developed a landscape of unexpected experimentation and innovation. Penn set out to create a community where religious freedom and entrepreneurial skills would work hand in hand. Unlike the religious philosophies that guided the formation of colonies in New England, his vision was to recruit advocates of diversity and multiculturalism from many parts of Europe to come to Pennsylvania. As a result, new immigrants began to flood into the city of Philadelphia. By the time of the American Revolution, Germans made up one-third of the total population of Pennsylvania. Most were Mennonite and Amish conservatives who came seeking religious freedom.

The British introduced the **metes and bounds** system that contrasted with the French long lot system of settlement and Spanish irregularly drawn land grants. In this system, the process of deciding on specific boundaries was often left up to individual land purchasers, not some central authority. As the desired quantity of land was purchased, buyers were then free to lay out their own boundaries based on tree markers or topographical features. A desire to secure the greatest amount of level, well watered, and fertile land means that individuals often scrambled to lay out boundaries. Later buyers could purchase the entire plot of land or only a part of it. The resulting changes in boundary features created a somewhat chaotic pattern of land ownership common to all the British colonies.

Ongoing Migration, Expansion, and Settlement

In the earliest years of the 19th century, approximately half the population of the United States lived south of the *Mason-Dixon Line*. This distribution pattern helps explain why this halfway point to be named the District of Columbia (Washington, D.C.) was selected in 1791 as the most appropriate location for the nation's new capital.

Identifiable European dominated areas, each with its own set of unique cultural and economic characteristics, had evolved in the United States and in various parts of Canada and Mexico by the early years of 19th century. Geographers refer to these culturally and economically distinct areas as **culture regions**, which often became a point of origin (**culture hearth**) for the spread of that set of cultural traits. In Mexico, a Spanish-influenced culture hearth had overpowered Native American culture in Tenochlitlan (Mexico City) and nearby towns, cities, and provinces. As a result of the mission system and military explorations, this Spanish-influenced culture hearth extended as far north as northern California and as far east as Florida. Use of the Spanish language and belief in the Roman Catholic faith became the official dogma of government and society. After Mexico won its independence from Spain in 1821, these Latino traditions continued.

Two distinctive culture hearths replaced Native cultures in Canada, one dominated by French-speaking Roman Catholics, and the other primarily by English-speaking Protestants. By 1865, most of the best agricultural lands within 100 miles (160 kilometers) north of the Canada-U.S. border had been settled. On the Atlantic Coast, English settlers had settled Nova Scotia and the coastlines of the Gaspé Peninsula and New Brunswick. By the latter part of the century, settlers expanded their developed land farther west despite more harsh environmental conditions in the Canadian prairies, west, and northern interior—and the almost impenetrable barrier of the Canadian Rockies. Winnipeg, Manitoba, Calgary, Alberta, and Vancouver, British Columbia, were the earliest important urban centers of early Canadian expansion into this part of the continent.

In what is now the United States, four major hearths emerged from the 13 original colonies. They were: (1) southern New England centered in the Boston area; (2) the Chesapeake Bay area; (3) the Middle Colonies centered in the city of Philadelphia; and (4) the South. By mid-century, the South's plantation system had spread from its core in Virginia, Maryland, and North Carolina to the Virginia-Piedmont in the interior. As soils were depleted and innovations like the cotton gin eased cotton harvesting, the region known as the South had expanded into Georgia and South Carolina first, and later into Florida.

As the westward migration of people and ideas began to unfold in earnest in the mid-19th century, the cultures and values of each of these subregions were carried along migration paths and to pioneer destinations. In Canada, territory dominated by the British expanded westward, while the French hearth remained confined to its original location. The territorial expansion of English Canada all the way to the Pacific Ocean (Figure 3.9) ultimately brought English culture and

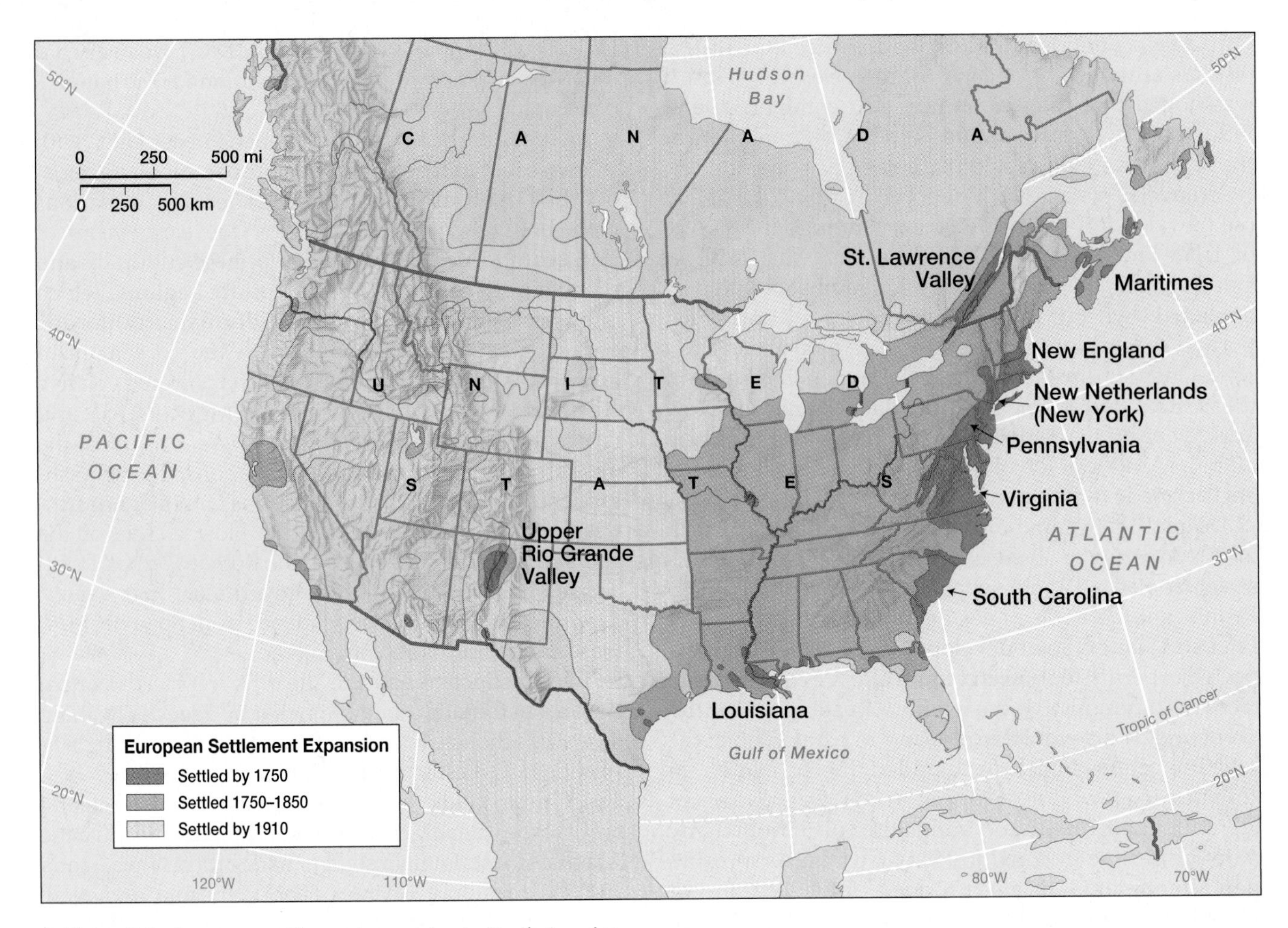

▲ **Figure 3.9** European settlement expansion in North America

economic systems as far west as Victoria on Vancouver Island and as far north as the Arctic Circle—while most of French Canada remained rural well into the 20th century. While colonial lands and boundaries were created through charters, patents, and grants from royalty, additional lands were added to the area of the young United States by treaty with other nations, by purchase, or by military conquest (see Figure 3.10).

When the United States gained its independence from England, formalized by the Treaty of 1783, the agreed upon western boundary of the new nation was the Mississippi River and the former colonists began to migrate to the other side of the Appalachians. Most of the new 13 states laid claims to these western lands, and one of the accomplishments of the new congress was to convince the States to cede these lands to the federal government for distribution to new settlers.

The area of the nation was doubled with the purchase of Louisiana in 1803. The now famous Lewis and Clark *Corps of Discovery* was directed in 1804 to locate the headwaters of the Mississippi-Missouri drainage so as to define the western boundary of American sovereignty (and to evaluate the possibility of commercial development of the lands west of the continental divide to the Pacific Coast). In 1819, a parcel north of Louisiana was purchased with the 49th parallel as its northern border. Florida was also acquired in 1819 by succession from Spain. The territory generally extended from the Mississippi River to the Atlantic Ocean and north to the 31st parallel, which was then the southern boundary of the United States.

The Republic of Texas, independent since 1835, was annexed in 1845. The Oregon Country was originally defined as everything from the Rocky Mountains on the east, the 42nd parallel (now the northern boundary of California) on the south to 54 degrees 40 minutes north latitude, the southern tip of the Alaskan Panhandle. Spain, Russia, Great Britain, and the United States all had some claim to the area. An agreement in 1846 gave British Columbia to Great Britain and the Oregon Territory to the United States.

The end of the Mexican War in 1846 resulted in the **Treaty of Guadalupe Hidalgo** in 1848 and the Mexican Cession. The half-million square miles included what would become the future states of Colorado, Utah, Wyoming, California, and most of Arizona and New Mexico. In 1853, the Gadsten Purchase was added for a railroad route.

Hawai'i was annexed in 1898 and became a Territory in 1900. Alaska was purchased from Russia in 1867. At the end of the Spanish–American War in 1898, Puerto Rico was ceded by Spain to the United States and is now administered as a commonwealth. In 1917, the United States bought the Danish West Indies in the

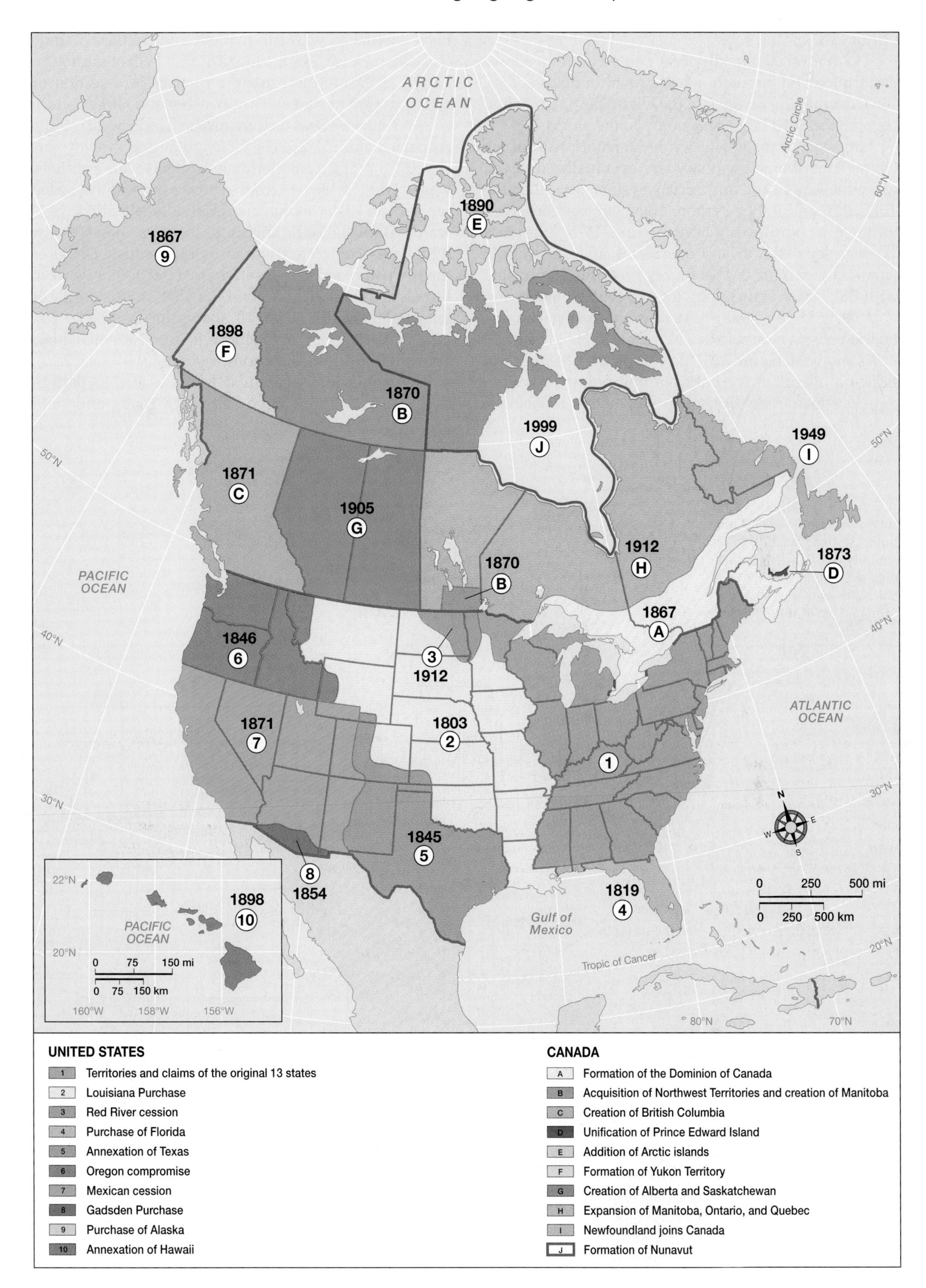
ARCTIC OCEAN
PACIFIC OCEAN
ATLANTIC OCEAN
Gulf of Mexico
Tropic of Cancer
Arctic Circle
1867 9
1898 F
1890 E
1870 B
1999 J
1949 I
1871 C
1905 G
1912 H
1873 D
1870 B
1867 A
1846 6
3 1912
1871 7
1803 2
1
1845 5
8 1854
1819 4
1898 10
0 250 500 mi
0 250 500 km
0 75 150 mi
0 75 150 km
22°N
20°N
160°W
158°W
156°W
80°N
70°N
60°N
50°N
40°N
30°N
20°N
UNITED STATES
1 Territories and claims of the original 13 states
2 Louisiana Purchase
3 Red River cession
4 Purchase of Florida
5 Annexation of Texas
6 Oregon compromise
7 Mexican cession
8 Gadsden Purchase
9 Purchase of Alaska
10 Annexation of Hawaii
CANADA
A Formation of the Dominion of Canada
B Acquisition of Northwest Territories and creation of Manitoba
C Creation of British Columbia
D Unification of Prince Edward Island
E Addition of Arctic islands
F Formation of Yukon Territory
G Creation of Alberta and Saskatchewan
H Expansion of Manitoba, Ontario, and Quebec
I Newfoundland joins Canada
J Formation of Nunavut

▲ **Figure 3.10** Territorial acquisitions in North America

Caribbean Sea and took possession, renaming them as the U.S. Virgin Islands. By 1860, United States territory had tripled and population had grown 15 times and by 1865, despite the ravages of the Civil War years, it had become the third largest economy in the world.

Almost all of the newly secured territory west of the Ohio-Pennsylvania boundary was eventually surveyed using the American Land Survey System also called the **Township and Range** system. This plan was conceived through the Northwest Ordinance in 1787 by President Thomas Jefferson, a land speculator and geographer (among other things). This survey procedure contrasted with the earlier French long lot, the English metes and bounds, and the Spanish land grant systems in the method of survey and in the exchange of title to land.

As shown in Figure 3.11, the surveyor's base line and principal meridian were the starting points for laying out 36 square-mile (93 square-kilometer) townships. Each township was subdivided into sections 1 mile (1.6 kilometers) on a side. In the 19th century an American farm was usually a quarter of a section or 160 acres (395 hectares). The township and range pattern of land survey was begun in the lands west of the Appalachians in 1785. The rectangular grid system, surveyed from principal meridians and base line parallels, was started in seven rows of townships ("the seven ranges") in eastern Ohio and gradually extended westward. All of the basic survey of the area shown in the figure was completed within half a century. Detail of the township and range pattern in a portion of western Illinois is shown in the second drawing as an example of the whole. The third drawing shows the six-mile-by-six-mile grid of a typical township, with the sequential numbering of the one mile sections.

The result is a rectangular land ownership pattern that is decidedly squared off. Since roads and field

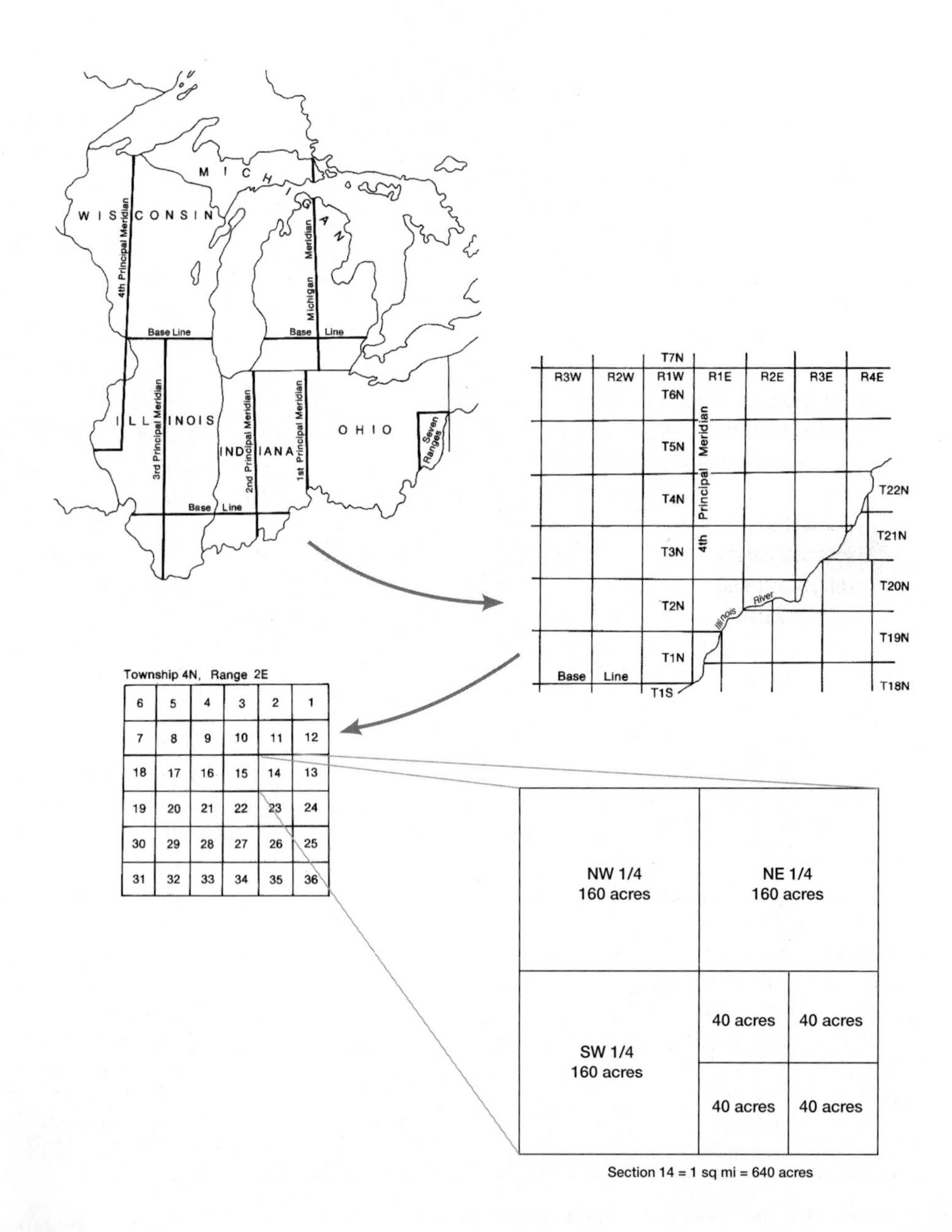

Figure 3.11 The U.S. land survey system. The U.S. land survey system was used to record land ownership wherever a previous system was not in place. Today many states use a combination of survey systems.

◀ **Figure 3.12** Center-pivot irrigated area overlayed on top of township and range pattern in eastern Colorado

patterns follow this same alignment, it's easy to notice the checkerboard patterns when flying over the Midwest and West. Only in places where center pivot irrigation systems make landscapes appear rounded, as in Nebraska and other Great Plains states, and in a few places where Spanish land grant boundaries still hold sway, such as in West Texas, is this regularity of squared-off townships interrupted (Figure 3.12).

By the mid-19th century, as agriculture, mining, lumbering, cattle raising, and other pursuits provided possibilities for economic survival in Native American territory, other new (and spatially disconnected) Euro-American culture hearths emerged in:

- Salt Lake City, where religious pioneers from the Church of the Latter Day Saints built an agricultural and urban empire in the Wasatch Valley;
- Northern California, as the Gold Rush added to settlement and cultural identity in the San Francisco and Sacramento area;
- Portland and south through the Willamette Valley, where fertile farmland was opened up and cultivated; and
- Virginia City, Montana, and other mining towns in the High Plains and northern Rockies, where minerals and other natural resources attracted new settlers and investors.

Between 1840 and 1920 new immigrants from Europe continued to arrive in North America in ever increasing numbers. Latin American and Asian newcomers also came during this period. Then, after this long period of uncontrolled immigration, legislation was passed in the 1920s that restricted the in-migration of selected groups. They might have been suppressed for other reasons, for example the depression, World War II, or tuberculosis. It was not until the 1960s that some of these laws were eased and immigration rates rebounded.

The settlement of most Euro-Americans in the central and western parts of North America after the 1860s came from two sources, the eastern part of the continent and from other countries. Motives for migration included push and pull factors such as the possibility of owning land, individual initiative, political decisions, escape from unpleasantness like the potato famine, and various other events occurring in Europe and other parts of the world. Between 1840 and the early 1920s, during the period of mass immigration into North America, the population of the United States increased from about 17 million to more than 105 million. Many individuals in the vanguard of the movement west wrote letters home encouraging their friends and relatives to join them. While some of the information about the glories of the new land presented in these letters no doubt was somewhat exaggerated, thousands of immigrants were encouraged to leave their homes and travel west by this mechanism. In an era before e-mail and televised news reports, information was also spread by church newsletters, railroad land companies, immigration bureaus, and newspaper advertisements. In addition, real estate and land development agents were sent directly to places like Stockholm, Hamburg, New York, and Atlanta in the hopes of luring new settlers to the West.

Political decisions affected the magnitude of movement to the West. Perhaps no legislation was as critical to opening up the area for settlement as the **Homestead Act** signed by President Abraham Lincoln in 1862. Thereafter, any adult citizen or person who intended to obtain citizenship could file for 160 acres (647 square meters) of public land, pay a nominal registration fee, and receive title after residing on the land for five years and showing proof of improvements. Prior to the passage of this act, most of the land had been purchased under the Preemption Act of 1841, which allowed settlers to buy land at $1.25 an acre. Homesteads could be, and were, often sold as soon as possible at market values, thereby creating the process of land speculation and establishing the infancy of the real estate business. A decade earlier, the **Donation Land Claim Act** awarded citizens free land if they were willing to travel the long distance to grow crops in the isolated and far-distant Willamette Valley of Oregon.

After the passage of various versions of the Homestead Act, settlers soon advanced westward along the Platte and Kansas Valleys and other river valleys, crossing beyond the 100th meridian. This is the line on the map that historian Walter Prescott Webb called the institutional fault line because, he claimed, it effectively divided the humid east from the arid west. Despite the impact of this climatic boundary, easterners, southerners, and new foreign-born immigrants seeking free land soon pushed well beyond this imaginary line of demarcation to build homes and set up farmsteads all the way to the Rockies. Even though many of the pioneers failed, more land was settled due to the Homestead Act and its enlarged versions passed in 1909 and 1916 (expanding the size of grants to 320 and 640 acres) between 1898 and 1917 than during the previous 30 years. In Montana, for example, in the peak year of 1910, almost 22,000 homestead applications were filed at the Montana Land Office with an average of more than 14,000 filed every year until 1919. Homesteading in most areas of the United States, except Alaska, was discontinued in 1934.

Construction of railroads joining the East and West coasts likewise changed the pace and nature of the westward flow of people in both Canada and the United States. In an aggressive effort to fill boxcars with grain and their government-granted land with people, settlers were actively, and at times quite aggressively, recruited by railroad companies. In partnership with immigration bureaus, new potential residents of midwestern and western states and territories were recruited at county fairs and other gatherings. In the "Boom of the Eighties," immigrants from the east, the south, and northern and western Europe responded in earnest. Tens of thousands boarded trains for the West. Once in place, many remained dependent on the railroads for transportation of their agricultural products to markets far away. Grain farmers on the Great Plains, for example, were limited to living within 20 miles of the nearest railroad because hauling their crop any farther would consume their profits. Later, county road systems would allow opening more untilled land.

Even the street patterns of frontier towns reflected the importance of the railroad. Main streets were laid out perpendicular or parallel to the railroad tracks, which were flanked by depots, grain elevators, and lumberyards. Companies laid out a dense network of new towns at regular intervals along the tracks to serve as water stops and grain collection points. There were so many speculative towns that they all could not successfully compete and many disappeared from the map by the early 20th century.

During this same time period, as new Euro-Americans boarded trains and settled the Midwest and West, many Native Americans struggled to adjust to reservation life. By 1890, the U.S. government had set up a system of small reservations on less valuable lands through out the Great Plains all the way to the Pacific. Here aboriginal people were isolated from new settlers and forced to sell or give away their most valuable lands. The existence of "Indian Country" had been threatened in 1887 after the **Dawes Act** was passed, requiring many to claim their own allotment within the reservation as individual land. Most of these parcels were then sold to non-Indians. All unclaimed surplus land was then made available for sale to Euro-American settlers. In the Far West, many tribes did all they could to hold onto their territory and to their cultures as the increasingly bloody impact of the nation's belief in the greater good of manifest destiny took root. When gold was discovered in California, and later in Idaho, Montana, the Yukon, and elsewhere, army posts were built to protect the routes to the mines and some to protect the Indians from the miners. Thereafter, conflicts between the original inhabitants of the continent and the newcomers intensified as warfare raged until the last Native American resistance was eliminated at places like Wounded Knee, South Dakota, and Chief Joseph's Wallowa Valley in Oregon.

An often overlooked aspect of the population diversity of the late 19th and early 20th centuries is the lingering presence of Latino families and workers in the American Southwest. Mexicans continued to occupy what geographer Richard Nostrand called their original *Hispano Homeland* in the upper Rio Grande Valley from Colorado to El Paso, in coastal California, and the Tucson area after the early arrival of new groups of Euro-American settlers, while many were associated with agricultural employment as they migrated farther north. Migrants from Mexico also relocated to the United States in the early 1860s to find employment in mining or railroad construction. Others moved to border cities such as El Paso and San Diego, where they formed some of the first ethnic enclaves in the American West.

Evolving Economic and Urban Development

The originally Spanish cities of St. Augustine, Los Angeles, San Antonio, and Santa Fe were frontier outposts. Likewise the Dutch settlement of New

Amsterdam was originally a fur trading post. The French built fur trading posts that were later to become the cities of Quebec, Montreal, Detroit, St. Louis and New Orleans. It could probably safely be said that more of the area of North America was explored in search of furs than for any other reason. However, most English colonial villages were primarily seaports and centers for trade and commerce. Williamsburg, Annapolis, Charleston, Boston, Savannah, Toronto, and Baltimore are examples. Philadelphia was the largest population center until 1790. In 1790 all cities with more than 10,000 people were seaports tied by commerce to Europe. All of these ports served **hinterland** areas from which commodities were collected and to which imports were distributed. The expansion and then economic dominance of New York was primarily due to the opening of the Erie Canal thereby increasing the city's hinterland. Transportation technology and accessibility of goods became important factors in the economic success or relative failure of towns and cities.

Coastal cities prospered during this early period of seacoast transportation dominated by sailing ships and wagon roads. Then, as canals and railroads were developed into the Atlantic coast interior in the 1830s through the 1850s, new cities were created and older settlements like Buffalo, Cincinnati, Louisville, and Pittsburgh grew rapidly. The sail wagon age gave way to the importance of steam power by rail and steamboat. St. Louis became the major assembly point and distribution center for those traveling into lands west of the Mississippi River. Likewise, New Orleans and Galveston became points of entry and exit in the Coastal South.

The role of railroads in the creation of the urban landscape can not be overstated. By 1890 they expanded the hinterland concept to a continental level by connecting almost all major towns and cities with a network of rail lines centered on Chicago. River towns, coastal towns, and more remote agricultural centers were now connected with industrial centers, government offices, universities, and with the entry points from Europe and Asia.

As the industrial core of the Canadian and American Northeast continued to expand after the 1890s, many of the immigrants and their children who originally settled in rural areas relocated to urban places, drawn by employment opportunities there. The growth of the steel industry at this time was an example of the need for labor in Pittsburgh, Chicago, and other such cities. These earlier immigrants were joined in large cities by most of the southern and eastern European immigrants who came after 1890. Boston and other smaller cities in New England and San Francisco in particular soon became home to large populations of Irish and Italian immigrants, while the port cities of the Great Lakes (e.g., Hamilton, Ontario, and the American cities of Milwaukee, Buffalo, Detroit, and Chicago) as well as Austin and San Antonio in Texas attracted ever larger populations of Germans. Meanwhile, the Dutch clustered in urban places on both sides of the Canadian–American border near the Great Lakes, while Scandinavian immigrants relocated to cities such as Minneapolis and St. Paul, Minnesota; Fargo, North Dakota; and Madison, Wisconsin. Between 1890 and 1920, large numbers of eastern European and Russian immigrants flocked into cities located in the industrial core such as Pittsburgh where they found employment in heavy industry. In contrast, many of the Jewish arrivals from the Russian Empire made their homes in midtown Manhattan where they worked in the garment industry, and later, in Toronto, Vancouver, Chicago, and Los Angeles.

With growth in the continent's manufacturing sector and ongoing arrival of new immigrants to work in it, cities became much larger, primarily due to more improvements in transportation. Cities developed electric mass transit systems that allowed workers to live further from where they worked. Land uses became more specialized into districts and larger population densities could be supported. By 1920, half of the population in the United States was urban.

Despite the common images of immigrants and their children having relative geographical mobility in North America, the then regional destinations remained fixed by industrial labor needs, at least up to the 1920s. Regional changes that did occur were the result of economic opportunities in developing areas. By far the most conspicuous change in the distribution of various ethnic and racial groups was the migration of African Americans to the cities of the Midwest, Northeast, and West. The regional distribution patterns of African Americans have shifted through time in response to the changing political and economic situation. In recent years, however, many African Americans in places like Oakland, Detroit, and Chicago are returning to the South in relatively large numbers as family ties and the relative stability of life in small towns and rural places draw them back. Today, as in earlier time periods, family connections and social networks both keep people in place once they settle in certain areas and help draw them home when they migrate away from these initial nodes of settlement.

Technological developments continued to make human mobility more possible. The interstate highway system built from the 1950s into the 21st century, once again created reduced travel times for people and products. Likewise, air transport in the late 20th century made nearly all aspects of life part of a global communications and transportation system. Some have suggested that globalization will mean the end of geographic differences among people and places. Others, your authors included, argue that technology offers more options to more people and that the geographies of places may change but will always carry unique and important attributes for continued interest in North American regions.

4 North American Political Economy

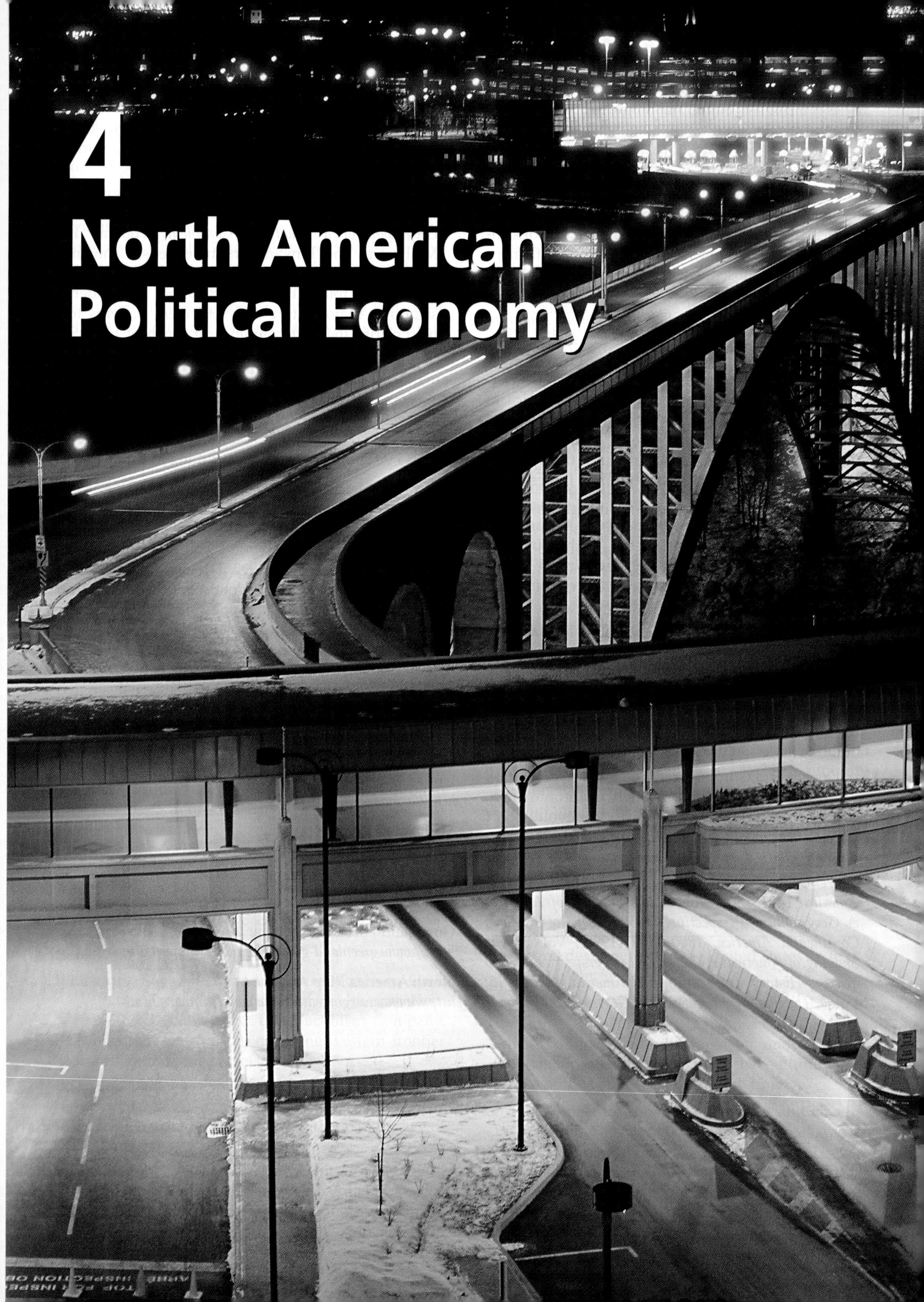

PREVIEW

- The study of political economy focuses on the related set of economic and political institutions that control the production, distribution, and consumption of goods and services.
- When geographers analyze the different levels of economic development, we divide them into the following four categories discussed in this chapter: primary, secondary, tertiary, and quaternary production.
- The technological treadmill is a term that refers to farmers (and other workers and business owners) being forced out of business because they can't keep up with current technological advances.
- A selected list of types of agricultural systems in the United States and Canada includes cash-grain farms, truck farms, dairy farms, mixed farms, and ranching.
- Beginning in the mid 1980s, increasing prosperity and mobility encouraged minority groups and new immigrants in North America to reside in the suburbs (rather than in high-density inner city neighborhoods) in what urban geographers call ethnoburbs.
- Increasing urbanization, suburbanization, and overall metropolitan development have resulted in the stagnation of many rural areas in Canada and the United States in recent decades.
- Since 1970, the number of women elected to the U.S. Congress has increased significantly. A female Majority Leader of the House of Representatives was elected in the November 2006 elections.
- Geopolitics refers to the relationships between geography and international relations and impacts upon how foreign policy is influenced by a nation-state's location, resource base, and other geographic factors relative to its competitors.
- Beginning in the cold war up to the post-Homeland Security era, U.S. foreign policy emphasized the importance of the Great Circle route. These policies have resulted in numerous (and at times unpopular) American military bases and missile tracking stations in Alaska, northern Canada, Greenland, and Iceland.
- Canada, the United States, and Mexico signed the North American Free Trade Agreement (NAFTA) in 1993 to link North American economic zones like never before in the history of the continent.

Introduction

Over the past four hundred years, the United States and Canada have become the wealthiest and most productive countries in the history of civilization. During this period, North Americans have created a unique **political economy**, or set of economic and political institutions governing the production, distribution, and consumption of goods and services. In this chapter, we examine the contemporary political economy of North America, focusing on the development of North America into a global economic and political power over the course of the four centuries. We examine the North American economic system, urban structure, political system, transportation, and communications in order to provide an overview of the North American political economy today. The themes developed within this chapter will be reinforced in discussion of specific regions of North America in subsequent chapters. All have an impact on the Canadian and American people.

◀ **Political geography on the landscape.** The Peace Bridge is an international connection between Canada and the United States at the east end of Lake Erie on the Niagara River above Niagara Falls. It connects the city of Buffalo, New York with the town of Fort Erie, Ontario

The Contemporary North American Economy

The evolving political economy of North America reflects not only changes in science, technology, and culture but also changing relationships between North America and other parts of the world. The United States and Canada were European colonies throughout most of the 17th and 18th centuries. During the 19th century, the political and economic power of the United States and Canada increased steadily. By the end of World War II, the United States was the leading political power in the world.

Economic Sectors

In examining the economy of any region on Earth's surface, it is useful to consider the relative importance of various stages in the process of producing and distributing goods and services. Economists divide a country's economy into four sectors: primary, secondary, tertiary, and quaternary. Each of these four sectors represents a different stage of production.

The **primary sector** of any country's economy includes activities associated with the identification and extraction of raw materials. Agriculture, forestry, mining, and fishing are important primary-sector occupations. The **secondary sector** includes activities associated with the transformation of raw materials into finished products. Thus the secondary sector is often called the manufacturing or industrial sector. Once a product is manufactured, it must then be distributed to the consumer and subsequently serviced and maintained. The distribution and servicing of finished products comprise the **tertiary sector**. Occupations in the tertiary sector include merchants, grocers, automobile dealers, physicians, attorneys, mechanics, and salespeople. The **quaternary sector** includes those activities associated with management, planning, technology, research, and development. Communication, financial services, scientific research, education, and government activity are also included in the quaternary sector.

Comparing the relative size and importance of the four sectors provides a rough measure of an economy's level of development. Generally speaking, more developed economies are characterized by larger and stronger tertiary and quaternary sectors. In North America, over 80 percent of all workers are employed in the tertiary and quaternary sectors, while the primary and secondary sectors are relatively small. In less developed countries, on the other hand, the percentage of persons employed in primary and secondary sector occupations is considerably higher.

Similarly, the relative importance of the tertiary and quaternary sector tends to increase over time as an economy develops. In North America, the percentage of persons employed in agriculture has declined steadily since 1900, and the percentage of persons employed in manufacturing has declined steadily since the 1950s. The history of the 20th-century North American economy has been one of steady expansion of the tertiary and quaternary sectors relative to the primary and secondary sectors. Every indication is that this trend will continue into the foreseeable future.

Changes in the relative size and importance of the primary, secondary, tertiary, and quaternary sectors also affect the economic bases of communities in which they are concentrated. The **economic base** of a community is that set of economic activities upon which the community relies in order to generate income from elsewhere. For example, the automotive industry is an important component of the economic base of Detroit. Government is the major economic base of Washington, D.C., and the entertainment and aerospace industries contribute heavily to the economic base of Los Angeles. The concept of an economic base is no less important to smaller communities; thus we associate Battle Creek, Michigan, with breakfast cereal; Akron, Ohio, with tires and rubber products; and Sudbury, Ontario, with nickel mining.

Frequently, the economic base of a community is reflected in local and regional cultures. Detroit, for example, is often called the "Motor City" or "Motown," reflecting its history as a center for automobile production. Likewise, many refer to Los Angeles as "Tinseltown." During the 19th century, Cincinnati was called "Porkopolis" in reference to its then-important economic base of meatpacking. The Midland-Odessa region of West Texas is sometimes called the "Petroplex," reflecting the importance there of the oil industry. The historic or contemporary economic base of a city or state has also been used frequently in nicknaming professional sports teams such as the Pittsburgh Steelers, Milwaukee Brewers, Seattle SuperSonics, Orlando Magic (from the nearby Walt Disney World "Magic Kingdom"), and Ottawa Senators.

Within any community, **basic** employment generates **non-basic** employment. Non-basic employees provide goods and services to employees in the basic industries of the community. For example, teachers, retailers, and physicians in the Detroit area provide services to the families of persons who work for General Motors, Ford Toyota, and other automobile companies. The political economy of any society, including North America, is comprised of places with different economic bases and the relationships among these places, as well as between that society and the rest of the world.

The Primary Sector

Few Europeans comparing North America with South America in 1600 would have predicted that four centuries later, North America would be the richest and most powerful region of the world, whereas South America would be less developed. After all, by 1600 such places as Mexico City, Lima, and Buenos Aires were already major cities, whereas few European Americans lived in North America: Only in Florida, New Mexico, and Quebec was there any meaningful amount of European settlement.

Why did North America prosper, while South America stagnated? The wealth of the Spanish and Portuguese colonies in Mexico and South America was associated with the mining of gold and silver. At that time, most Europeans associated wealth with possession of gold and silver; that person or country possessing the most gold and silver was the wealthiest. However, those subscribing to this point of view neglected the fact that gold and silver have little practical value in meeting life's necessities. Gold and silver cannot be eaten, worn, or used to build shelters. North America had relatively little gold and silver, but it did have abundant quantities of resources that could be converted to food, clothing, and shelter, including agricultural land, fish, and timber. As the value of these resources in producing useful goods and services came to be appreciated, North America's resource base became better appreciated and the continent prospered.

Many of the seemingly humble but ultimately valuable resources of North America were associated with the primary sector. During the Colonial era, and for the first century of American independence, the United States and Canada were primarily agricultural societies. Well over 90 percent of Americans earned their livings by farming or agricultural labor (including slave labor) at the time of American independence in 1776. With the development of the secondary, tertiary, and quaternary sectors, the relative importance of the primary sector began to decline. As late as 1900, nearly 40 percent of all Americans lived on farms. Today, barely 1 percent of North Americans earn a living by farming. Even in "farming" states and provinces such as Iowa, North Dakota, and Saskatchewan, only small minorities of the population are actually employed in agriculture.

Despite continued declines in the number of farmers, North America is the world's breadbasket. In 2003, Canada and the United States together produced more than 88 million metric tons of wheat, or 15 percent of the world's total. The United States alone produced over 256 million metric tons of corn, or about 40 percent of the world's total. Vast quantities of food and food products are exported from the United States and Canada throughout the world. For example, in 2002 the United States led the world by exporting 24 million metric tons of wheat; Canada was fourth with 12 million. The United States led the world by exporting 47 million metric tons of corn, and it was third behind Thailand and India in rice exports with more than three million metric tons exported.

Vast quantities of food and food products are exported from the United States and Canada throughout the world. Relative to 1900, there are far fewer farms and far fewer farmers, but the per-acre productivity of each farm has more than offset these declines.

Why has the productivity of individual farms increased so dramatically, while the number of people employed in agriculture continues to decline? To a considerable extent, the increased productivity of North American agriculture is the result of technology. Fertilizers, pesticides, improved tillage and crop rotation practices, new crop varieties, and hybrids have generated per-acre yields undreamed of a century ago. While technological improvement has caused great increases in the per-acre productivity of North American agriculture, the continued introduction of new technology has placed economic stress on large numbers of individual farmers. A farmer's profit is determined by the difference between the cost of producing a crop and the price that the farmer receives by selling it. In order to increase profit, farmers invest in new technologies that increase per-acre yields or decrease the per-acre cost of production. However, increasing yields cause an increase in the overall crop supply, and increased supply depresses prices. To maintain profit margins in the face of lower profits, the farmer must continue to increase production. Further increases in production require the adoption of additional technology. Thus technology improves productivity while lowering prices. This in turn generates incentives to develop more technology, and the process repeats itself.

This process is sometimes called the **technological treadmill**. The technological treadmill forces farmers who are unable to keep up with technological advances out of production. In marginal agricultural areas, their farms may be abandoned or converted to other land uses. In profitable areas, farmers who have been driven out of production usually sell or rent their land to more successful neighbors. Thus the number of farms decreases, while the size of the average farm increases and production becomes concentrated on the most profitable land. Even in highly productive agricultural regions, many are concerned that the continued operation of the technological treadmill may continue to drive farmers out of production or will promote corporate as opposed to individual or family ownership or operation of farmland. As we will see in chapters on the Corn Belt and Great Plains, local farmers and public officials have turned to government policy to counteract or mitigate the impacts of the technological treadmill on local communities.

The effects of the technological treadmill are evident in examining the changing distribution of farming across North America. In areas most highly suitable for crop production, such as the Corn Belt and the Great Plains, the percentage of land under cultivation has not declined very much in recent years. In other areas, such as parts of New England and the South, agriculture has all but disappeared. Despite the declining number of farms and farmers, agriculture continues to play an important role in the economies of many

▶ **Figure 4.2** Nike sportsware manufacturing plant in Vietnam

In recent years, high-technology industries have sprung up in a variety of locations. High-technology industries have been especially prevalent in areas near major research universities. Some, such as Apple and Dell Computers, were founded by university students and later expanded into major corporations. The Silicon Valley of California is located in easy proximity to the University of California at Berkeley and Stanford University in Palo Alto. Other major high-tech centers include the Boston area (Harvard and the Massachusetts Institute of Technology), Seattle (University of Washington), and Austin (University of Texas). In some cases, very specialized high-tech firms arise near universities well known for associated research. For example, Norman, Oklahoma—the home of the University of Oklahoma, the National Weather Center, and the National Severe Storms Laboratory—has become the headquarters for companies that undertake meteorological research, process weather-related data, or manufacture meteorological instruments.

In the contemporary world, a large number of manufacturing firms are multi-locational. Multi-locational firms maintain production facilities in different communities. Many are transnational, with production facilities in many different countries. Because the cost structure of different attributes of the production process differs from one aspect of production to another, multi-locational firms can save money in different aspects of production. For example, the actual work of manufacturing is more labor-intensive than is the research and development that goes into developing new products. Thus actual production plants are located in areas where labor is cheap, while other factors determine the location of corporate headquarters. Nike Shoes, for example, is headquartered in Beaverton, Oregon, but most of the actual manufacturing occurs in China, Indonesia, Vietnam, and Malaysia (Figure 4.2).

In general, corporate headquarters tend to be located in large metropolitan areas (Figure 4.3). Historically, corporate headquarters were generally found in the downtown areas of major cities in the Northeast. In recent years, many have relocated to suburban locations or to the Sun Belt. Either way, location in a large metropolitan area provides convenient access to transportation and communications.

The Tertiary Sector

As we have seen, a very important component of the development of the North American political economy in the 20th century has been the shift in employment from the primary and secondary to the tertiary and quaternary sector. Tertiary-sector or service-sector employment includes such activities as personal services such as health care and retailing (see Figure 4.4). Historically, such employment was distributed relatively evenly across North America relative to the overall distribution of population. Most services were provided by individual entrepreneurs and small firms with predominantly local clienteles. Every small community had its own locally owned bank, hotel, grocery store, dry-goods store, and other retail enterprises whose customers were generally local residents and nearby farmers. Physicians, dentists, lawyers, barbers, and other tradespersons provided various personal services to local inhabitants as well. Especially outside the large manufacturing cities of northeastern North America, a major raison d'etre for small urban communities and towns was the provision of services to local farmers

◄ **Figure 4.3** New York City is a center of corporate headquarters as are many other large cities in North America

and other local residents. The service sector generally consisted of non-basic employees whose income was generated indirectly from the basic economic activities associated with the communities in which they lived and worked.

The increasing dominance of tertiary-sector employment in the 20th century was accompanied by a shift from individual entrepreneurship to multi-locational, corporate service provision. Many retail enterprises are still owned by individual operators or families, but a large majority of retail sales in North America today are recorded at outlets of large retail shopping chains. Corporations such as Wal-Mart, K-Mart, and Dayton Hudson (Target) maintain thousands of large department stores in communities throughout the United States, Canada, and elsewhere. As well, large corporations dominate retail markets for various specialty items such as Tru-Value, Ace, Home Depot, and Lowe's in hardware; Albertsons and Safeway in groceries; Eckerd and Walgreens in pharmacies; and many others. Corporate management has also pervaded the provision of many personal services, as is evident to anyone eating at McDonald's or Kentucky Fried Chicken, drinking coffee at Starbucks, getting a haircut at Supercuts, buying eyeglasses at Pearle, or staying overnight in a Motel 6 or Holiday Inn.

▲ **Figure 4.4** Health care is a large tertiary sector industry

The growth of large retail outlets in the provision of goods and personal services has allowed certain communities to specialize in the tertiary sector. For example, Wal-Mart's emergence as the world's largest retailer has created hundreds of jobs in and near its Bentonville, Arkansas, headquarters, making northwest Arkansas one of North America's leading centers for retail-based employment. In northwest Arkansas, therefore, retailing has become a basic rather than a non-basic activity. No longer are individual retail outlets tied to local communities.

Another major component of growth in the tertiary sector has been the development of tourism and the tourism industry (Figure 4.5). In 1900, few North Americans had the time or money to travel very far from home. Only the wealthy had sufficient leisure or sufficient income to take pleasure trips. By the 1920s, however, leisure travel had become more commonplace among middle-class North Americans. More and more Americans held non-farm jobs. By the end of World War II, most employers granted paid vacation time to their employees. Moreover, the mass diffusion of the automobile gave people the opportunity to take long-distance trips. In many parts of North America, tourism became a major contributor to the local economy.

Many different types of places emerged as tourist destinations. These included large cities, national parks and national monuments, beaches, and other resorts.

▲ **Figure 4.5** Remote fishing resort in Alaska

Gambling became a focus of tourism in Las Vegas, Atlantic City, and more recently in small communities such as Tunica, Mississippi. Other communities have constructed amusement parks in order to attract tourists. A prime example, of course, is Orlando, where Walt Disney World has become one of the world's most popular tourist attractions and where numerous other theme parks and tourist sites have been developed.

Ironically, retailing has itself become an important locus for tourist activity. The Mall of America outside Minneapolis is one of the Upper Middle West's leading tourist attractions. Factory outlet malls, many of which are located outside major metropolitan centers, have become popular tourist destinations as well. The factory outlet mall near San Marcos, Texas, in fact, is the Lone Star State's fourth leading tourist attraction.

The Quaternary Sector

The quaternary sector includes government, research and development, education, journalism and the media, and similar activities. Unlike tertiary-sector employment, quaternary-sector employment has long been concentrated in relatively few places. Government functions are concentrated in Washington and Ottawa at the federal level, in state and provincial capitals, and in county seats. The growth of government activity during the 20th century has created economic booms in Washington and Ottawa as well as in many state capitals.

Of course, not all government expenditures are concentrated in state capitals. Especially during and after World War II, military and defense spending formed a significant component of federal spending in the United States and to a lesser extent in Canada. Much of these funds were spent on military bases. In the United States, the lion's share of military installations has been located in the South and in the Sun Belt. The concentration of military bases in the Sun Belt is due to a number of factors, including the influence of powerful Southerners in Congress, a culture more favorably disposed to military activity as a way of life, and the presence of infrastructure supporting military needs. Military base communities generate substantial levels of civilian employment. For example, Tinker Air Force Base outside Oklahoma City is Oklahoma's largest single-site civilian employer. In addition, the presence of military personnel and civilian employers generates substantial demand for teachers, physicians, merchants, and other non-basic employees.

Research and development forms a significant share of the economy in many places. Many research and development activities are funded by government contracts. Especially during the Cold War, a large share of government-funded research and development activities were associated with military and defense-related issues. These tended to be concentrated heavily along the East and West Coasts and in the Sun Belt. Indeed, the term "Defense Perimeter" has been used to describe the concentration of United States military spending along the Atlantic Coast, westward across the South to California, and north along the Pacific to Seattle and Alaska. Relatively little government research and development funds are spent in peripheral areas outside the Sun Belt.

Other research and development activity is associated with the private and non-profit sectors. Large corporations maintain extensive research and development operations, which in many cases are located near corporate headquarters. Research and development activity is especially critical to the development of high-technology industry, and areas like California's Silicon Valley are well known for the importance of research and development to local economic bases. Rochester, Minnesota's Mayo Clinic has long been recognized as an important center for medical research (see Figure 8.5).

Over the course of the 20th century, the role of formal education in North American culture and society has increased dramatically. In 1900, only a small minority of persons in North America graduated from high school or attended college. The large majority of persons left school as teenagers and entered the labor force. During the early 20th century, however, business and the professions began to expect or require higher levels of formal education. For example, by the end of World War II nearly every state required its teachers to graduate from college. Lawyers, who prior to the 20th century could be admitted to the bar following private study in the office of an experienced attorney, were now expected to complete college and formal legal training in law schools. As a result of these trends, the number of years of schooling completed has increased dramatically over the course of the 20th century, while income gaps between those who have completed college and those who have not have widened substantially.

The result of this increased emphasis on higher education has been economic growth and development in college and university communities. Places such as Austin, Tallahassee, Columbus, and Madison have grown in response to the combined impacts of state universities and state governments. In small, peripheral states, college and university communities have become growth centers even when their states have not grown. For example, the college town of Iowa City, Iowa, ranked ninth in population in Iowa in 1960 but ranks fourth in the state today. West Virginia, Wyoming, South Dakota, and Arkansas are other peripheral states where growth has been concentrated in college towns relative to the rest of their states.

Another important aspect of quaternary-sector growth has involved journalism, publishing, the media, and entertainment. For many years, the cities of New York and Los Angeles have been major media centers in the United States. In recent years, cities such as Atlanta, Vancouver, Orlando, and Nashville have become increasing important media centers. The entertainment industry is often linked to tourism. Tourists visiting southern California often take trips to Hollywood as well as to the homes of movie and television stars and to production studios. Las Vegas, Atlantic City, and Branson, Missouri, are other places whose tourist industries are closely linked to the entertainment industry.

The Changing Urban System of North America

The development of North America's political economy has also had dramatic impacts on settlement patterns. Prior to 1900, as we have seen, a large majority of North America's people worked in the primary sector and lived in rural areas. Urbanization, which was already evident before 1900, became much more commonplace in the 20th century. Changes in the economy of North America during the 20th century have contributed to a dramatic restructuring of the continent's system of settlement since 1900.

This is evident from examining lists of the ten largest cities in the United States in 1850, 1960, 1980, and 2004 (Table 4.1). In 1850, all but one of the three of the largest

TABLE 4.1 Total Population of the Ten Largest Cities in the United States: 1850, 1960, 1980, 2004

1850		1960		1980		2004	
New York City	515,547	New York City	7,781,984	New York City	7,071,639	New York City	8,104,079
Baltimore	169,054	Chicago	3,550,404	Chicago	3,005,072	Los Angeles	3,845,541
Boston	136,881	Los Angeles	2,479,015	Los Angeles	2,966,650	Chicago	2,862,244
Philadelphia	121,376	Philadelphia	2,002,512	Philadelphia	1,688,210	Houston	2,012,626
New Orleans	116,375	Detroit	1,670,144	Houston	1,595,138	Philadelphia	1,470,151
Cincinnati	115,435	Baltimore	939,024	Detroit	1,203,339	Phoenix	1,318,041
Brooklyn	96,838	Houston	938,219	Dallas	904,078	San Diego	1,263,756
St. Louis	77,860	Cleveland	876,050	San Diego	875,538	San Antonio	1,236,249
Spring Garden, PA	58,894	Washington	763,956	Phoenix	789,704	Dallas	1,210,393
Albany	50,763	St. Louis	750,026	Baltimore	786,775	San Jose	904,522

Source: U.S. Bureau of the Census, Census Population, 1850 and census tabulations in 2004 (*Top 50 Cities in the U.S. by Population and Rank*).

between the two countries has remained the world's longest unfortified boundary. Canada was centrally located relative to United States objectives in the Arctic and along air routes between the Western and Eastern Hemispheres. The foreign policy of Canada has been tied closely to that of the United States throughout the 20th century, although at times many Canadians have expressed opposition to what they regard as American heavy-handedness and over-commitment in world politics.

After World War I, the United States retreated from ongoing involvement in global geopolitics. Such was not the case after World War II, however. After six years of war, the economy, population, and landscape of Europe was devastated, and the United States and the Soviet Union had emerged as the world's leading superpowers. Soon after World War II ended, the "cold war"—pitting the United States and her democratic, Western allies against the Soviet Union and her communist, Eastern allies—began. The United States and Canada moved to stabilize relationships with Western Europe and to increase their influence in other parts of the world.

The cold war ended abruptly in the late 1980s and early 1990s with the collapse of communism in Eastern Europe and the Soviet Union. The end of the cold war left the United States as the world's remaining superpower. By century's end, there was no doubt that the United States was the leading political and economic power in the world. The collapse of communism and the demise of state economic planning in favor of the market economy of the United States and her allies spurred continued globalization of the world economy, although many persons in both the United States and Canada remained skeptical about whether globalization was in North America's interests.

Trade, Transportation, and Communications

The establishment and expansion of both internal and external trade networks, the development of surface and air transportation, and the establishment of global communication networks were all critical to the development of North America into the world's leading global, economic, and political power. Over the course of the 20th century, trade between North America and the rest of the world increased exponentially. This increase in international trade was critical to North America's rising influence in the global economy.

Throughout the century, the United States and Canada have been major trading partners of each other. For the most part, this symbiotic trade relationship has benefited the economies of both countries. However, Canada has frequently undertaken policy steps to ensure that its economy remains independent and not totally dominated by its much larger southern neighbor. The question of how much to encourage trade between North America and other areas of the world has been a matter of ongoing political debate. There are two major schools of thought concerning international trade policy. Those advocating protectionism argue for high tariffs, or taxes on importing and exporting goods across international boundaries. The intent of such tariffs is to protect domestic producers from foreign competition. Advocates of free trade, on the other hand, argue that protective tariffs should be reduced or eliminated, and that international exchange is beneficial rather than harmful to the domestic economy.

In general, the northeastern industrial region has been supportive of protective tariffs throughout the 20th century, while the South and the West have been more supportive of free trade. This is evident from looking at the distribution of votes in the House of Representatives on the Smoot-Hawley tariff bill of 1929. The Smoot-Hawley tariff bill provided for import duties of as much as 40 percent on imported industrial products from outside the United States. The bill was opposed strongly by Democrats in the South and West, but the Republican majority in the House of Representatives passed it with especially strong support in the northeastern industrial core.

During the late 1920s, other industrial powers in addition to the United States enacted high protective tariffs similar to the Smoot-Hawley Tariff. The resulting slowdown in trade reduced demand for industrial goods, and many economists regarded this slowdown as a cause of the worldwide depression of the 1930s. After World War II, economists throughout the world took steps to discourage high protective tariffs in the future. As a result, the world's leading powers signed the General Agreement on Trade and Tariffs (GATT). This agreement, which is actually a series of bilateral trade agreements among countries, is designed to discourage protective tariffs and promote free trade.

In the late 20th century, many countries throughout the world joined with their neighbors to establish regional trade blocs. The United States, Canada, and Mexico signed the **North American Free Trade Agreement (NAFTA)** in 1993. The NAFTA agreement called for the free movement of goods between the three countries.

The NAFTA proposal was highly controversial in both Canada and the United States. In both countries, legislative approval of NAFTA occurred only following intense, bitter political debate. The NAFTA agreement was opposed strongly by northeastern Democrats in the United States. Many northeastern Democrats represented highly unionized and minority-dominated districts, and their constituents feared that NAFTA would cause significant loss of jobs as employers moved their operations to Mexico or other low-wage areas. The Sun Belt, on the other hand, strongly supported NAFTA. For example, all 30 members of Congress from Texas, both

Democrats and Republicans, supported the NAFTA bill. More than a decade after NAFTA was initiated, it is evident that the agreement has had significant impacts on local economies and landscapes, as we will see in detail in some of the regional chapters. The NAFTA vote, like other political debates involving trade and foreign policy, also symbolizes the ongoing changes in the internal political geography of North America.

Political and Economic Geography of North America

Changes in North America's position in the global economy had considerable influence in domestic politics in both countries as well. The political geography of both countries has evolved in response to both international and domestic events throughout North America's history.

To put these comments into context, let us briefly examine the political structure of the United States and Canada. Both countries are **federal states**, in which governmental power is shared formally between the federal governments based in Washington and Ottawa respectively, and the state or provincial governments. The U.S. Constitution spells out the relationship between federal and state authority. Of particular importance is the 10th Amendment, which specifies that all powers not explicitly granted to the federal government in the Constitution "are reserved to the States or to the people."

These considerations influence the structure of electoral competition in both countries. Both have explicitly defined legislative, executive, and judicial branches of government. In both countries, the people elect representatives to federal, state or provincial, and local legislative bodies. In both cases, geography is a significant basis of representation in that districts used to elect representatives to the U.S. House of Representatives and state legislatures, as well as to Canada's House of Commons and provincial legislatures, are delineated territorially, with each representative accountable primarily to the residents of the territorially defined district from which he or she has been elected. However, the relationships between the branches of government vary considerably between the United States and Canada. The Canadian political system is modeled after that of the United Kingdom. The chief executive, or Prime Minister, is simultaneously a legislator and a member of the House of Commons. The Prime Minister is also the leader of the party or coalition of parties holding a majority of seats in the House of Commons. Should that party lose its majority, or should the coalition collapse, the Prime Minister resigns and is replaced by the leader of the new majority. In the United States, in contrast, the legislative and executive branches are separated. The President is elected independently of the House of Representatives and the Senate, and the President may or may not be a member of the party holding a majority in either or both houses of Congress.

The U.S. President is elected through a system known as the Electoral College. Seats in the House of Representatives are apportioned among the states on the basis of population, and each state has two members of the Senate regardless of size. Each state has representation in the Electoral College in accordance with its representation in Congress. Thus Kansas, for example, with two Senators and four Representatives, has six electoral votes. The total number of electoral votes is 538, with the Constitution requiring a majority to secure the Presidency.

These considerations influence the electoral geography of North America. In both countries, parties compete with each other for electoral advantage across territorially defined states, provinces, and districts. The party that is successful in splicing together a majority coalition wins control of the Presidency, of either house of Congress, or in the case of Canada's parliamentary system, of both.

The first half of the 20th century was characterized by a remarkable stability in the geographic structure of partisan politics in the United States. The "Solid South" was controlled by the Democratic Party, while most of New England and the upper Middle West were heavily Republican. As a result, elections were won and lost in the Middle Atlantic and Great Lakes states. In all but one Presidential election between 1900 and 1956, at least one major-party candidate for President came from New York, Ohio, or Illinois. The West was also volatile politically, but its sparse population left this section with little influence on national politics before the second World War.

Since the 1970s, an entirely different pattern has emerged. The Democrats lost their grip on the once-solid Democratic South, while New England and the upper Middle West became strongly Democratic. The West, except for the Pacific Coast states, became heavily Republican. In general, the Democrats are strongest in central cities, in areas with large numbers of minority voters, in places oriented to the primary and secondary sectors of the economy, and in government-oriented and academic communities. The Republicans, on the other hand, are strongest in traditionally Republican rural areas such as the Great Plains and Appalachia, in outer suburbs, in areas oriented to the tertiary sector, and in military base communities. As a result, the key battlegrounds in national elections today are suburbs and the Sun Belt, especially the South. Strong support in these areas was critical to the electoral success of Republican Ronald Reagan in the 1980s and Democrat Bill Clinton in the 1990s, whereas in 2000 and 2004 George W. Bush swept the entire South, narrowly winning the

Presidency. Indeed, not since John Kennedy in 1960 has a President without roots in the Sun Belt been elected.

Conclusion

The 20th century was a remarkable period in the history of North America. The United States and Canada emerged as the strongest, most productive, and wealthiest places in world history. The transition of North America to global political and economic leadership was associated with profound changes in the economic, demographic, and cultural characteristics of both countries. As we begin to examine individual regions of North America in subsequent chapters, we will look at how these changes have affected individual regions and how the impact of such changes is reflected in the regions' cultural landscapes and in the lifestyles of their residents.

Suggested Readings and Web Sites

Books and Articles

Borchert, John R. 1987. *America's Northern Heartland: An Economic and Historical Geography of the Upper Midwest*. Minneapolis, MN: University of Minnesota Press. Historical economic geography of the region focusing on the Twin Cities and their relationship to the upper Midwest.

Bureau of the Census. 2000. *Census of Population*. Washington, D.C.: United States Bureau of the Census. Provides decadal information on economic, social, population, and other variables useful in doing research and teaching about the United States.

Castells, Manuel. 1996. *The Information Age: Economy, Society, and Culture*. Vol. I, *The Rise of the Network Society*. Cambridge: Blackwell. A global perspective on the role of the United States, Canada, and other nation-states in the shaping and reshaping of today's information age.

Cronin, William. 1991. *Nature's Metropolis: Chicago and the Great West*. New York: Norton. Historical geography focusing on relationships between Chicago and the agricultural and mineral-producing areas of the West.

Dreidger, L. ed. 1987. *Ethnic Canada: Identities and Inequalities*. Toronto: Copp Clark Pittman. Analysis of ethnic settlement in Toronto and other parts of Canada.

Hudson, John C. 1993. *Crossing the Heartland: Chicago to Denver*. New Brunswick, NJ: Rutgers University Press. Historical examination of transportation, settlement, and culture in the Corn Belt.

Johnson, Hildegard Binder. 1976. *Order Upon the Land: the U.S. Rectangular Land Survey and the Upper Mississippi Country*. New York: Oxford University Press. Analysis of impacts of the Northwest Ordinance upon the land settlement patterns in the Great Lakes and Corn Belt regions.

Lockridge, Ross, Jr. 1947. *Raintree County*. Boston: Houghton Mifflin. Historical novel about a rural county in Indiana during and after the Civil War.

Meyer, David R. 2003. *The Roots of Industrialization*. Baltimore: The Johns Hopkins University Press. A well-written analysis of the beginnings of industrialization in North America and its impact on subsequent landscapes and economic systems.

Stanford, Q.H., ed. 1998. *Canadian Oxford World Atlas*. 4th ed. Toronto: Oxford University Press. Comprehensive atlas with mapped data showing Canada's economic, cultural, and political patterns, population distribution, and environmental features.

Warkentin, J. 1997. *Canada: A Regional Geography*. Scarborough: Prentice Hall. A well-written geography text on Canada focusing on the nation's various regions, peoples, and landscapes.

Weller, Phil. 1990. *Fresh Water Seas: Saving the Great Lakes*. Toronto: Between the Lines Publishers. Environmental history and geography of the Great Lakes.

Statistics Canada. 2006. *Census of Population*. Ottawa: Statistics Canada. A rich source of information on Canada including population data, economic and social statistics, and other variables (comparable to the U.S. Census Bureau in scope and depth).

Wheeler, James, Yuko Aoyama, and Barney Warf, eds. 2000. *Cities in the Telecommunication Age: The Fracturing of Geographies*. London: Routledge. An excellent overview of the structures, functions, and linkages of today's cities as key players in the information age.

Zukin, Sharon. 1991. *Landscapes of Power: From Detroit to Disneyland*. Berkeley: University of California Press. A must-read for all students interested in the geography of North America; one of the most cited sources on urban, economic, and cultural landscapes in North America.

Web Sites

CIA World Factbook
http://www.cia.gov/cia/publications/factbook/

The Library of Congress: American Memory Railroads
http://memory.loc.gov/ammem/gmdhtml/rrhtml/rrintro.html

NAFTA Information
http://www.fas.usda.gov/itp/Policy/NAFTA/nafta.html

U.S. Employees by Economic Sector
http://www.census.gov/econ/census02/

5
The Atlantic Periphery

PREVIEW

Environmental Setting

- Much of the Atlantic Periphery is part of the Appalachian Mountain range. The Appalachians reach the sea in most parts of the Atlantic Periphery, and the coastal plain is very narrow or absent. The western portion of the Atlantic Periphery is part of the Canadian Shield.
- The topography of the Atlantic Periphery was strongly affected by glaciation over the past two million years.
- The climate of the Atlantic Periphery is cool and wet, with cold, snowy winters and mild, humid summers.

Historical Settlement

- The original inhabitants of the Atlantic Periphery were part of the Northeast Area Cultural Complex. Many of the original Native American cultures of the Atlantic Periphery were wiped out following exposure to European diseases.
- The Atlantic Periphery was the first part of North America to be settled by Europeans. Control over the area was disputed between the British and the French, and later between the British and the Americans.
- The population of the Atlantic Periphery has been dominated by persons of British Isles origin since the mid-18th century.

Political Economy

- The Atlantic Periphery is an economically poor and impoverished area, and for centuries many of its residents have struggled to achieve adequate living standards.
- The cool climate and poor soils have meant that agriculture in the Atlantic Periphery is generally marginal. In a few areas, specialized crops continue to be commercially produced.
- Forestry and fishing have long been important to the economy of the Atlantic Periphery, but their importance has declined in part because of excessive logging and overfishing. Industrial activity has been limited because of the region's peripheral location and resource-oriented economy.
- In recent years, tourism and recreation have become important contributors to the Atlantic Periphery's economy.

Introduction

In a small rural community along the coast of Maine, an elderly man celebrated his hundredth birthday surrounded by family and friends. One guest asked the old gentleman if he had lived in the same town all of his life. "Not yet!" he replied.

This apocryphal story illustrates the very strong sense of place felt by many persons living in a region that for many North Americans is off the beaten path. The Atlantic Periphery—the northeastern corner of North America, including Canada's Atlantic Provinces and northern New England—is a region that is rich in history and diverse in character but is little known to many. The Atlantic Periphery was the first part of North America to be explored and settled by Europeans. Leif Ericsson established a settlement in present-day Newfoundland a thousand years ago. Five hundred years later, British and French explorers saw the Atlantic Periphery as a gateway to the riches of North America, and they contested vigorously for control over the region.

Despite this long and colorful history, the Atlantic Periphery is an area highly dependent on outside economic and political forces, both in Europe and in North American cities to the south and west. Its cool and damp climate, rugged topography, and isolation have made it difficult for many parts of the Atlantic Periphery to sustain themselves. Like the Great Plains (Chapter 11) and the Appalachians (Chapter 9), the Atlantic Periphery has exported people for decades. Many parts of the region have experienced steady population losses, as natives of the region depart for western Canada, the Sun Belt, or large metropolitan areas. Yet the attachment to place shared by many natives of the Atlantic Periphery is strong. President Calvin Coolidge, a native of Vermont, is said to have boasted that no member of his family ever moved west. Like the flinty and stubborn New Englander beginning his second century of life on the Maine coast, many a native of the Atlantic Periphery maintains a strong allegiance to his homeland, despite economic and environmental hardships. The Atlantic Periphery—romanticized in classic stories for children such as *Rebecca of Sunnybrook Farm* and *Anne of Green Gables*—may experience a renaissance in the 21st century as people opt out of the urban rat race and rediscover their attachments to local communities.

◀ An aerial view of a lighthouse on Cape Tryon, Prince Edward Island, Canada

Environmental Setting

The Atlantic Periphery contains the Canadian provinces of Newfoundland and Labrador, Nova Scotia, Prince Edward Island, and New Brunswick. Collectively, these four provinces are known as the "Atlantic Provinces." The province of Newfoundland and Labrador includes the island of Newfoundland along with Labrador on the mainland to the northwest. In 2001, the province changed its name officially from "Newfoundland" to "Newfoundland and Labrador." In this chapter, we use the term "Newfoundland and Labrador" to refer to the province, and "Newfoundland" to refer to the island. On the American side of the international boundary, the Atlantic Periphery includes most of Maine and New Hampshire, all of Vermont, and northeastern New York State (Figure 5.1). The region's rugged and sometimes spectacular scenery, its proximity to the ocean, and its cool, humid climate have all influenced the area's long and colorful history and continue to influence the region today.

Landforms of the Atlantic Periphery

Much of the Atlantic Periphery is part of the Appalachian Mountain region, which stretches northeastward from Alabama to eastern Canada. As we saw in Chapter 2, the Appalachian chain is a very old mountain range that has been affected by millions of years of erosion. During the ice ages of the past two million years, all but the highest of the region's mountains were covered with ice. As a result, even the highest peaks of the Atlantic Periphery are not nearly as high as those in the western United States and Canada. Mount Washington in New Hampshire is the highest peak in the Atlantic Periphery at 6288 feet (1920 meters) above sea level.

The retreating ice also scraped away much of the region's soil cover, leaving most of the area with thin, rocky soil that is not very useful for agriculture. In inland areas, there are numerous lakes and ponds, many of which are very popular with campers, hunters, fishermen, and tourists. To the north and west of the Appalachian region is the Canadian Shield, discussed in more detail in Chapter 18. The Adirondack Mountains of New York are a large plateau region that is part of the Canadian Shield, which also includes the uplands of northern Quebec and Labrador. The topography of this region was also affected considerably by glaciation.

The coastal plain that is prevalent farther to the south is largely absent in the Atlantic Periphery, and therefore there is little flat land near the coast. The coastlines of the region are rugged and spectacular, with numerous harbors, some of which are submerged river valleys that were modified by glaciation. Many of these harbors became the sites of cities that have been settled for several centuries.

The rugged coastal topography which created these fine harbors has also impeded overland transportation. Until the advent of the automobile and the airplane, most long-distance transportation was by sea. Even today, residents of small coastal communities tend to rely on boats for local transportation. In general, relatively few people and few human activities are found in the mountains. Most people and most of the region's larger cities and towns, including Burlington, Vermont; Portland, Maine; St. John's, Newfoundland; and Halifax, Nova Scotia, are located along the coast and in the valleys of the Connecticut, Penobscot, Saint Lawrence, and other rivers.

Climate and Hazards

The climate of the Atlantic Periphery reflects the region's northern location and rugged topography. Generally speaking, the region has a climate characterized by cool summers and cold, snowy winters. Precipitation levels are quite consistent from month to month throughout the year, and snow covers the ground for several months each winter in many locations.

Because weather systems in North America tend to move from west to east, the ocean has relatively little moderating influence on large-scale weather patterns. However, the ocean has considerable local influence on temperature and precipitation. Temperatures near the coast are cooler in summer and milder in winter than at inland locations. The complex pattern of ocean currents off the coast of the Atlantic Periphery also influences the region's climate. The cold Labrador Current drifts southward immediately offshore, and the warm Gulf Stream flows to the east of the Labrador Current. Offshore winds blowing across the Gulf Stream and the Labrador Current contribute to frequent maritime fogs. Fog is prevalent in most coastal areas of the Atlantic Periphery, especially in Newfoundland and along the New England coast. These fogs result in cooler temperatures, contribute to cloud cover, and sometimes cause hazards to shipping and aviation.

Despite these fogs, generally speaking the Atlantic Periphery does not experience catastrophic natural hazards of the magnitude of the earthquakes of California, the tornadoes of the southern Great Plains, or the hurricanes of Florida. Nevertheless, residents of the Atlantic Periphery must contend with several distinctive types of natural and climatic hazards. Heavy snowfalls, sometimes accompanied by blizzard conditions, often occur in winter. Many communities in northern New England and in Canada's Atlantic Provinces average more than a hundred inches (250 centimeters) of snow each winter. Along the coast, some of this snow is brought in by *nor'easters*, or coastal storms associated with winds moving from the northeast to the southwest. Nor'easters can bring as much as two or three feet (0.6 to 0.9 meters) of

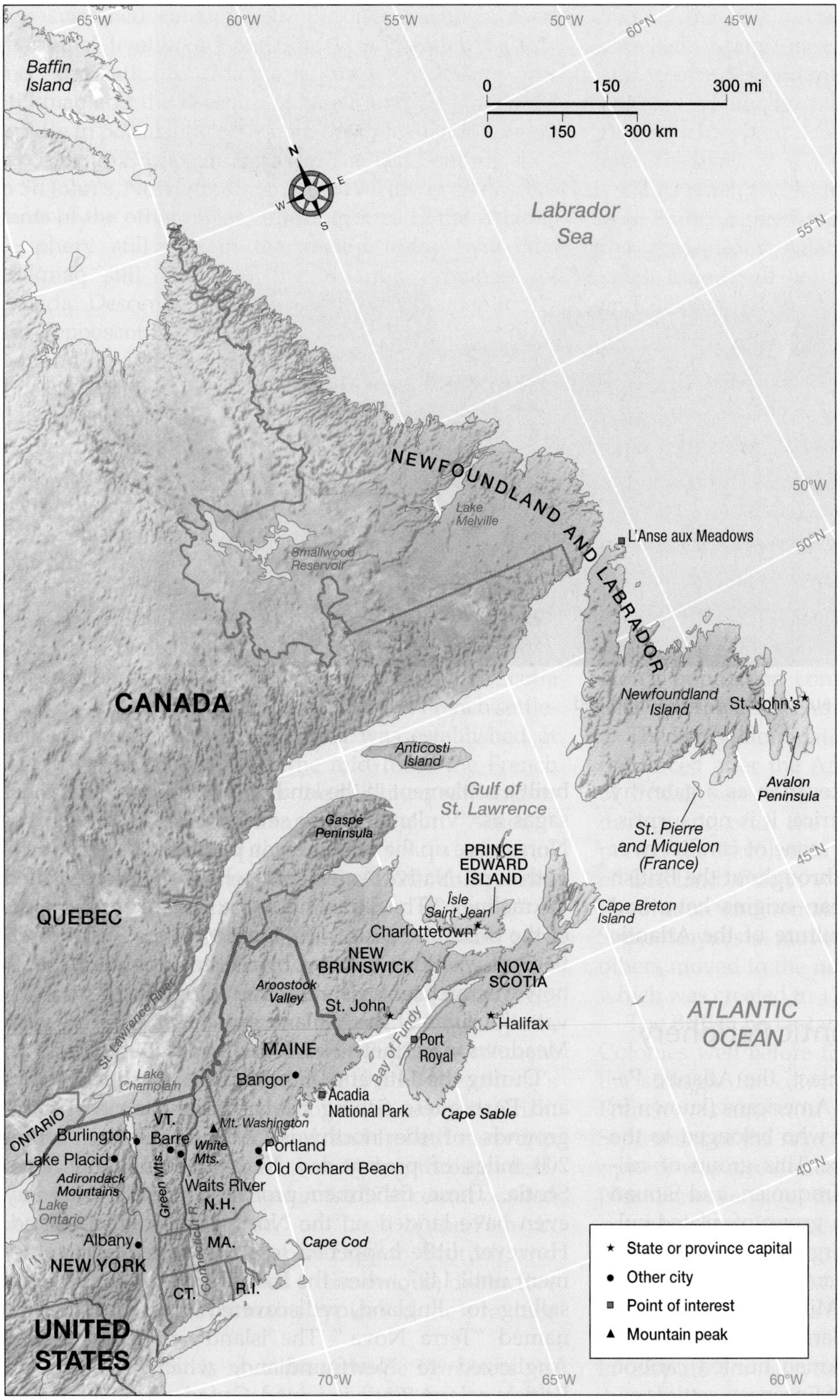

Figure 5.1 The Atlantic Periphery region

snow along with high winds and blizzard conditions. Hurricanes and tropical storms moving northward from the Caribbean and the south Atlantic coast occasionally bring high winds and heavy rains to the coasts of Maine, New Brunswick, and Nova Scotia in late summer and early fall.

Historical Settlement

The settlement history of the Atlantic Periphery is long and colorful. The region was home to a variety of Native American cultures prior to European contact. It was the first part of North America to experience

▲ **Figure 5.5** Landscape of Aroostook County, Maine

in high demand in England and western Europe, whose forests had long since been cleared. The tall, straight trunks of white pine trees, some of which grew to heights of nearly 200 feet (60 meters), were converted into masts for military and commercial ships. Commercial logging has been important to the Atlantic Periphery's economy for hundreds of years.

Other settlers turned to the sea in order to make a living. The fishing grounds off the coasts of Newfoundland and Nova Scotia had been discovered and exploited decades before Columbus crossed the Atlantic. Fisherman caught huge quantities of cod, haddock, and many other species from the shallow continental shelf. By the 17th century, Newfoundland had become well established as a center for commercial fishing.

Although forestry and fishing remain important components of the Atlantic Periphery economy, the future for both of these primary-sector activities is now uncertain. The huge stands of virgin forest discovered by the region's 17th-century settlers have been almost entirely cut down. Much of the land remains forested, but the quality of lumber in the second- and third-growth forests that remain is much lower than what was once the case. Timber production for pulpwood remains significant in northern Maine and parts of New Hampshire. However, trees grow slowly in the cool climate of the region. Timber production in the Southeast, where the warmer climate allows trees to grow faster, is outpacing that of the Atlantic Periphery. Forestry is relatively more important in the Atlantic Provinces of Canada, with extensive logging operations and paper milling taking place, especially in New Brunswick and Newfoundland.

Fishing has also been a major contributor to the Atlantic Periphery's economy for hundreds of years. Returning from Newfoundland to England in 1497, John Cabot reported that codfish off the Newfoundland coast were so thick that they could be caught by hanging wicker baskets off the sides of ships. In the words of another commentator, "there were six- and seven-foot long codfish weighing as much as 200 pounds (90.7 kilograms). There were great banks of oysters as large as shoes. At low tide, children were sent to the short to collect 10-, 15-, even 20-pound (4.5, 6.8, 9.1 kilogram) lobsters with hand rakes for use as bait or pig feed." Before modern refrigeration was invented, dried and salted codfish could be kept for long periods and were relatively light and easy to transport. Salted codfish from the waters off the Atlantic Periphery became an important source of animal protein throughout Europe.

Residents of the coastal Atlantic Periphery harvested codfish and other marine animals in great numbers for centuries, with little apparent impact on fish populations. In the 1950s, however, large and highly mechanized fishing trawlers as large as small ocean liners began to appear in the Grand Banks and other fishing areas. These ships weighed as much as 8000 tons (7280 metric tons) and towed nets as long as two-thirds of a mile in circumference. Radar, sonar, echograms, and other fish-finding technologies were implemented. These trawlers—in effect floating fish factories—could haul up as much as 200 tons (182 metric tons) of fish an hour—twice as much as a 16th-century fishing ship could catch in an entire year. While earlier fishing was limited to daylight hours in the summer, these highly

BOX 5.2 The Outports of Newfoundland

As early as the late 1600s, Newfoundland was known throughout North American and Europe as a major fishing center. Many Newfoundland fishermen lived in small fishing villages known as outports (Figure 5.A). Outports were located wherever there were suitable harbors along with sufficient space to construct houses and collect wood for building construction and the drying of salted fish. Some outports were inhabited year-round; others were abandoned in winter, with local residents moving to less exposed locations. The outports' economies were sustained almost entirely by fishing. Transportation between outports was largely by boat, and outports were therefore isolated from the outside world during storms and when harbors became icebound in winter.

By the middle of the 20th century, government officials became concerned that the isolation and small size of many outports made the provision of education, health care, and other services inefficient and uneconomical. Between 1954 and 1972, the Newfoundland government encouraged consolidation by inducing people to move from smaller to larger outports, abandoning the smaller ones. By the 1980s, nearly half of Newfoundland's more than 1200 outports had been emptied.

The economies of the remaining outports, however, were threatened by continued overfishing. In response, the Canadian government imposed a temporary ban on cod fishing off the Atlantic coast in 1992. Given the importance of fishing to the economies of Newfoundland, this ban proved to be devastating to outports and their residents. In an effort to alleviate hardship, the Canadian government provided fishermen, cannery workers, and others who lost their jobs as a result of the ban with payments equivalent to their previous wages for two years, provided that they would seek retraining in other fields. The government's intention was to encourage resettlement of outport residents to Ontario and western Canada, but many persons retained strong place ties and declined to move.

▲ **Figure 5.A** Newfoundland outport of Petty Harbour

mechanized ships hired rotating crews and were operated 24 hours a day year-round. By the 1970s hundreds of factory trawlers registered in Britain, Japan, Spain, France, West Germany, Iceland, and other countries were harvesting fish off the Atlantic Periphery coast. In effect these factory ships strip-mined the seas. Not surprisingly, by this time the annual harvest of cod and other fish began to decline significantly. In response, the U.S. and Canadian governments began to impose limits on fishing, but these limits had severe impacts on coastal fishing communities, especially in Newfoundland (Box 5.2).

Mineral resources in the region include granite and marble in Vermont and iron ore in upstate New York and Labrador. Historically, however, mining and mineral exploitation have never been of more than local significance to the economy of the Atlantic Periphery. This may change, however, in the years to come. Oil and gas exploration has taken place off the coast of Nova Scotia and Newfoundland for decades. By the turn of the 21st century, three natural gas reservoirs with a combined productive capacity of over 400 million cubic feet went into production near Sable Island, located 200 miles (321.9 kilometers) east of the Nova Scotian mainland. The $2 billion Sable Island project is complemented by a $3.7 billion dollar oil production project off the Newfoundland coast. Some 150,000 barrels of oil a day are currently pumped through this Hibernia oil project. Also, the provincial government of Newfoundland has signed an agreement to exploit one of the world's largest nickel deposits at Voisey's Bay in Labrador. This development is expected to provide hundreds of permanent jobs and as much as 10 billion dollars in revenue to the provincial government.

Manufacturing, Innovation, and Trade

For the most part, the Atlantic Periphery did not share in the large-scale development of industry in North

America. The region's rugged terrain impeded transportation, and its relatively small population reduced available markets for industrial products. Many of the few industries that did develop in this region were oriented to natural resources, notably forestry and fishing. Although many of the once-commonplace pulp and paper mills of northern New England and Atlantic Canada have been closed, paper production continues in places such as northern New Hampshire, western Maine, and New Brunswick. Fish processing has long been an important activity as well. In the outports of Newfoundland and in fishing villages along the Maine and Nova Scotia coasts, for centuries fishermen have dried, salted, and sold cod and other fish. Canneries, in which fish products are produced for sale on commercial markets, are also common. Firms doing business on the Internet ship lobsters from Maine to gourmet restaurants and individual customers throughout the world.

Other businesses and industries located in New England in recent years are associated with the region's culture, place image, and resource base. For example, many associate Vermont with Ben and Jerry's ice cream, reflecting the importance of dairy farming to the Green Mountain State. One of Maine's most familiar businesses is L.L.Bean, which began as a small company that provided rugged outdoor clothing to local fishermen and loggers. Today, L.L.Bean manufactures and sells clothing and equipment on a worldwide basis to campers, cyclists, skiers, hikers, and recreational fishermen (Figure 5.6). The company actively promotes awareness of ecologically sensitive use of the outdoors, particularly in Maine and other parts of the Atlantic Periphery.

Spillovers from Megalopolis

Today, the once-isolated economy of the Atlantic Periphery has been affected increasingly by its relative proximity to Megalopolis, a region addressed in Chapter 7. The population and wealth of the Boston through New York City to Washington, D.C. area has spilled over into the Atlantic Periphery and is evident in the location of branch plants, tourism, second home development, and permanent in-migration.

▲ **Figure 5.6** L.L.Bean headquarters, Maine

With increasing mobility, residents of the crowded urban areas of Megalopolis have long associated the Atlantic Periphery with scenic beauty, recreational opportunities, a lower cost of living, and a more relaxed pace of life. Since the 19th century, urban dwellers from large metropolitan areas in the United States and Canada have vacationed in the Atlantic Periphery. Resorts such as Bar Harbor, Maine, and Campobello, New Brunswick have long attracted wealthy residents of Boston, New York, and other East Coast cities. Old Orchard Beach on the Maine coast is a favorite vacation destination of residents of Quebec where shopkeepers and motel owners post signs in French (Figure 5.7).

The lower cost of living has also attracted business opportunities. Executives of large corporations have recognized that cheap housing and scenic amenities are attractive to their employees. In 1963, IBM opened a major facility in Burlington, Vermont. The plant soon became Vermont's largest single-site employer, attracting a white-collar workforce and paving the way for additional high-tech employment in the Lake Champlain valley. Other branch plants followed suit. Even more people have moved to southern New Hampshire and southern Maine, which are within easy commuting range of the Boston metropolitan area. New Hampshire, which has no income tax and no sales tax, has been especially attractive to those wishing to avoid the high taxes in nearby Massachusetts.

As middle- and working-class Americans and Canadians began to spend more money on vacations and travel, the Atlantic Periphery emerged as an important venue for tourism. Abundant winter snows and scenic mountain ranges made upstate New York and Vermont important ski resorts. Lake Placid, New York (the home of the 1980 Winter Olympics), as well as Killington and Mount Snow in Vermont, are among the most popular ski resorts in the Atlantic Periphery. They attract thousands of skiers from throughout the northeastern United States and Canada every winter (Figure 5.8). In the autumn, thousands of people come to the Atlantic Periphery to view the spectacular fall foliage (Figure 5.9). Package fall foliage tours attract visitors from as far away as Texas, California, and Europe. Tourism is also popular in summer when mild temperatures along with opportunities to participate in hiking, camping, fishing, boating, and other water sports attract vacationers. The coastal areas of Maine, Nova Scotia, and Prince Edward Island, along with the scenic mountain resorts of northern New England, are especially attractive destinations for tourists in the summer.

Many people who experience scenic rural areas while vacationing eventually decide to purchase second

◄ **Figure 5.7** Old Orchard Beach, Maine

homes in these rural areas. Some are occupied seasonally, and others are eventually occupied year-round by their owners or by tenants. The Atlantic Periphery is one of the most important areas reflecting this process, along with such places as the Sierra Nevada of California, the Ozarks of Arkansas and Missouri, the southern Appalachians of North Carolina and Tennessee, and northern Minnesota, Wisconsin, and Michigan. Throughout northern New England, Nova Scotia, and other parts of the Atlantic Periphery, thousands of people who originally visited the area during winter, fall, or summer vacations bought property in the region, living in these second homes for several weeks or months each year. Many decide eventually to retire to these communities. At the same time, some younger people anxious to avoid the fast pace of urban life in Megalopolis decide to settle down in the Atlantic Periphery. Since 1980, the states of New Hampshire and Vermont, which are closest to Megalopolis, have had the highest population growth rates of the entire northeastern United States, although population growth in more isolated Maine slowed considerably in the 1990s.

◄ **Figure 5.8** Skiing in Vermont

▲ **Figure 5.9** Fall foliage in New England

Cultures, People, and Places

People throughout the Atlantic Periphery share problems associated with a sometimes harsh physical environment, isolation, and economic dependency. Yet the region contains a large and fascinating variety of individual places and cultures. Although interaction across the international boundary is considerable, for purposes of discussion we divided the region into its Canadian and American components.

Canadian Places in the Atlantic Periphery

The Canadian portion of the Atlantic Periphery, as we have seen, includes the provinces of Newfoundland and Labrador, Nova Scotia, Prince Edward Island, and New Brunswick. Nova Scotia and New Brunswick joined with Quebec and Ontario to become the Dominion of Canada in 1867, and Prince Edward Island joined the dominion six years later. Today's province of Newfoundland and Labrador was a separate British colony until 1948, when it voted to join Canada.

As discussed earlier, the island of Newfoundland was the first place in North America to be visited and settled by Europeans. The economy of Newfoundland has been oriented to the sea for hundreds of years. Most residents of Newfoundland live along the coast, with few people inland. As we have seen, transportation between rural settlements along the coast was largely by boat, contributing to the island's isolation as well as its sense of separate identity. Although relatively few people live in or near the interior forests, timber production has long been a major contributor to the economy of the province as well.

The island of Newfoundland along with Labrador on the mainland was governed by the United Kingdom as a separate colony until 1948, when residents of the colony voted narrowly to join the Dominion of Canada. This long history of separate political administration has affected Newfoundland's culture, and bonds between local residents and the national government are sometimes tenuous. Even today, public opinion polls show that a large majority of Newfoundland's residents identify more closely with the province of Newfoundland and Labrador than they do with the country of Canada.

The most densely populated part of Newfoundland is the Avalon Peninsula in the southeastern part of the island. The province's capital and largest city, St. John's, is located on the Avalon Peninsula (Figure 5.10). The city was established in the early 1600s as a center for the fishing industry. British and French interests frequently battled for control of the city and its lucrative fishing trade throughout the 17th and 18th centuries. The city was made the provincial capital upon entry into Canada in 1948. Nearly a third of Newfoundland and Labrador's population lives in the greater St. John's area.

Figure 5.10 St. John's, Newfoundland

Nova Scotia consists of Cape Breton Island, which is located 70 miles west of Newfoundland, and the mainland area to the south and west. After the Great Expulsion, thousands of Scots settled in the colony, whose name means "New Scotland." Nova Scotia has the largest population of African descent in the Atlantic Provinces. During the American Revolution, British officials guaranteed freedom to any slave who emigrated to the colony. During the 19th and 20th century, several thousand persons of African ancestry moved to Nova Scotia from various British colonies in the Caribbean. The economy of Nova Scotia, like that of Newfoundland, is oriented to the sea and has long emphasized fishing and trade.

Like Newfoundland, Nova Scotia is dominated economically and politically by one major city. Halifax on Nova Scotia's east coast is not only the largest city in Nova Scotia but the largest in the Atlantic Periphery. Halifax, which was founded in 1749, is the eastern terminus of Canada's major railroads and highways, and is also a major port (Figure 5.11). The Canadian Royal Navy also maintains extensive facilities in Halifax.

Figure 5.11 Halifax, Nova Scotia

No
which
popul
settle
securi
name
Victor
tia, P
agricu
fruit,
and e
In
mainl
the C
Borde
New
lane b
ice-fil
lar fe
mainl
minu
provi
Islanc
the ir
Amer
what
Ne
Scotia
Abou
Frenc
lation
Bruns
speak
and e
from
durin
provi
Provi
of na
In
New
in a s
three
ton, a
arily
St. Jo
and
other
porta
garde
Franc
capit

Am
Peri

Sout
Perip

6 Quebec

PREVIEW

Environmental Setting

- Quebec straddles three physiographic regions: the Canadian Shield, Appalachia, and the St. Lawrence Lowlands.
- The Canadian Shield and the Appalachian portions of Quebec are divided by the Saint Lawrence River, with its adjacent lowlands. The Saint Lawrence flows from southwest to northeast across the province.
- Quebec's weather and climate can be extreme with blizzards and ice storms common during winter months. Summers are relatively short and mild, with ample precipitation throughout the year.

Historical Settlement

- Jacques Cartier, who arrived from France in 1534 on the first of several trips to search for the Northwest Passage to Asia, was one of many European explorers and settlers to devastate First Nations peoples and their cultures in Quebec.
- The French also had closer and more collaborative relationships with some of the groups of First Nation people than did the English. The exploitation of furs was particularly dependent on maintaining relationships with the region's first inhabitants, who were very experienced hunters.
- Quebec's early settlement patterns are dominated by the Saint Lawrence River. The *rang* or long lot system of land division was designed to give everyone access to the river.

Political Economy

- Quebec remains an integral part of the Canadian and larger North American economy despite cultural differences.
- The traditional economy of Quebec emphasized agriculture along with trapping, forestry, fishing, and mineral extraction.
- Since World War II, Quebec has become a predominantly urban society. Quebec City, the administrative center, and Montreal, the financial center, dominate Quebec's political economy.
- The urbanization and industrialization of Quebec fueled a sense of isolation among many Quebecois. In the 1990s, Quebec voters narrowly rejected a secession referendum. Recently, the Canadian government has enacted impediments to further secession efforts in the future.

Culture, People, and Places

- One of the reasons Quebec has such a distinctive sense of place is the central role of religious expressions that are visible in its local landscapes. Roman Catholic religious structures such as cathedrals, shrines, educational centers, cemeteries, and retreat houses are found everywhere, especially in rural areas.
- About two-thirds of the residents are Francophones, making Montreal the second largest French-speaking city in the world.

The Future of Quebec

- Quebec's economy, like that of the rest of North America, continues to move toward a post-industrial structure dominated by services and information technology.
- In future years, Quebec's leadership may continue to push for cultural autonomy along with greater control of the province's economic future, regardless of whether Quebec achieves full political independence.

Introduction: A Francophone Island?

English-speaking Americans traveling in Quebec may feel like they are in Europe, rather than in North America. Despite the dominance of French road signs; a preponderance of French shops and restaurants; street, city, and town place names; and literally thousands of beautiful Catholic churches; a visit to Quebec is very different than a visit to France in many ways. This place is distinctly Canadian as well as French, and thus is a unique region of North America (see Box 6.1).

Quebec is Canada's largest province in land area, with 600,000 square miles or 1.6 million square kilometers. If Quebec was an independent country, it would be the world's 18th largest country in area—more than twice the size of Texas and three times the size of France. Quebec's history and geography clearly set it apart as a well-defined political and cultural North American region because it is the only large cultural hearth of French language, customs, and heritage on the continent. And its influence stretches well beyond its provincial boundaries since French Canadians migrated to many other parts of Canada and the United States during the past two and a half centuries. France's historic impact on Quebec remains dramatically visible

◄ Casino de Montreal, a province-owned casino on Ile Notre-Dame overlooking Vieux Port

BOX 6.1 Who Are the French Canadians?

The use of the word *Canadians* has a fascinating past. The term emerged during the early 1700s to refer to people of native-born French or mixed French and Indian ancestry, most of whom were trappers, traders, or peasant farmers. Until the British conquest, there were only two groups of people living in New France: *les francais* and *les Canadiens*. After the British takeover, no common terminology evolved to describe the two founding European groups in Canada. *Canadien* was associated with French-speaking people and all other inhabitants (British settlers, loyalists, and indigenous English) were called *les Anglais* (the English) or the British. But English-speaking residents of Canada were never referred to as *Canadiens*.

After confederation, new terms were needed to describe each of the two groups. Canadians of French origin became known as *les Canadiens francias* (French Canadians) and the remainder of the population were called *Canadians* (or English Canadians).

In the 20th century, several major changes happened again. Today groups are defined according to their language, territory, and ethnicity as either *Francophones* or *Anglophones*. The term *Canadiens* has become an anachronism in Quebec (except for the Montreal Canadiens hockey team) and may disappear altogether when the older generation passes away.

Source: From "Quebec, Past, Present and Future," (in) *Introducing Canada: Content Backgrounders, Strategies, and Resources for Educators*, William W. Joyce. Washington, D.C.: National Council for the Social Studies, pp. 116–117.

today—in religious beliefs, customs, food habits, language, and the cultural landscape.

The Canadian census of 1996 reported that more than 82 percent of Quebec's 7.1 million people declared French as their native language. These French-speaking Canadians are known as **Francophones**. The rest of Quebec's people speak English (**Anglophones**), and/or an aboriginal or a foreign language. English speakers tend to be concentrated in Montreal, the Eastern Townships, and the Ottawa Valley. Aboriginal speakers like the Cree and Inuit live in the far north, while the Mohawk, Montagnais, Abenaki, and Micmac speakers live in the central and southern parts of the province. Many more recent immigrants speak Portuguese, Spanish, Russian, or one of many African or Asian languages. In fact, Montreal, the largest city in Quebec and second largest in Canada, is today one of the most cosmopolitan and diverse cities in the world due to the recent and ongoing arrival of immigrants and refugees from many other parts of the world.

Quebec's position as a French-speaking community located in a continent dominated by English speakers has contributed to its unique political economy. Historically, Quebec's economy was oriented to resource extraction, agriculture, and transportation along its major artery, the Saint Lawrence River. The British took control of Quebec in 1759. The area then became known as "Lower Canada."

Under British administration, the province was largely French-speaking and remained so after Canada achieved its independence in 1867. Today many Francophone **Quebecois** are increasingly aware of their status as a linguistic minority within Canada and within North America. In recent years, many have argued for seceding from Canada and making Quebec an independent country. Some of the complex and emotionally charged reasons that French-speaking Quebecois continue to debate the issue of whether to remain a part of Canada or become a sovereign state are discussed later in this chapter. The location of the Quebec region in eastern Canada is shown in Figure 6.1.

Environmental Setting

Prior to Canadian independence, Quebec was known as "Lower Canada." The name "Lower Canada" refers to Quebec's location *downstream* along the Saint Lawrence from its neighboring province, Ontario (or "Upper Canada").

Landforms

Much of Quebec is part of the Canadian Shield. This landform region contains some of the oldest rocks on the North American continent. Much of the land consists of relatively flat to rolling terrain. Given the region's northerly location, it is not surprising that Quebec was covered with ice during the recent Ice Ages. The advance of the glaciers some eight to ten thousand years ago scraped away much of the soil and left a landscape with innumerable bogs, marshes, and lakes (Figure 6.2).

Southeastern Quebec, along with the Gaspé Peninsula, is part of the Appalachian chain. The highest peaks in Quebec are found in the Gaspé Peninsula, but these are barely 4000 feet (1200 meters) above sea level. This region has also been subject to extensive glaciation, creating beautiful ski areas.

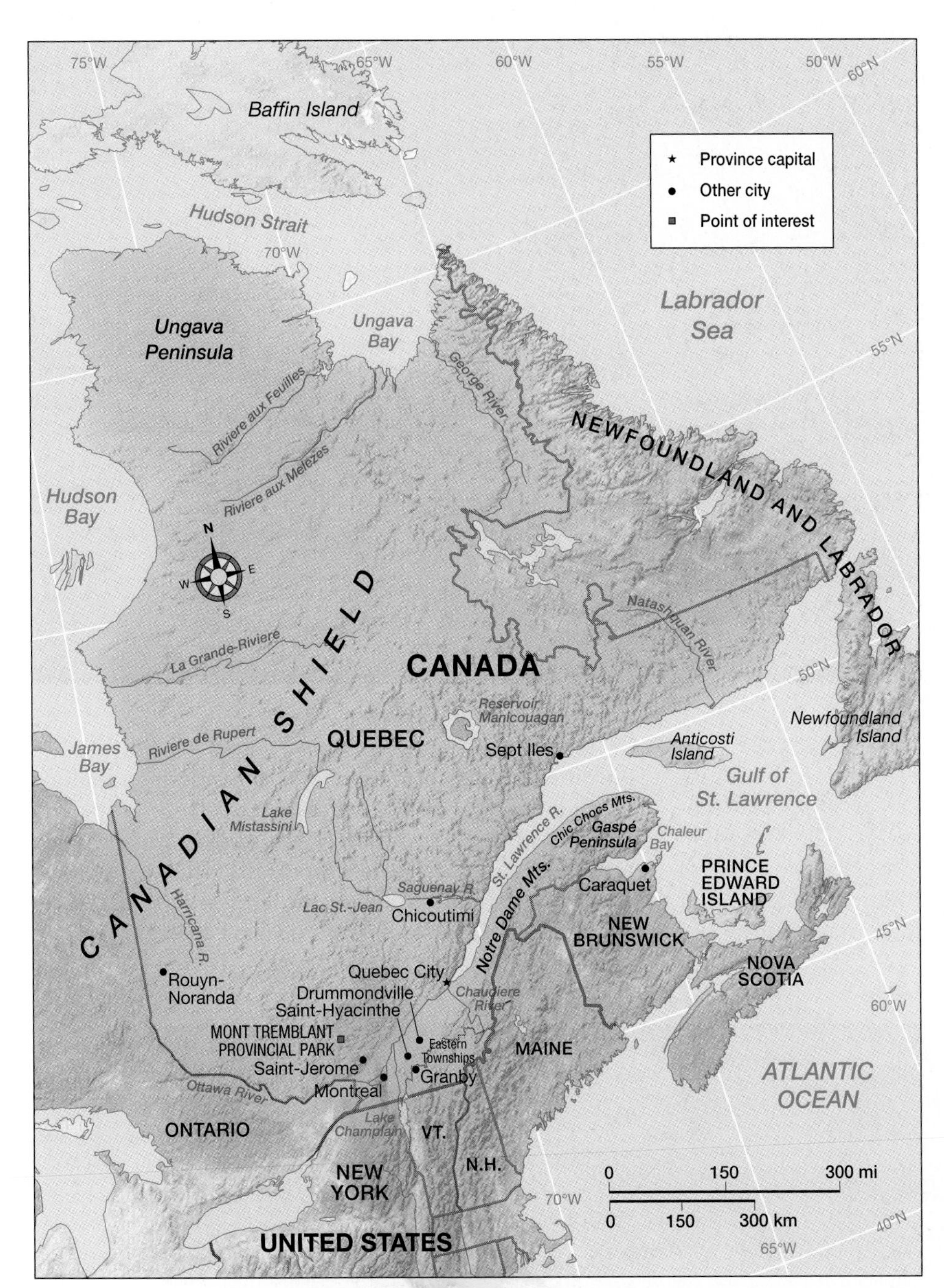

Figure 6.1 The Quebec region

The Canadian Shield and the Appalachian portions of Quebec are divided by the Saint Lawrence River, with its adjacent lowlands. The Saint Lawrence flows from southwest to northeast across the province. The two major cities of the province, Montreal and Quebec City, are located at key points along the river. Montreal is at the head of navigation for oceangoing ships including coal and iron ore ships heading east from interior minefields. Downstream from Quebec City, the river widens into an estuary. Within a hundred miles (160 kilometers) downstream from Quebec City, the estuary of the Saint Lawrence is so wide, in fact, that it takes more than two hours to cross the river by ferryboat (Figure 6.3).

Climate and Biodiversity The climate of Quebec, like that of the neighboring Atlantic Periphery, tends to be cool and damp. Winters are long and cold, with devastating ice storms and blizzards common occurrences. Summers are relatively short and mild, with ample precipitation throughout the year. Quebec's location on the eastern margin of the North American continent ensures that continental influences on the province's climate outweigh oceanic influences, given that most weather systems traverse Quebec from west to east.

Quebec's environmental setting has had a considerable influence upon its economic geography. The combination of poor soil, rugged terrain, and a harsh climate has made much of Quebec unsuitable for

▶ **Figure 6.2** Bogs and lakes in Quebec

agriculture. Most of the province is covered by dense boreal forests, and logging has been an important economic activity for hundreds of years. The Canadian Shield portion of the province has extensive mineral deposits including iron and aluminum. The lakes and rivers, while impeding agriculture have, however, encouraged transportation. The meandering, shallow streams could not be navigated by large ships but could easily be used by explorers, traders, and trappers with their pirogues, canoes, and rafts. This water transportation network allowed the French easy access to the North American interior, and since the 16th century,

▲ **Figure 6.3** View from the Saint Lawrence River

the Saint Lawrence and its tributaries, such as the Saguenay, have provided convenient access to the Great Lakes, the Mississippi Valley, and other parts of what is now the American Midwest.

Historical Settlement

At least 500,000 First Nations people were living in what is now Canada when John Cabot, representing the British crown, arrived in 1497. Cabot was followed by Jacques Cartier, who arrived from France in 1534 on the first of several trips to search for the Northwest Passage to Asia. Cartier failed to establish a permanent settlement in the name of France, but he did carry news back to Europe about the fishing grounds off the Atlantic shore and in coastal rivers. This news encouraged follow-up expeditions from Western Europe to exploit the resources of the region's forests, rivers, and seas.

During his second visit to the Saint Lawrence Valley, Cartier captured one of the tribal leaders of the Iroquois and several women who lived in the area. Returning from France on a third trip (on a mission to find gold and silver), Cartier somehow "forgot" to bring the chief and Iroquois women back home—even though he brought several hundred colonists with him. During a major uprising that followed, many French settlers were killed. Over the years, disease and warfare took a terrible toll on the indigenous residents of Quebec. In less than 150 years, some estimates report that the aboriginal population in the area had decreased by at least 90 percent.

Not until 1608 did the French attempt to establish a permanent settlement in Quebec. At that time, Samuel Champlain founded Quebec City. In establishing a permanent settlement, Champlain had three goals: to find a route to China and the East Indies, to develop the fur trade, and to convert the indigenous people to Catholicism. Once permanent settlements had been established, the French population had increased to more than 60,000 by 1750. Most of this population increase by the end of the colonial era, however, was due to the high birthrate of early colonists. Relatively few persons migrated from France to Quebec after 1700. French identity and culture in the region have persisted despite this very limited migration during the colonial era. Comparatively, English colonists, who were Protestant, had smaller families once they arrived in North America, but thousands more settled in the United States and Canada as immigrants who came from Great Britain.

During the late 16th to the 18th centuries, France established its power over a huge area of North America (Figure 6.4). Centered along the Saint Lawrence Valley,

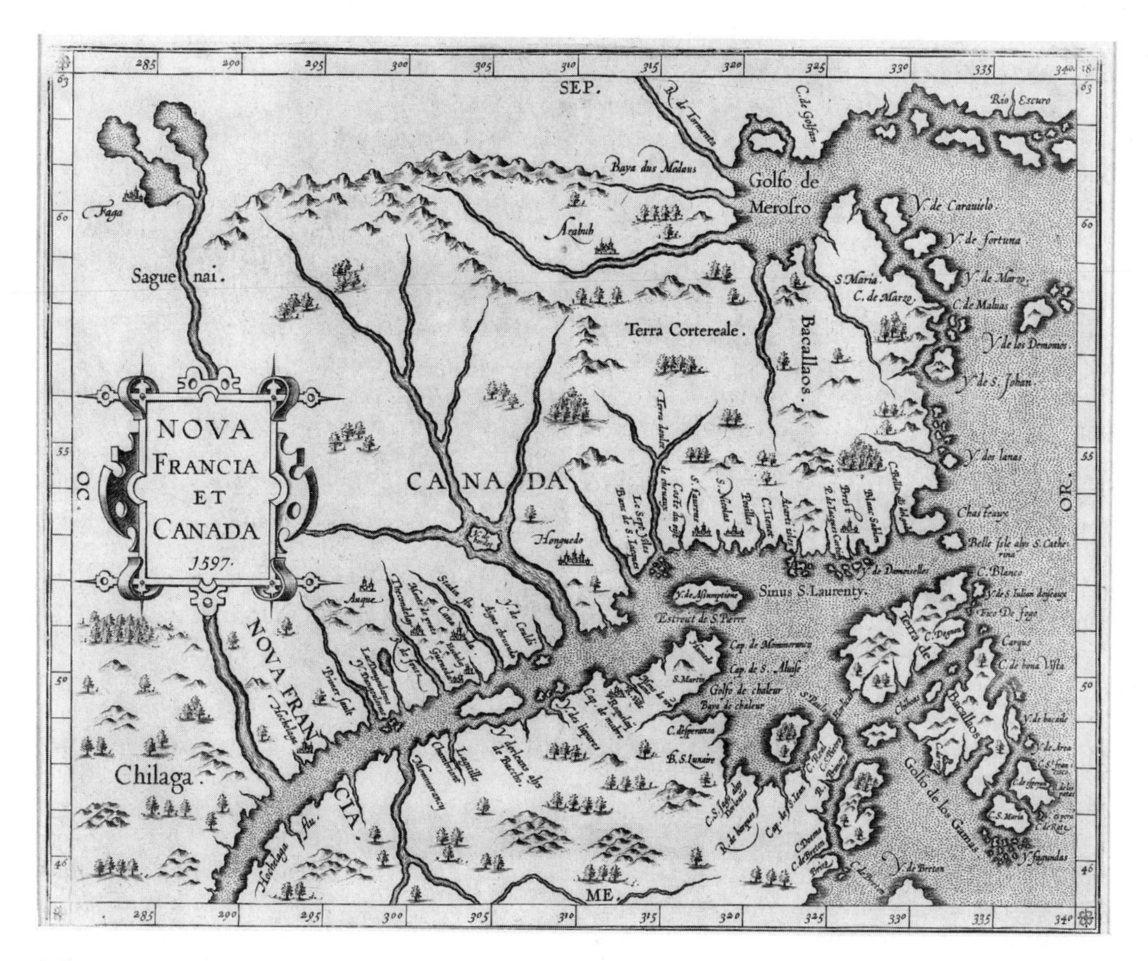

▲ **Figure 6.4** New France in 1597

New France soon became a globally connected fur trading empire. The Saint Lawrence River provided access to interior resources by way of the Great Lakes and interior river systems. The explorer La Salle tested the waters of this interior route in 1682 by making his way from the Saint Lawrence across the Great Lakes and down the Mississippi to New Orleans. Others used Lake Superior as a jumping-off point for fur hunting and trading expeditions into the interior following rivers to trading sites as far away as Manitoba. This effective French network of trade with native peoples overtook the English, who had chosen to construct their fur-trading posts at faraway Hudson Bay. It also planted the initial settlements that would later grow into cities with French place names like St. Louis and Detroit. Daring French *voyageurs*, in fact, were the earliest Europeans to explore most of the Great Lakes area and the Mississippi Valley.

The French also had closer and more collaborative relationships with some of the groups of First Nations people than did the English. The exploitation of furs was particularly dependent on maintaining relationships with the region's first inhabitants who were very experienced hunters. Beginning with Champlain's trip up the Saint Lawrence, the French went to great pains to cultivate positive relationships with indigenous aboriginal groups like the Algonquin. They were effective political allies, and they had wide-ranging mobility with their birchbark canoes, snowshoes, and knowledge of the geography of rivers, lakes, and trails through the northern woods. By 1750, the relationship that developed between the French and the Algonquins had become the most enduring and extensive bond between Europeans and native peoples in all of North America. According to Meinig (1986, p. 113): "The success of the fur trade grew out of mutual acculturation and interdependence between the Algonquins and the French. . . . the great convoys of furs and of trading goods moving up and down the Ottawa and the easy comings and goings of hundreds of Indians and voyageurs to Montreal were seasonal exhibits of this relationship."

A critical element of this alliance was a shared antagonism against other native peoples, particularly the Iroquois who lived south of Lake Ontario. Most abhorred and feared were the members of the Iroquois Five Nations (Seneca, Cayuga, Onondaga, Oneida, and Mohawk), who were allies of the Dutch and later the English. These sometimes-deadly counter allegiances encouraged the French fur traders to expand their territory to the west instead of to the south. Eventually the French also were able to foster good relationships with other groups such as the Hurons who helped them open up new fur hunting territory to the north.

Since a great deal of wealth could be made from furs and fishing, few early settlers from France wanted to clear the thick forests to become farmers. However, more peasants were eventually recruited from France to Quebec. Once they recognized that the Saint Lawrence Lowland was fertile, the valley of the Saint Lawrence became a magnet for rural settlement. Unlike the British system of settlement in **colonies**, the French mostly came as individuals who were participants in the feudal **seigneurial system**. This system was dictated by Jean Talon, who was sent to manage New France by King Louis XIV in 1665. Talon's goal was to create a rural society modeled after the one that existed at that time in France. He recruited his landless peasants in France with the promise of land. He sent for young women (most of whom were orphans or daughters of impoverished rural families in France) to be brides for the settlers, and imposed the French feudal system of land ownership known as the seigneurial system. Tracts of land were given to certain favored people such a military officers, politicians, and the Roman Catholic Church. These awardees had to swear allegiance to the king, pay for workers to come to New France, and promise to have their fields cultivated only by their tenants. Peasants were required to cultivate the fields, pay annual dues, and pay rent to the seigneur for use of his grinding mill and ovens.

By 1760, there were approximately 200 seigneuries in Quebec, each consisting of 3 to 9 miles (5 to 15 kilometers) of land. These landholdings dominated the landscape along the rivers, especially the Saint Lawrence, which was the region's primary transportation artery. In order to ensure access to transportation, the French seigneurs developed a rang ("row") or long lot system of land tenure (Figure 6.5). Land was allocated to settlers in long, narrow strips or rows, with each strip fronting on a river or, in inland areas, a road. Farmers built their houses along the river or along the roads. The rang system ensured that each farmer had access to transportation, and it also prevented a small minority of farmers from monopolizing the best lands. According to historical geographer Donald Meinig: "There was so much frontage available along the St. Lawrence that in 1750 settlement area could, with little exaggeration, be described as two riverine strips over two hundred miles long and a mile deep on either side of the river . . ." (1986, p. 110).

The 18th century was characterized by ongoing tension between France and Britain, both of which were competing for dominance in the global economy. They contested for domination of colonies in North America and elsewhere. In 1756, war broke out between France and Britain. During this "Seven Years War," known in the United States as the French and Indian War, both France and Britain relied on their First Nations allies. The Algonquins and Hurons supported France and the Iroquois supported Great Britain. British victories in Quebec were critical to their victory in the war, and in 1763, the victorious British forced France to cede Quebec to Great Britain. The British agreed to allow

▲ **Figure 6.5** Example of the French long lot system as viewed from above

Quebec's Francophones to continue to speak French and to practice Catholicism. Birthrates were so high during this pre-industrial period that land became scarce along the Saint Lawrence and its tributaries by the mid-1800s. Unable to obtain land, large numbers of French-speaking Canadians emigrated to areas of the Canadian Shield, where they tried to find ways to maintain their agricultural lifestyles on the thin, rocky soils. They also moved to the Appalachian Upland region between the Saint Lawrence and the U.S. border, where they purchased farms from British emigrants. Others moved to New England, where many found work as loggers or in factories.

This migration helped spread the extent of French influence during a period when conflicts between Anglophones and Francophones were once again heating up. Local rule by English governors was filled with elitism and corruption. In addition, the economy was depressed and tensions between the French-Canadian majority and the British minority were running high. By 1837, rebellions against British rule had flared in both Upper and Lower Canada, aimed at gaining more of a voice for local French-speaking residents. These rebellions were quickly silenced by the British Army. Nearly 300 rebels were killed in six different battles and the leader of the resistance escaped to the United States. Afterward, the Act of Union was passed in 1841. This Act created a single elected assembly and merged the two parts of Canada into one Province of Canada.

With this kind of tight control by the British in place, how did the French find ways to maintain their culture, ethnic identity, and linguistic separation after the passage of the Act of Union? Factors that influenced this cohesion included the French citizens' own strong desire to remain Catholic and French; the institutional support of the Roman Catholic Church that provided them with educational and spiritual support; the high birthrate that insured an ever expanding population in a concentrated area of settlement; and the isolated rural lifestyle of the majority of the Francophones, which separated them from Anglophone culture and influence. With this separation of cultures as a fact of life, a new political economy of the region evolved.

Political Economy of Quebec

Primary-Sector Activity

Even before the British took control of Quebec, local residents realized that the lowlands along the Saint Lawrence River represented the best farmland in the province. To the north, the climate is too cold and the soils of the Canadian Shield are too poor. Likewise, the uplands of southern Quebec near the U.S. border, like adjacent northern New England, are covered with thin, rocky soils.

In the 18th and early 19th centuries, wheat was the primary crop grown by Quebecois farmers. Once the prime wheat-growing areas of western Canada and the Great Plains of the United States had been opened for settlement, farmers in Quebec could no longer produce wheat competitively. Instead, they began to produce hay, potatoes, apples, oats, dairy products, sugar beets, and vegetables. Marginal farmland was abandoned.

Other Quebecois made their livings through alternative primary-sector activities including trapping, fishing, mineral extraction, and forestry. Cities such as Thetford Mines and Sherbrooke grew up around the

▶ **Figure 6.6** Aluminum smelter at La Baie, Quebec

production of asbestos, but recent evidence linking exposure to asbestos to cancer has largely eliminated the global market for asbestos. Aluminum is another important mineral resource found in Quebec, with the major production areas located north of Quebec City in the Canadian Shield near La Baie (Figure 6.6). Like the nearby Atlantic Northeast, Quebec is heavily forested and has ample quantities of timber. Many persons farmed in the summer and logged during the winter. Logs were floated down streams into the Saint Lawrence River. Lumbering remains an important contributor to Quebec's economy, especially downstream from Quebec City on both sides of the Saint Lawrence. Pulpwood production and paper milling continue to contribute substantially to Quebec's industrial economy (Figure 6.7).

The Urban and Industrial Economy of Quebec

Although a large majority of 18th- and 19th-century Quebecois were farmers, many others opted for urban life. Montreal and Quebec, both of which were founded

▶ **Figure 6.7** Paper mill in Quebec

▲ **Figure 6.8** Walled portion of Quebec City

by early French settlers, became thriving urban centers. Quebec City is located at the head of the Saint Lawrence estuary, and Montreal is located at the head of navigation on the Saint Lawrence. Quebec City became Quebec's administrative center (Figure 6.8), while Montreal, whose location allowed it to control access into the interior, became its financial and economic center (Figure 6.9). Montreal grew to be Canada's principal seaport and also a major shipping, transportation, and distribution center. On the fringe of the Francophone core, the city of Ottawa was selected as Canada's capital because of its location between the cores of Francophone Lower Canada and English-speaking Upper Canada. Although Ottawa itself is

◄ **Figure 6.9** Montreal

located in Ontario, the Ottawa River adjoining the city forms the Ontario–Quebec border and suburbs such as Hull are located in the province of Quebec.

Quebec's urban population began to grow rapidly after World War II. The province had several advantages to industrial employers, including a large labor force (following decades of high birthrates), ample water and hydroelectric power, and effective transportation by road, rail, and sea. Montreal in particular grew as a major industrial and commercial center. Factories were established in large numbers in Montreal, Quebec City, and other cities. Many were oriented to local resources, including food production, aluminum smelting, and the manufacturing of wood and paper products. Others produced consumer durables such as cars and trucks, aircraft, ships, appliances, and furniture. Iron and steel and chemical production also became important to the Quebec economy. However, multinational corporations based in English-speaking Canada and in the United States owned most of the factories. The fact that many of the profits from the industrialization of Quebec flowed to English-speaking areas would eventually fuel Quebecois nationalism.

The Development of Quebecois Nationalism and the Quiet Revolution

In recent years, the issue of maintaining French-Canadian cultural identity in a continent dominated by English speakers has become a major political issue. Today, Quebec residents are deeply divided as to the desirability of secession from Canada.

Canada, like the United States, is a **federal state**. In a federal state, power is formally shared between the national government and local governments—in the case of Canada, those of the provinces. Each of Canada's provinces has considerable autonomy. The British North America Act of 1867, which provided for Canadian independence, declared that Canada would be officially **bilingual** and that French and English would have equal status in Canada's national parliament as well as in the province of Quebec. Government business was to be conducted in both French and English in all provinces.

Although Quebec remained predominantly French-Canadian, the percentage of Canadians who spoke French as a first language declined steadily during the late 19th and early 20th centuries. The declining influence of French speakers across Canada was the result of several factors. These included the opening of western Canada to settlement. Many persons immigrated from Europe and the United States into Ontario and western Canada, and most of these immigrants learned English rather than French. In addition, as we have seen, many French Canadians moved to the United States. It has been estimated that as many as 40 percent of the province's residents left Quebec between the 1830s and the 1940s. Cultural differences between Quebec and the rest of Canada were reinforced by the fact that a large majority of Quebecois are Catholics, whereas the majority of non-Francophone persons in Canada are Protestants.

The decades immediately after World War II are known in Quebec as the **Quiet Revolution**. During this period, not only was Quebec transformed from an agrarian to an urban industrial society, but also the intellectual underpinnings of Quebecois nationalism were transformed. Pre–World War II Quebec was a heavily rural society, dominated by the Roman Catholic Church. As Quebec became more urbanized, Quebecois became less concerned about maintaining the rural and religious character of their society. Rather, they developed increased awareness of their status as a cultural and linguistic minority within a predominantly Anglophone North America.

The Quebec Secession Movement

Today, although 82 percent of Quebec residents speak French as their first language, most of Canada's other provinces have only small minorities of French speakers. Eighty-five percent of Canada's French speakers live in Quebec. A large majority of people in all of the other provinces of Canada speak English as their "mother tongue," except in New Brunswick, where about a third of the population is Francophone. The "other" languages spoken by many are predominantly European languages such as Italian and Ukranian in Ontario and the Prairie Provinces; European languages and Chinese in British Columbia; and Inuit and indigenous Native Canadian languages in Yukon, the Northwest Territories, and Nunavut. Moreover, the global domination of the United States in the world economy after World War II intensified the status of English as the world's predominant language. The ability to speak English became more and more critical in the world economy. Within Canada, English-speaking Toronto surpassed Francophone Montreal as Canada's dominant population, commercial, and industrial center. In the 1950s and 1960s, some experts predicted that French-Canadian culture and the French language would disappear from Canada, especially in urban areas, as Quebecois became **assimilated** into the English-dominant majority culture of Canada.

In response to these threats, political activists began to promote efforts to maintain the French language and French Canadian culture. Many Quebecois saw themselves as a distinct **nation**—that is, a group of people bound together by a common cultural heritage, language, and history within a place—relative to the rest of Canada. Activists identified "three pillars" of

▲ **Figure 6.10** Bilingual commercial sign in Old Quebec City

Quebecois nationalism: the French language, the Roman Catholic Church, and French legal and political traditions. Based on these pillars, efforts were made to prevent assimilation into the majority English-speaking culture. A sharp debate soon arose over whether this goal could be better achieved within the Canadian constitutional framework or through a politically independent Quebec. Those holding the latter view argued that Quebec should secede politically from Canada.

In the late 1960s, supporters of Quebecois nationalism founded the Parti Quebecois (PQ). The party soon won widespread support, and in 1976 it won a majority of seats in Quebec's parliament. The PQ government aggressively promoted the use of French throughout the province. One law mandated that road signs be posted only in French. Although this law was declared unconstitutional, most signage in Quebec continues to be in French only (Figure 6.10). The province of Quebec also sponsored a law requiring all children residing permanently in the province to be educated in French-speaking schools, regardless of the ethnic or linguistic backgrounds of their families. The PQ's activities encouraged French speakers to seek positions as educators and civil servants, and they promoted the activities of small businesses owned by and catering to local French-speaking communities. Today most business, especially outside Montreal, is conducted in French (Figure 6.11). The PQ sponsored a referendum on possible secession, but this was rejected by 60 percent of the electorate.

The success of the PQ reinforced the ethnic and political divisions between Quebec and the rest of Canada. Many English speakers decided to leave Quebec and move to other parts of Canada. For example, it has been estimated that more than half of the 25,000 English speakers in the Gaspé Peninsula (about a quarter of the region's total population in the 1970s) left Quebec between 1975 and 1991. By the mid-1990s, more than half of Gaspé's English speakers were senior citizens.

In the 1980s, the Canadian federal government attempted to respond to Quebecois nationalism by redefining the relationships among the provinces within Canada's federal system. In 1987, provincial leaders signed the Meech Lake Accord. This agreement was intended to ensure that Quebec's efforts to preserve its

▲ **Figure 6.11** Street scene in Quebec City showing business signage in French

BOX 6.2 Notes on the Referendum: An Update from a Canadian Correspondent to the American Geographical Society

Dear Hilary,

After scrounging through my files, it appears that the last time I wrote to you was in 1994! Nothing like spending five years in administration to really put one out of touch with the wider world. Well, spring prom season came and went, successfully I think, and the skateboarders are back at it. For the rest of us that must mean the snowy season has taken a recess and that Quebec politics, which has no off-season, is still consuming a big part of our time. Actually, there have been a few recent events that might interest your politically minded readers.

The referendum of 1995, as you remember, was not just a real squeaker—the NO side carrying the day by the narrowest of margins (54,000 votes)—it has turned out to be a sort of road to Damascus experience for most of us. *We came perilously close to losing the whole country on October 31, 1995.*

Just exactly how close we came was illustrated in May, 1997 by the revelations of Jacques Parizeau, the ex-premier of Quebec, on the occasion of the launching of his new book, *Pour un Quebec Souverain*. Before the referendum took place, Quebecers were assured by Premier Parizeau that after a majority YES vote there would be an offer of partnership, both economic and political, made to the rest of Canada, and that only if negotiations on this offer were to fail over a period of a year or so, would Quebec move to declare independence unilaterally. But in his book, it was revealed that Parizeau had intended no such lengthy waiting period at all. In fact, he writes that a unilateral declaration of independence could have taken place "within a week or ten days" following a victory for the YES side in the referendum. It seems that this was all part of his "grand plan" for a quick move to sovereignty and that the U.S. would have been placed in an interesting position in all this. . . . Parizeau was following former French president Valery Giscard d'Estaing's advice to declare independence more or less immediately, in return for which France would have recognized the new state of Quebec and promised to put great pressure on the U.S. to do likewise.

It would have been hard to imagine what it would have been like to wake up, say, on Remembrance Day, 1995 (November 11) to find out that my Canadian passport was useless . . . and that somewhere in Quebec City a new currency was being minted that would have replaced all of the Canadian money left in my wallet. Very scary thoughts, all of them. My wife and children happen to hold U.S. citizenship and so would never have become "stateless" but on the assumption that I would have become a migrant back to Canada, presumably I faced the prospect of applying for citizenship to my own country! Not a pleasant thought at all.

Source: Adapted from "Quebec Politics" by Curt Rose. *Focus* (1997) Vol. 44 (3): 36–38.

language and culture would be given constitutional protection. However, Manitoba and Newfoundland declined to ratify the Accord. Five years later, the Charlottetown Agreement, which would have recognized Quebec as a "distinct society" within Canada's constitution, was proposed but it was voted down in a national referendum. The failure of the Meech Lake Accord and the Charlottetown Agreement sparked a resurgence of Quebecois nationalism. In 1995, the provincial government sponsored a referendum on secession. The referendum barely failed, with 50.6 percent voting to remain part of Canada and 49.4 percent voting for secession (Figure 6.12). The story of what would have happened if the referendum had passed in 1995 is told in Box 6.2. Today, the idea of gaining full and complete independence from the rest of Canada remains a salient and often emotional issue for many Quebecois.

The fact that the secession vote was so close illustrates the level of division on this important issue within Quebec. Most English-speaking Quebec residents oppose secession. In Montreal, many felt that secession would further erode the city's influence in international trade and commerce. First Nation communities were especially opposed to secession. In referenda, more than 85 percent of Cree and Inuit in Quebec voted in favor of the province's remaining a part of Canada. Even among the Francophone population, a substantial minority of the population has been skeptical of secession.

How secession might affect long-standing economic ties with other parts of Canada, particularly Ontario, was a matter of concern. Some also considered that Quebec's secession would isolate the Atlantic Provinces geographically. The secession debate began to cool as the turn of the 21st century approached. In 1998, Canada's Supreme Court ruled that Quebec could not legally secede on its own. Two years later, Canada passed a federal law requiring not only that secession be supported by a clear majority, but also that negotiations take place to resolve issues involving borders, taxation, and the seceding province's responsibility for Canada's national debt before any province could secede legally. In addition, some felt that Quebec would continue to prosper within the North American Free Trade Agreement and through establishing trade relationships with the European Union, and that this economic growth might be imperiled by attempting to form an independent state.

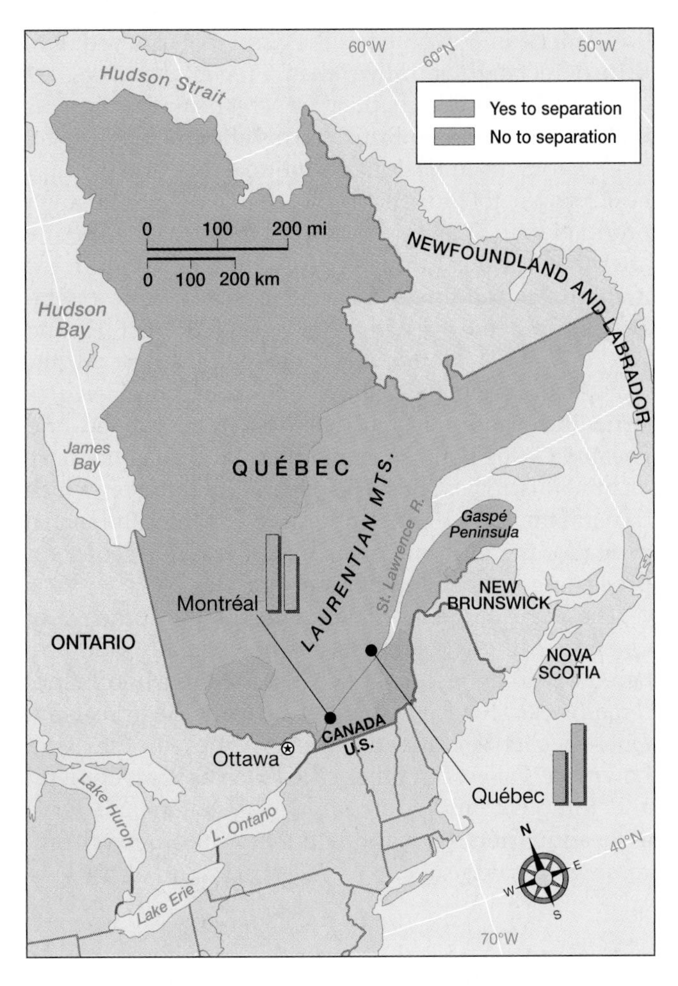

▲ **Figure 6.12** Referendum on secession, 1995

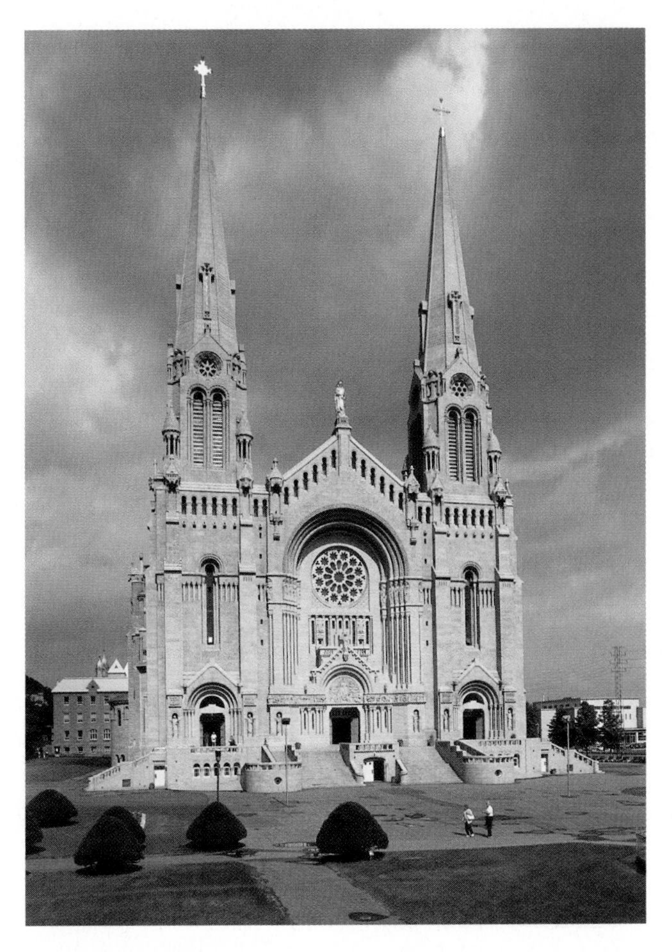

▲ **Figure 6.13** Basilique Ste.-Anne-de-Beaupre in Quebec City, part of a French Catholic religious landscape

Culture, People, and Places

As we have seen so far in this chapter, Quebec's identity as a region is defined by its linguistic and religious distinctiveness, its rural settlement patterns, and its intense attachment to being French. Quebec is clearly the heartland of ethnic identity and political power for Francophone Canadians. French-speakers now constitute about a fourth of the Canadian population.

One of the reasons Quebec has such a distinctive sense of place is the central role of religious expressions that are visible in its local landscapes. Roman Catholic religious structures such as cathedrals, shrines, educational centers, cemeteries, and retreat houses are found everywhere, especially in rural areas (Figure 6.13). Although the moral and ethical influence of the church has waned in recent years, most of the residential and commercial districts of Quebec's small towns and villages are grouped around a large stone Roman Catholic Church in the center of town.

Place names are another reminder of the enduring influence of French culture in this part of North America. Most noticeable on a map are all the towns, cities, mountains, and other geographic features that are named for saints. From the Saint Lawrence River to the town of St. Maurice, Ste. Anne de Beaupre, and St. Jerome, place names appear everywhere as reminders of the French Catholic heritage of this region.

Southern Quebec

Southern Quebec is the social, economic, and political center of the Francophone region. Southern Quebec has more than 90 percent of the total population of the province but only a small percentage of the land. As we have seen, this area is made up of two distinctive subregions: the Appalachian Uplands and the Saint Lawrence Lowlands (Figure 6.14).

The Appalachian Uplands are located just west of the Atlantic Ocean and north of the United States to the south and east of the Saint Lawrence Lowlands. This region includes the Gaspé Peninsula, which was first settled by the Vikings in the 11th century and later by Basque fishermen. The beautiful Chic Choc Mountains surprisingly rise up out of the interior and are an extension of the northern Appalachians (Figure 6.15).

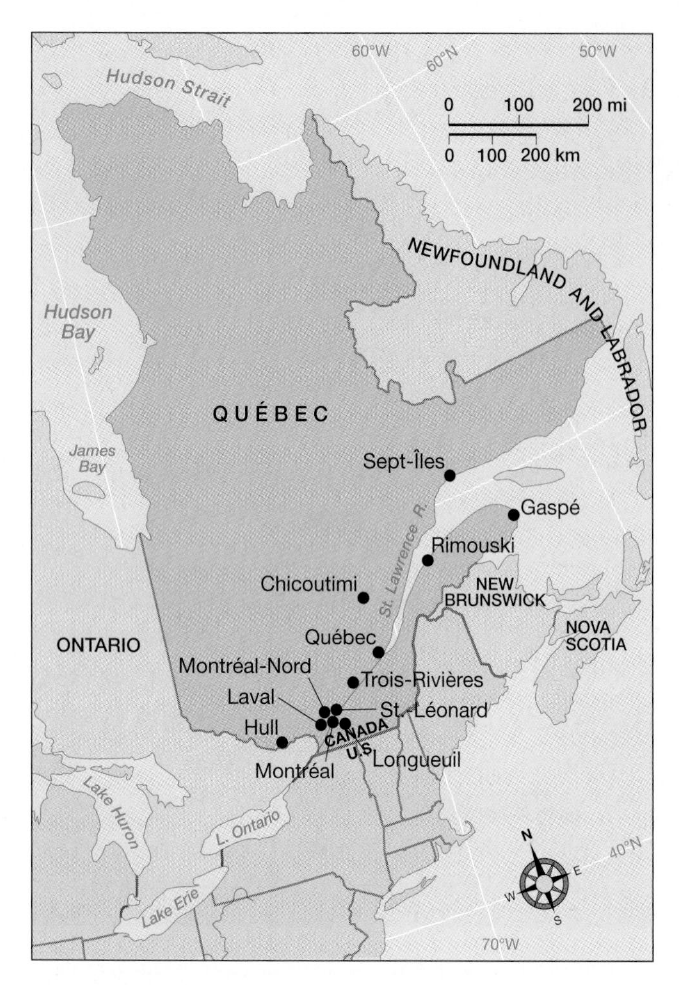

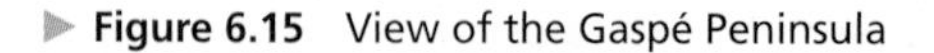
▲ **Figure 6.14** Quebec's primary places of settlement

The Appalachian Uplands also include Estrie (Eastern Townships), which are a cluster of communities in the rolling hills of the Appalachians just east of Montreal. Settled originally by British Loyalists after the American Revolution, the land is surveyed into British rectangular rather than French long lots. As noted earlier in this chapter, French-speaking migrants took over abandoned British farms after the English left to look for jobs in Boston or Montreal. Estrie is perhaps best known to tourists as the center of maple sugar production in Quebec. Every year in late March, sunny afternoons and cold nights cause the sap to run in the maple trees sending local "sugar shacks" into operation boiling the sap in large vats (Figure 6.16 and Box 6.3). This is the most prosperous part of the Appalachian Uplands since it is more conducive to agriculture and forestry than the more rugged and isolated Gaspé Peninsula, and it also has an abundance of minerals for export. The city of Sherbrooke, which has an emerging high-tech industry, is the commercial capital of the Eastern Townships. It is also a center for processing lumber and minerals.

The Saint Lawrence valley remains the core area of Quebec. Although farmers are now greatly outnumbered by urban residents, the historic French influence remains clearly visible on the landscape. Montreal and Quebec City are the two major cities in the Saint Lawrence Valley, but other cities such as Trois-Rivieres and Rimouski are local commercial centers and serve as headquarters for various lumber and mineral processing operations.

Northern Quebec

Northern Quebec contains over 90 percent of the land in the province but only a few people. Geologically, the land is part of the Canadian Shield (described in more detail in Chapter 18 on the Far North) but has few people and few opportunities for economic development. Following the completion of the Grand Trunk

► **Figure 6.15** View of the Gaspé Peninsula

BOX 6.3 Making Maple Syrup in Quebec's Eastern Townships

Every March, the combination of sunny days and cold nights causes the sap to run in the maple trees in Quebec's Eastern Townships ("Estrie") region. *Cabanes a sucre* (sugar shacks) go into operation, boiling the sap collected from the trees in buckets. At some modernized locations, a series of complicated looking tubing and large vats do the job. The many commercial enterprises scattered across the landscape host "sugaring offs" and tours of the process for locals and visitors alike, including tapping the maple trees, boiling the sap in vats, and *tire la neige*, pouring hot syrup over cold snow to give it a taffy consistency just right for pulling and eating. A number of cabanes also serve hearty meals of ham, baked beans, and pancakes, all drowned in maple syrup.

Railway (now the Canadian National Railroad connecting Quebec City to Winnipeg), French farmers from the growing agricultural area in the Saint Lawrence Lowland began migrating into the Clay Belt of northwestern Quebec in search of open land. Both the Quebec government and the Catholic Church encouraged settlers to take up "free" land. Such movement would allow these persons to maintain their French culture, rather than lose it following a move to the United States. Eventually, about 15,000 small, non-commercial farms were established. Because of the rigorous climate and poor soils, however, most proved unprofitable and many were eventually abandoned. Today, mineral resources are much more important to northern Quebec's economy. Forestry, mining, and hydroelectric power production are important ways that people in northern Quebec make their living.

▲ **Figure 6.16** Sugar Shack in Saint Casimir de Pontneuf, Quebec

Urban Places

Today, a large majority of Quebecois live in cities that are growing at a rate considerably faster than that of the rural population. Montreal is the largest city in Quebec and the second largest in Canada. Established as a religious outpost in 1642, Montreal is located at the confluence of the Saint Lawrence and Ottawa Rivers. The original settlement was located on an island where shipments needed to seek portaging around the Saint Lawrence River. The presence of the rapids also meant that ocean-going ships could not penetrate farther upstream than Montreal. These considerations encouraged early commercial development making Montreal the center of the fur and lumber trade and connecting it with a vast hinterland in the interior. Montreal's central commercial location was reinforced in the 1850s, when a railroad bridge was completed across the river, eventually connecting it by rail to Portland, Maine. (Figure 6.17).

Today's towering skyscrapers, high-density downtown and suburban residential districts, and international ambience mark Montreal as a truly global city. About two-thirds of the residents are Francophones, making Montreal the second largest French-speaking city in the world. The other third are a combination of Anglophone, American, and other foreign-born immigrant groups. Here one finds a bicultural city with two principal languages, two traditions, and two school systems. Two distinctive historical eras are also celebrated here: One honors the distant past by preserving the landmarks of New France while the other celebrates the post-Expo 67 and World's Fair "New Quebec" era of national identity and provincial pride.

▲ **Figure 6.17** Montreal's central business district

▲ **Figure 6.18** Montreal's underground city

▲ **Figure 6.19** Cobbled streets in Lower Town, Quebec City

During the 20th century, Montreal's growth has been outpaced by growth in Toronto and Vancouver. Montreal is working hard to improve the efficiency and livability of the city. The visionary quality of the city's downtown rejuvenation plan is most noticeable to visitors when they see Montreal's massive covered city, a multilevel complex of more than 80 acres that links business establishments, hotels, museums, high rise apartments, and the metro station (Figure 6.18).

Quebec City, the second largest urban place in Quebec, is located at head of the Saint Lawrence River estuary. It has fewer geographical advantages for growth than Montreal and therefore never achieved the commercial status and influence of Montreal. Nevertheless, Quebec City is Quebec's provincial capital and it has grown into a commercial, cultural, and government center with an intensely French sense of place. In contrast to Montreal, the large majority of Quebec City's people are Francophones. The old part of Quebec City includes *Upper Town*, which is the old walled part of the city and is the center of government and religion. Upper Town's cobbled streets and walls give it a distinctly Parisian feel (Figure 6.19). *Lower Town* is the commercial center along the wharf-lined river. It also has a rather large area of narrow cobbled streets and small shops. One of the city's most important economic functions is travel and tourism since it is one of the most unique and enviably charming cities on the continent.

The Future of Quebec

Quebec occupies a unique niche within North America. In a continent dominated by English-speaking cultures, Quebec's French-influenced culture stands out. Since the 17th century, many aspects of Quebec's culture including its dominant language, religious traditions, architecture, and cultural landscape show clear signs of long-standing French influence.

How might Quebec's unique French heritage and culture impact its future? A century ago, it was commonly assumed that Quebec's French heritage would disappear as Quebecois assimilated into the dominant English heritage of Canada. During the late 20th century, however, Canada's Francophone minority took steps to preserve its cultural heritage. Such efforts were successful, and there is now little likelihood that French culture in Quebec will disappear. However, these efforts to maintain a distinctive French culture in this unique province of Canada have also reinforced the cultural

7 Megalopolis

PREVIEW

Environmental Setting

- Megalopolis is located astride the three major physical regions of the eastern United States—the Atlantic Coastal Plain, the Piedmont, and the Appalachians.
- Many of the region's major cities are located along the Fall Line, where the Atlantic Coastal Plain meets the Piedmont.

Historical Settlement

- Megalopolis has been a magnet for immigrants from all over the world for hundreds of years. Prior to the American Revolution, numerous settlers arrived from the British Isles and continental Europe. The early settlers included large numbers of Puritans, Quakers, and Catholics who wished to escape religious prejudice in England.
- The cities of Megalopolis have attracted millions of immigrants throughout American history. During the late 19th and early 20th centuries, newcomers from southern and eastern Europe moved to the region.
- Today, large numbers of immigrants from Asia, Latin America, Russia, Africa, and elsewhere are continuing to contribute to the unparalleled cultural diversity of the region.

Political Economy

- Megalopolis has been the economic and political core of North America since the 18th century.
- During colonial times, the area comprising Megalopolis was North America's major agricultural region and contained its largest ports. During the 19th century, the region became part of the major industrial region of North America. Today, the region is a leader in the contemporary post-industrial economy.
- The region is highly urbanized, but far more people live in suburbs than in central cities. Since the 1970s, the suburbs of Megalopolis have become increasingly diverse and have generated many employment opportunities.
- Cultural diversity has been central to the region's economic core position for more than 200 years.

Culture, People, and Places

- Although the five major cities of Megalopolis dominate the region's economy and culture, millions of persons live and work in smaller cities such as Providence, Hartford, Albany, Atlantic City, Richmond, and Norfolk.
- Despite the region's large population and high density, parts of the Megalopolis region are lightly populated and retain an undeveloped character. Specialized agriculture and tourism are important to the economies of many such places.

Introduction

In 1961, the French geographer Jean Gottmann coined the term "Megalopolis" to describe the densely populated urban corridor of the northeastern United States. In coining the term "Megalopolis," Gottmann was referring to the fact that the five major cities that comprise Megalopolis—Boston, New York, Philadelphia, Baltimore, and Washington—have expanded so much that they have, in effect, become a single, very large metropolitan area.

To be sure, the area that we consider Megalopolis—that is, these five cities, their suburbs, nearby smaller cities, and surrounding rural territory—is the most densely populated part of North America. Yet by no means is Megalopolis a homogeneous area. Each of the major cities of the region retains its distinctive character and culture characteristics. Nor is the region uniformly urbanized. Within the region are many places that are little developed and retain their traditional rural characteristics.

Relative to other regions of North America, Megalopolis is fairly small. The entire region, which extends from Maine to Virginia, includes only about 50 thousand square miles—an area smaller than Illinois. Yet the region contains the greatest concentration of wealth and power in the history of the world. For more than 200 years, the cities of the Northeast have comprised the economic and political core of North America. This relatively small area contains the continent's largest and densest population, its largest and most important centers of financial and governmental power, and its greatest concentrations of wealth.

Why did Megalopolis emerge very early as the political-economic core of North America? Much of the region was part of the Middle Atlantic colonies during colonial times. As we saw in Chapter 4, the Middle Atlantic colonies had numerous geographical advantages over eastern Canada, New England, and the South. These advantages included fine harbors, favorable

◄ The Empire State Building in Manhattan, New York, the heart of Megalopolis

opportunities for land transportation, agricultural prosperity, and a central location relative to the other British colonies in North America. Well before the American Revolution, commercial and political leaders in the Middle Atlantic colonies took advantage of these favorable conditions and established cities that would emerge as the economic core of the United States after independence. During the 19th century, Megalopolis became an important industrial area. Today, it is one of the leading centers of post-industrial economic activity in the United States and indeed throughout the world.

Another hallmark of the northeastern corridor has been its ethnic and cultural diversity. Even in colonial times, the Middle Atlantic region was characterized by far more diversity than any other part of North America. For more than 200 years, diversity in the northeastern corridor has continued to increase as persons from all over the world have flocked to this area in search of economic opportunity and political freedom. Immigrants and newcomers from other parts of North America continue to contribute to the region's vitality and economic health.

Environmental Setting

Relative to the rest of North America, population densities in the area we call Megalopolis have been high for several hundred years. Megalopolis has been the most densely settled region of North America since the 18th century. Not surprisingly, humans have dramatically modified the ancestral landscape of the region. The fact that this region has been modified so much by human activity means also that its boundaries are imprecise and are defined more by cultural than by natural phenomena. For our purposes, we define Megalopolis as the region including the five major urban centers of the Northeast—Boston, New York, Philadelphia, Baltimore, and Washington—along with nearby smaller urban centers and rural areas whose economies and cultures are closely linked to the five major cities (Figure 7.1). Thus, Megalopolis extends from southern Maine to the tidewater of Virginia. It also includes the capitals of states in which these cities are located along with other significant cities, such as Hartford, Albany, Harrisburg, Richmond, and Norfolk, along with their suburbs and exurban territory.

Landforms

The area that we call Megalopolis extends across three major physiographic regions—the Atlantic Coastal Plain, the Piedmont, and the Appalachians (see Chapter 2). Within Megalopolis, the Atlantic Coastal Plain extends along the Atlantic Coast southward from New York City. It is only a few miles wide in northern New Jersey, but widens from north to south. Southern New Jersey, the Delmarva Peninsula between the Delaware River and Chesapeake Bay, and the Tidewater region of Virginia are all part of the Atlantic Coastal Plain. The land comprising the Atlantic Coastal Plain is low and flat, with sandy soils.

West of the Atlantic Coastal Plain is the Piedmont, with its rolling topography. The Appalachians, as we have seen, are ancient mountains that have been worn down by millions of years of erosion and glaciation. The Catskills of southeastern New York, the Poconos of Pennsylvania, and the Shenandoah Mountains of northern Virginia are part of the Appalachians. Today, these areas contain crowded campgrounds and resorts visited every year by millions of tourists wanting relief from the region's warm, humid summers.

The entire region is traversed by numerous rivers, most of which flow from northwest to southeast into the Atlantic. These rivers rise in the Appalachians and flow across the Piedmont into the coastal plain and eventually into the ocean. The major river systems of Megalopolis, from north to south, include the Charles, Narragansett, Connecticut, Hudson, Delaware, Susquehanna, Potomac, and James rivers. Harbors are often found where the rivers enter the ocean. Indeed, the Delaware and Chesapeake Bays are **estuaries**, or river valleys that became submerged when sea levels rose following the melting of glaciers at the end of the most recent ice age. Delaware Bay is the estuary of the Delaware River; to the southwest, Chesapeake Bay is an estuary formed by the Susquehanna and Potomac rivers.

Many of the major cities of Megalopolis are located along the region's major rivers. In particular, many are located along the fall line. The fall line is located where the Piedmont meets the Atlantic Coastal Plain. The term "fall line" refers to the fact that many rivers descend from the Piedmont onto the coastal plain over rapids and waterfalls. These falls were important to settlement for two reasons. First, their presence prevented oceangoing ships from penetrating any further upstream. Second, the falls provided water power to run the region's early industries. Sites along the fall line were therefore very attractive sites for urban development. Trenton, Philadelphia, Baltimore, Washington, and Richmond are all located on or near the fall line.

Climate and Natural Hazards

The region's climate is temperate, with precipitation throughout the year. The southern portion of Megalopolis has a humid subtropical climate, with hot, humid summers and mild winters. (During the 19th

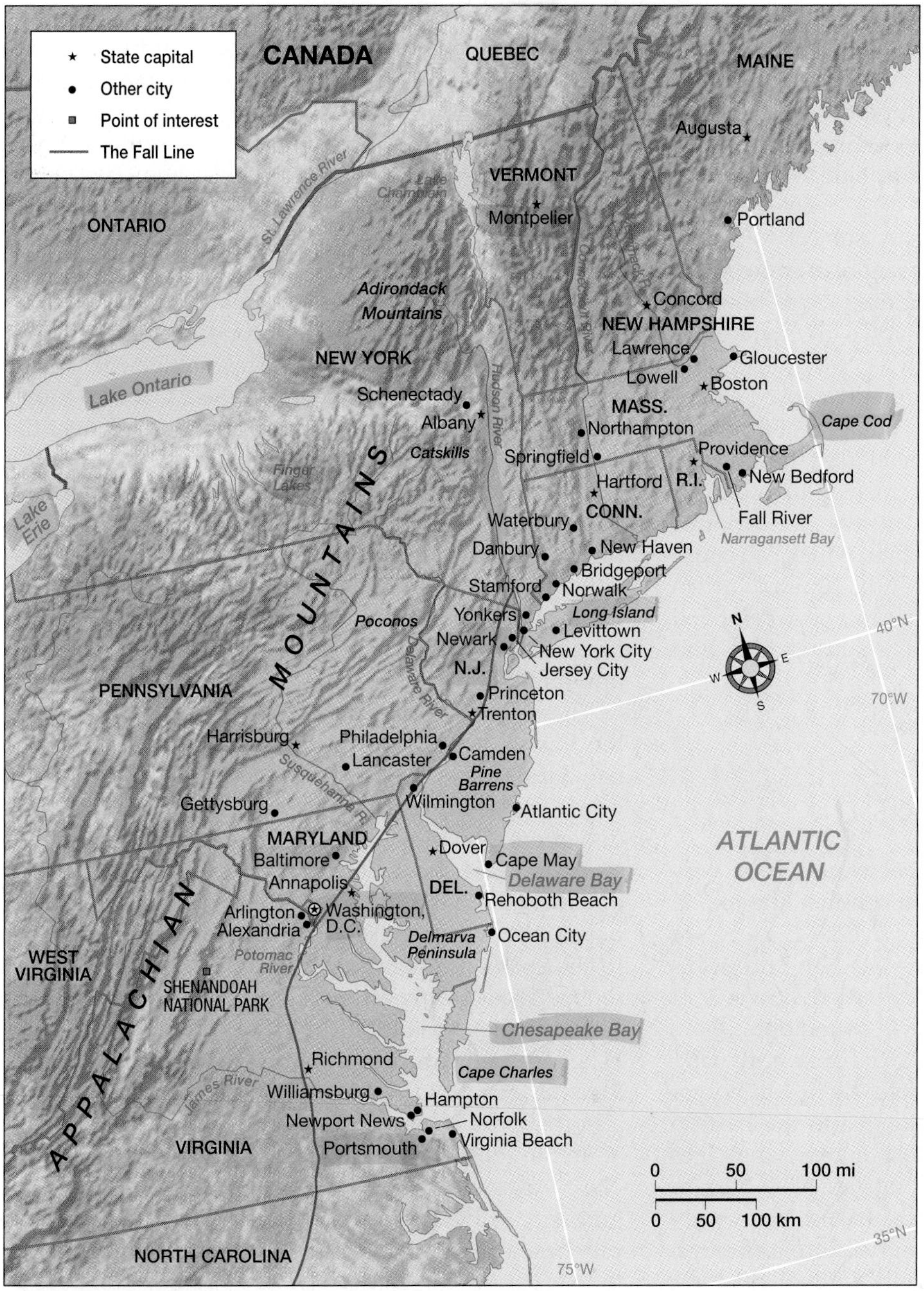

Figure 7.1 A region called Megalopolis

century, the British diplomatic corps identified Washington as a "tropical" outpost, and employees in the British diplomatic service stationed in Washington were given higher pay for serving under more tropical conditions than were ordinarily experienced in Europe.) To the north and further inland, the climate can be classified as more continental. Winters are longer and colder and summers are cooler and shorter. Average winter snowfalls vary considerably from south to north across the region, ranging from an average of 16 inches (41 centimeters) per year in Washington, D.C., to 86 inches (218 centimeters) per year in Boston.

Megalopolis is occasionally threatened by hurricanes that sweep northward along the Atlantic Coast. On September 21, 1938, a major hurricane raced across Long Island, New York, with wind gusts of more than 120 miles (193.1 kilometers) per hour. Seven hundred persons lost their lives, and more than 15,000 homes were damaged or destroyed. Half of Long Island's lucrative apple crop was destroyed. The damage in Long Island alone was estimated at more than six million dollars, which by today's standards would make this storm one of the 20 costliest natural disasters in recorded history. The storm raced across Long Island Sound and caused more than 500 additional fatalities and extensive damage in eastern Connecticut, Rhode Island, and Cape Cod. Tides of 18 to 25 feet (5.5 to 7.6 meters) were recorded in coastal communities in Connecticut and Rhode Island, causing major damage to thousands of houses and other buildings. Nearly 6000 ships and boats were destroyed or heavily damaged. Fortunately, with more sophisticated warning systems today, hurricanes are more likely to cause property damage than large numbers of fatalities; yet the events of 1938 serve as a reminder that the region is susceptible to occasional tropical storms.

The entire region is occasionally subject to heavy snowfalls that can bring two feet or more of snow along with high winds, cold temperatures, and blizzard conditions. Nor'easters (see Chapter 5) sometimes affect coastal regions as far south as New Jersey and Delaware. These storms have frequently caused extensive property damage, power outages, numerous injuries, and occasional fatalities in part because of increased numbers of traffic accidents. In summer, thunderstorms can bring brief, heavy downpours. Heavy rainfall contributes to flooding, which can cause major problems, especially in poorly drained urban areas whose natural pervious cover has been removed and replaced by buildings, roads, parking lots, and other impervious surfaces.

Historical Settlement

The Atlantic Periphery and Quebec were settled very early in Euro-American history. In contrast, the region between the Hudson and Potomac rivers—the area we now know as Megalopolis—was settled by Europeans somewhat later. Once the area was settled, however, it quickly attracted a large and diverse population.

Prior to European colonization, the area now comprising Megalopolis was inhabited by a variety of Native American groups that belonged to the Eastern Woodland culture complex. Most spoke languages related to those of the Iroquois of upstate New York or the Algonquin of the eastern Great Lakes. Many subsisted through a combination of agriculture and hunting and gathering. Women typically cultivated beans, corn, squash, and other plants while men hunted deer and other game animals. Some of the original forests were cut down by Native Americans, who removed trees in order to clear land for farming and to harvest timber for fuel, constructing houses, and making bows and arrows, tomahawks, and other artifacts. Thus the landscape of present-day Megalopolis had already been modified well before the arrival of the Europeans, who then changed the original landscape far more dramatically. Although most of the original inhabitants of the region were killed or driven away with the encroachment of invading settlers, the names of several cultures live on in names of communities and natural features such as the Susquehanna and Potomac rivers, and Narragansett and Chesapeake bays.

Among the first permanent European residents of present-day Megalopolis were Dutch fur traders, who established Fort Nassau at the site of today's city of Albany, New York. The Fort Nassau site was at the head of navigation on the Hudson River, meaning that it was as far upstream as oceangoing ships could travel. Thus it was an ideal location for a new settlement dependent on resource extraction and trade. The Dutch traded with Iroquois fur hunters and traders who had access to the riches of the "West" on the other side of the Appalachian Mountains by way of the Mohawk Gap. Building on their success, a few years later, in 1625, the Dutch constructed a second fort on the southern tip of Manhattan Island. This settlement, which they called New Amsterdam, became a walled, feudal-looking, multicultural city of about 800 people over the course of the next 30 years or so (Figure 7.2). Nearby Dutch settlements were named Bronck's, Breuckelen, and Harlem—place names that linger today as the Bronx, Brooklyn, and Harlem. Other Dutch settlers established farms along the banks of the Hudson, including the home of President Franklin D. Roosevelt, whose first paternal ancestor in North America was a Dutch immigrant.

These Dutch settlers' neighbors to the northeast were the English who were busily engaged in establishing new farms and small towns in New England. The "Pilgrims" of southeastern Massachusetts were religious dissenters who moved across the Atlantic in 1620 after several years of exile in the Netherlands. They established Plymouth Colony, comprised of what is now southeastern Massachusetts, including Cape Cod. In the 1630s, a much larger group of English Puritans moved from southeastern England to Massachusetts. These Puritans founded the city of Boston and established settlements throughout central and southern New England. In 1691, the Plymouth Colony was absorbed into Massachusetts.

To the southeast of the Dutch settlers of present-day New York were Swedes, who set up a colony under the guidance of Peter Minuit, the man who had originally purchased Manhattan from the Native Americans. By the 1660s, the Swedes, assisted by larger numbers of Finns, had built two forts on land along the Delaware River. However, by that time, trouble with the English broke out over trading rights up the Hudson, which ultimately led to the capture of both Dutch forts by the English settlers. After taking possession of New Amsterdam, the English changed the community's name to New York in honor of the English King's brother, the Duke of York.

To the southeast, the English Quaker leader William Penn was given lands along the Delaware River as repayment for a debt. His territory became the main focus for the settlement of English, Welsh, and German immigrants. More than 10,000 new immigrants arrived in a four-year period in the late 17th century. Only about 18 years after it began, this colony had at least 18,000 residents. Penn's policies were quite radical by the standards of the day. He set up liberal immigration

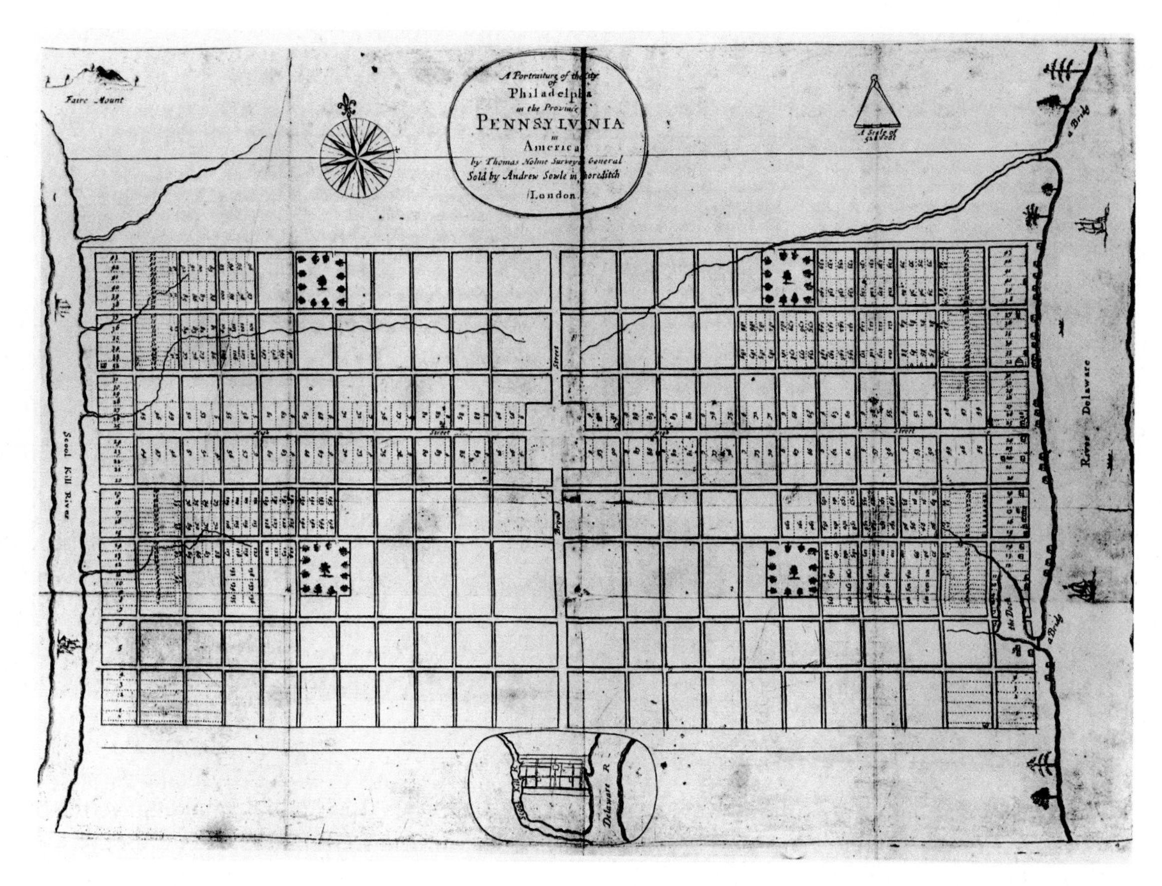

▲ **Figure 7.2** William Penn's 1682 map of Philadelphia, Pennsylvania, showing a pattern of gridiron streets, a central square, parks, and residential lots

and settlement policies and actively recruited people to come from different parts of Western Europe by advertising in German-language publications in Europe. This encouraged Germans to settle in a large area extending from Philadelphia to Lancaster and German Baptists, Moravians, and Mennonites to settle on the fertile farmland along the Delaware Valley. These early Pennsylvanians were soon joined by Scots, Irish, Welsh, and English farmers.

Penn's town planning innovations also made this place different from other parts of the Middle Atlantic and New England regions (see Figure 7.2). With the help of a fertile agricultural hinterland, and the productivity of hard-working and entrepreneurial residents, the city of Philadelphia soon grew into the largest and most cosmopolitan urban place in the colonies. Nearby southeastern Pennsylvania soon emerged as the nation's breadbasket. Maryland, meanwhile, was settled by English Catholics who, like the Quakers in Pennsylvania and the Puritans of New England, were often eager to escape religious persecution in the British Isles. Like Pennsylvania, Maryland welcomed settlers of all religious faiths.

The diversity of population in what would become Megalopolis increased steadily during the 17th and 18th centuries, and continued after the United States achieved independence. English, Irish, Welsh, Scots, and German immigrants dominated foreign-born migration flows into this North American region. During the 1830s and 1840s, in fact, millions of new residents migrated here with more than 70 percent coming from just two places of origin: England and Germany. By the late 19th century, immigrants from southern and eastern Europe joined these earlier arrivals. Families and individuals from Poland, Russia, Scandinavia, and the Austro-Hungarian Empire along with tens of thousands of others from Italy, Greece, and other parts of the Mediterranean survived the long ocean voyage and passed through the immigration processing center on Ellis Island (Figure 7.3). Some moved on quickly after arrival to seek employment in the mines and factories of central and western Pennsylvania or the Great Lakes states.

Other new immigrants stayed in the Boston, New York, Philadelphia, and Baltimore metropolitan areas, fueling new population growth and adding to the economic potential and cultural richness of the region. Massive immigration into this part of North America continued steadily until the outbreak of World War I and, soon thereafter, the passage of restrictive immigration laws in the 1920s. These laws limited the numbers of new foreign-born in-migrants in the United States according to a strict quota system. However, flows of African Americans from the American South soon

▶ **Figure 7.3** Ellis Island Historic Monument

replaced the flows of foreign-born immigrants. And since the 1960s, millions of immigrants and refugees born in Latin America, Asia, and Africa have settled in Megalopolis.

Political Economy

Well before the American Revolution, what is today's Megalopolis was clearly recognized as the core area of British North America. The region contained the Thirteen Colonies' largest and densest populations, its greatest ethnic diversity, its wealthiest settlement, and its most significant agricultural and industrial products. Not surprisingly, after independence the area emerged as dominant within the newly independent United States. Today, the region remains an important core area within North America and within the entire world.

The Establishment of Megalopolis

As we have seen, the name Megalopolis implies a conurbation, or a continuously settled collection of urban areas. Travelers driving along Interstate 95 or riding Amtrak trains between Washington and Boston can certainly observe nearly continuous urban development throughout the region.

Since colonial days, the focus of Megalopolis has been its major cities. Four of these—Boston, New York, Philadelphia, and Baltimore—were established as port cities. At the time of independence, these were four of the five largest cities in the United States (the other was Charleston, South Carolina, which was also a port city). Washington, in contrast, was not founded until after the United States became independent, and it was built specifically as a government center.

Why was the development of Megalopolis focused so heavily on port cities? The British and other European colonial powers established colonies in order to extract resources for transformation into finished products (see Chapter 4). The colonies of the present-day Middle Atlantic region had abundant quantities of resources, including fish, timber, furs, and agricultural land. These resources, along with products made from processing them, were shipped to Europe. Exports were carried out through port cities that were instrumental in promoting trade between the colonies and England. As we have seen, the Northeast is a well-watered region with many large rivers flowing from the interior to the coast. Not surprisingly, port cities were developed along most of the major rivers of the region (Figure 7.4). Indeed, each of the major cities became the economic center of its surrounding colony, with Boston, Providence, New Haven, New York, Trenton, Philadelphia, Baltimore, and Richmond playing this role for the colonies of Massachusetts, Rhode Island, Connecticut, New York, New Jersey, Pennsylvania, Maryland, and Virginia respectively.

The first U.S. census, taken in 1790, reveals the importance of port cities to the economy of the fledgling United States (Table 7.1). All of the ten largest cities in the United States in 1790 were port cities. Some, including New York, Philadelphia, and Boston, remain major cities today; others such as New Haven and New Bedford have been eclipsed in size and importance.

Figure 7.4 Baltimore's Inner Harbor

TABLE 7.1 The Largest Cities in the United States in 1790

Rank	City	Population
1	New York City	33,131
2	Philadelphia	28,522
3	Boston	18,320
4	Charleston	16,359
5	Baltimore	13,503
6	Northern Liberties, PA	9913
7	Salem	7921
8	Newport	6716
9	Providence	6380
10	Marblehead, MA	5661
10	Southwark, PA	5661
12	Gloucester, MA	5317
13	Newburyport, MA	4837
14	Portsmouth, NH	4720
15	Sherburne (Nantucket), MA	4620
16	Middleborough, MA	4526
17	New Haven	4487
18	Richmond	3761
19	Albany	3498
20	Norfolk, VA	2959

Source: U.S. Bureau of the Census.

The Emergence of New York and the Development of Washington, D.C.

Once the United States became independent, Americans began to pay more attention to the development of the interior. Even before independence, many American colonists ignored the Proclamation of 1763 and began to settle west of the crest of the Appalachians. The westward movement accelerated after independence, especially after the Louisiana Purchase secured New Orleans and the Mississippi for American shipping.

The settlement of the interior proved critical to the emergence of New York as the dominant economic center of the United States. In 1790, Philadelphia, with 28,522 people, was more than two-thirds the size of New York City (Table 7.1). Some thought that Philadelphia's more central location and its selection as the temporary capital of the United States would allow it to rival, or perhaps even surpass, New York in size and economic importance. However, physical geography played a crucial role in New York's rise to predominance. A glance at the map of the physical geography of the Northeast shows that New York, unlike Philadelphia or Boston, is blessed by relatively easy access to the Great Lakes and other interior locations along the Hudson and Mohawk rivers. New York financiers recognized this locational advantage and were quick to provide funds to construct the Erie Canal along the Mohawk River Valley. These advantages were reinforced after railroads came into common use, as the Mohawk River Valley also became the site of the New York Central Railroad. Access to the interior from Philadelphia, in contrast, was blocked by the rugged Appalachians, which were much more difficult and expensive to traverse. Thus New York combined the advantage of a fine harbor and easy access to foreign countries with ready access to the North American interior. By 1850, New York had more than twice as many people as Philadelphia and Boston combined, and it had consolidated its position as the country's dominant city—a position which it has held ever since.

The newly independent United States also had to address the question of where to locate a permanent capital city. Philadelphia, which was the site of the Constitutional Convention of 1787, had been selected as the capital temporarily, for a ten-year period between 1790 and 1800. However, leaders of the new government, particularly those from outside the Northeast, did not want Philadelphia, New York, or other established financial centers to become seats of political power as well. Thus the first Congress in 1789–90 devoted considerable attention to identifying a capital city that would be separate geographically from these seats of financial power.

Eventually, Congress decided among locating the city along the Delaware, Susquehanna, or Potomac rivers. Northern representatives preferred the two northern alternatives, whereas those from the South argued for the Potomac. Eventually, the North agreed to allow the federal government to assume the states' war debts, which were greater in the South than the North, in exchange for the more southerly location. The states of Maryland and Virginia ceded 100 square miles (260 square kilometers) of land to the federal government, and the capital city of Washington was soon constructed. In 1843, however, the government decided that it did not need the Virginia portion of the District of Columbia and returned this land, which contains the present-day city of Arlington, to Virginia.

Industrialization and Deindustrialization in the Northeast

Independence and the settlement of the interior coincided with the Industrial Revolution, and the cities of the Northeast were well positioned to take advantage of the opportunity to industrialize. The northeastern corridor had several geographical advantages that encouraged industrial development in the 19th century. As we have seen, many northeastern cities were ports, facilitating international trade. As well, many had developed access to raw materials and markets in the interior. This accessibility was reinforced after the railroad came into common use, as cities such as New York, Philadelphia, and Baltimore became the eastern termini of railroads connecting the East Coast with the Great Lakes and the Corn Belt. The Northeast also had abundant water power, especially along the fall line, which is located in the zone of transition between the Atlantic Coastal Plain and the Piedmont, as well as the head of navigation of many eastern rivers. Most of the major cities of the Northeast and other major industrial centers such as New York, Trenton, Philadelphia, Baltimore, Washington, and Richmond are located along the fall line.

The Northeast also had the country's largest concentration of capital to finance industrial development. Moreover, the densely populated region contained an ample supply of labor, which was augmented by the arrival of large numbers of European immigrants. Not surprisingly, the Northeast, along with the nearby cities of the Great Lakes, soon became North America's preeminent industrial center and one of the major industrial regions of the world. After World War II, heavy industry began to decline. More efficient and technologically advanced production processes reduced the need for industrial labor, and many firms took advantage of the opportunity to hire cheaper labor and relocated to the South and West or to other countries. However, the deindustrialization of the northeastern states had less of an effect on the population and economies of the Northeast than was the case in cities such as St. Louis, Cleveland, or Detroit.

Why did the northeastern corridor escape the negative consequences of deindustrialization felt in many Great Lakes and midwestern communities? In part, the northeast prospered relative to the Great Lakes because of the greater diversity and locational advantages of the northeastern corridor cities. The industrial base of the Northeast was much less dependent on heavy industry, and contained a much larger variety of lighter and more technologically sophisticated industries. The northeastern corridor also took advantage of opportunities to move from industrial to post-industrial employment. By the 1980s the Northeast, along with California, had emerged as a leading center of high-tech industry. Medical research and services, entertainment, communications, financial services, insurance, and publishing also remained concentrated in the Northeast. The Northeast remains North America's leading financial center, and continues to contain a disproportionate share of the country's corporate headquarters.

Cities and Suburbs

The shift from an industrial to a post-industrial economy in the 20th century was accompanied by spectacular growth in the region's suburban population. Today, far more residents of the northeastern corridor live in suburbs than in central cities. Suburbanization in North America dates back to the late 19th century, when developers began buying and constructing homes on land on and near railroads emanating outward from Philadelphia, New York, Boston, Washington, and other major cities. Trains to accommodate commuters were scheduled, and developers extolled the virtues of suburban life, combining fresh air and space with easy access to downtown workplaces—indeed, the sales pitches of developers in the 1890s were not that different than those a century later. Until World War I, however, suburbanization was primarily confined to the very wealthy. After the war, and after the automobile came into common use by the middle class, the pace of suburbanization began to increase. Well-known northeastern suburbs such as Ardmore, Pennsylvania; Greenwich, Connecticut; Rye, New York; and Bethesda, Maryland, were established and grew rapidly between the two world wars.

The pace of suburbanization increased dramatically after World War II. The GI Bill of Rights included provisions that encouraged returning military veterans to purchase houses. Many had young families and wished to raise their children away from the noise, crime, and grime of the central city. During the 1930s and 1940s, few houses were built in the United States because of the Depression and World War II. Once the war was over, however, millions of returning veterans and their families now needed low-cost housing. In response to these demands, developers built millions of suburban houses throughout the northeastern corridor and elsewhere in North America.

The city of Levittown, New York, typifies the suburbanization of the mid-20th century (Figure 7.5). Levittown is located on Long Island some 20 miles east of New York City. World War II veteran William J. Levitt, who worked in constructing prefabricated housing for the Armed Forces during the war, decided to use his expertise to mass-produce homes for the burgeoning civilian market. Purchasing hundreds of acres (1 acre equals 4050 square meters) of potato fields on Long Island, Levitt mass-produced thousands of identical houses using crews of workers with highly specialized skills. His firm owned or hired all links in the production chain. The company owned lumberyards and appliance wholesalers, and hired teams of electricians, carpenters, plumbers, and other skilled workers. Each house had four rooms and a bathroom on a 25 feet × 30 feet (7.6 meters × 9 meters) slab foundation. These mass production techniques made the houses uniform and subject to later ridicule, but also made them cheap: In the late 1940s, a veteran with a Fair Housing Administration loan could buy a home in Levittown for less than $8000. A second Levittown was built outside Philadelphia beginning in 1953.

The last third of the 20th century was characterized by two significant developments in the suburbanization of the United States. Before and shortly after World War II, the suburbs were "bedroom" communities: Most residents commuted to work in central cities. By the 1970s, however, more and more people worked, as well as lived, in the suburbs. Suburban employment began to increase much more rapidly than did employment in the central city. Suburban locations were attractive to business interests for several reasons. The price of land per acre was much cheaper than in the city, and business owners could construct much larger, less expensive factories and offices in the suburbs. Suburban locations also allowed business interests to take advantage of accessibility to highways, avoiding the traffic congestion associated with narrow, crowded city streets and reducing transport costs. Suburban

◀ **Figure 7.5** Levittown, New York, 1954

BOX 7.1 Heterolocalism: Vietnamese Americans in Northern Virginia

In 1975, the war in Vietnam ended with the collapse of South Vietnam's government, as the communist government based in Hanoi assumed control of the entire country. After the fall of the South Vietnamese government in the mid-1970s, more than 700,000 refugees from South Vietnam moved to the United States. Many of these refugees were government officials, military officers, or business executives.

Like other immigrant groups throughout North American history, Vietnamese refugees tended to select their new homes within North America on the basis of contact with relatives, friends, and acquaintances in Vietnam—a process that we know as **chain migration** (see Chapters 3 and 4). The largest community of Vietnamese-Americans became established in southern California, with another large concentration along the Gulf Coast of Texas and Louisiana, including Houston and New Orleans. Another substantial community of Vietnamese Americans became established in the northern Virginia suburbs of Washington, D.C. Many Vietnamese refugees in northern Virginia had contacts with American military personnel associated with the Pentagon, and some were former officers in the South Vietnamese armed forces. As these people brought friends and relatives to the Washington area, a thriving ethnic community began to emerge.

The establishment of a Vietnamese community in the Washington suburbs is in some ways reminiscent of ethnic settlement in Megalopolis during the late 19th and early 20th centuries, but there are some major differences. Migrants from Italy, Poland, Russia, and other origins who moved to the cities of Megalopolis tended to inhabit tightly packed ethnic neighborhoods. Within each neighborhood, most residents belonged to the dominant ethnic group. The primary focus of social interaction was within the neighborhood. Residents socialized with one another, worshipped in the language of the "old country," and read newspapers and magazines published in their native languages.

In contrast, Vietnamese Americans moved into suburban homes scattered throughout several counties in Northern Virginia. Few had Vietnamese-American neighbors, and their children attended school with children of many other ethnic groups. Yet community leaders made concerted efforts to maintain ethnic ties and to preserve Vietnamese culture. A shopping mall in Arlington became recognized as an important Vietnamese-American center; Vietnamese-American entrepreneurs bought and renovated many of the stores and restaurants, which attracted Vietnamese-American consumers from throughout the metropolitan area. Using cell phones and the Internet, community leaders kept in touch with local residents of Vietnamese ancestry across the region. Gatherings were organized for religious activities and for celebrating Vietnamese holidays. In these ways, members of Virginia's Vietnamese-American community retained ties with one another and with their homeland, despite the fact that their community, unlike ethnic communities a century ago, was very much spread out geographically.

Source: Joe Wood, 1997.

locations were especially attractive to high-technology industries. By the 1980s, suburbs such as the Route 128 corridor outside Boston and the Rockville Pike corridor northwest of Washington, D.C., were well known as centers of high-tech industry.

The population of the suburbs also began diversifying rapidly. In the 1940s, 1950s, and early 1960s, suburbs were overwhelmingly white, with very few African Americans and other racial minorities. Many suburbs had formal or informal codes prohibiting the sale of property to minorities. Levittown, for example, forbade the sale of houses to African Americans until racial covenants in housing were declared unconstitutional by the U.S. Supreme Court in 1948. In the late 1960s, the federal government enacted and began to enforce fair housing laws that prohibited discrimination in housing throughout the United States. These laws, along with increased prosperity among minority homeowners, resulted in the rapid suburbanization of African Americans and other minorities. Many suburban jurisdictions attracted large numbers of minorities. Today a majority of residents of Prince Georges County, Maryland, east of Washington, are African Americans. This county is the largest and most prosperous minority-dominated suburb in the United States. In recent years, Hispanics and Asian Americans outside Washington, New York, and other major cities have also suburbanized in large numbers.

During the late 19th and early 20th centuries, members of ethnic groups often settled in densely packed neighborhoods whose populations were dominated by members of the same ethnic group, who not only lived near one another but also tended to work, socialize, attend worship services, and participate in social activities within the neighborhood. New York's Little Italy and heavily Irish South Boston are examples of such neighborhoods. Today, however, suburbanizing members of ethnic groups maintain close personal ties without living in the same immediate area—a process known as **heterolocalism** (Box 7.1).

Over the past three decades, more and more people have taken advantage of continued improvements in transportation to enjoy suburban lifestyles. Commuter trains transport thousands of people every day from once-rural outposts such as Harper's Ferry, West Virginia, the Pennsylvania Dutch country west of Philadelphia, and the Hudson Valley to work in the cities of Megalopolis and their suburbs. As in other urbanizing areas on the fringes of large metropolitan centers, Starbucks coffee, the *New York Times* and

Washington Post, and other upscale accoutrements of 21st-century city life are increasingly evident in the once-isolated and bucolic countryside.

Unity and Diversity in the Northeast

The suburbanization of minority populations in Megalopolis over the past few decades underscores the importance of cultural diversity as a factor in the development and maintenance of the region as an important economic, cultural, and political center. The region was the most diverse region in North America long before the American Revolution, and along with California and Texas is one of the most diverse areas of North America today in terms of ethnic and racial diversity.

Since the 1960s, other destinations—especially California—have eclipsed New York in the number of new immigrants. Yet the cities of Megalopolis, especially New York and Washington, remain magnets for immigrants from all over the world. New York remains one of the world's most polyglot and diverse cities: A visitor walking through Manhattan or Brooklyn can hear conversations, buy and read newspapers and magazines, and view signs in dozens of languages. Of course, the contributions of immigrants and their descendants to the Northeast have enriched North American and global culture immeasurably. Frank Sinatra, Babe Ruth, Lou Gehrig, John F. Kennedy, Jon Bon Jovi, Tom Clancy, John Travolta, Sylvester Stallone, Bruce Springsteen, Colin Powell, Jennifer Lopez, and Carmelo Anthony are but a few of the children and grandchildren of immigrants into the communities of Megalopolis who have enriched American and global culture and achieved worldwide renown.

Cultures, People, and Places

Given its large population, economic diversity, and importance, it is not surprising that Megalopolis contains a large variety of distinctive places. The five major urban centers of the region—Boston, New York, Philadelphia, Baltimore, and Washington along with their suburbs—contain the preponderance of the population, but large numbers of people live in smaller metropolitan areas or in non-urban areas. Here, we discuss the five major cities, smaller urban settlements, and less urbanized rural areas in sequence.

The Major Cities of Megalopolis

The five major cities of Megalopolis can be found along a nearly straight line from northeast to southwest, yet they are each unique and distinctive. The northernmost of the five cities is Boston, which has been the dominant financial, commercial, and political center of New England for more than 300 years (Figure 7.6). The original site of Boston is a narrow peninsula jutting into Massachusetts Bay. During the 19th century, much of the land between the peninsula and the mainland was filled in, creating the "Back Bay" district.

▲ **Figure 7.6** Downtown Boston, Faneuil Hall Marketplace, 1842

▲ **Figure 7.10** Quincy Market in Boston

recent years following the legalization of gambling (Box 7.2). Further to the south, residents of Baltimore and Washington cool off at Ocean City, Maryland, and Rehoboth Beach, Delaware, which is sometimes called "The Nation's Summer Capital" because so many government officials and employees vacation there (Figure 7.11). Offshore islands such as Martha's Vineyard, Nantucket, Block Island, and Kent Island in Chesapeake Bay also draw many vacationers.

To the west, parks and resort areas such as Virginia's Shenandoah National Park and Skyline Drive, Pennsylvania's Pocono Mountains, and New York's Catskill Mountain region also draw millions of visitors annually. Historic sites such as the Civil War battlefields of Antietam and Gettysburg, Colonial Williamsburg, and various Revolutionary War battle sites also draw large numbers of visitors. Tourism is the major economic activity and source of revenue in these communities.

▶ **Figure 7.11** Rehoboth Beach, Delaware

BOX 7.2 The Revitalization of Atlantic City

Atlantic City, New Jersey, has long been one of the United States' prime beach resorts, with a long and colorful history. It is built on one of many barrier islands that are located along the Atlantic seaboard throughout the eastern United States. In the mid-19th century, Atlantic City became a popular resort for vacationers wishing to escape the summer heat of New York, Philadelphia, and other East Coast cities. One early settler, Richard Osborne, was a civil engineer who supervised the construction of a railroad bridge to connect Atlantic City with the mainland. Osborne also named the streets of Atlantic City; those parallel to the ocean were named after major bodies of water, and those perpendicular to the ocean were named after U.S. states. These street names were used to name properties in the board game *Monopoly*.

The railroad, which was completed in 1854, brought thousands of tourists each year to Atlantic City. Numerous hotels and restaurants were opened. Many were located on piers built along the Boardwalk, which was first completed in 1880. The original Boardwalk was heavily damaged by a hurricane in 1889, but it was replaced by a new, larger, and more substantial boardwalk that is still used today. The heyday of Atlantic City's prosperity was in the early 1900s, when it was visited by hundreds of thousands of tourists annually. Many tourists attended vaudeville shows and concerts given by the day's most popular performers.

After World War II, Atlantic City began to decline. The number of tourists dropped in part because commercial air travel allowed tourists to visit the Caribbean, Florida, and Hawai'i more easily and inexpensively. Many who did continue to visit Atlantic City were day trippers and elderly persons, who contributed relatively little to the city's economy. Crime rates increased and many buildings deteriorated.

To counteract this decline, in 1976 the voters of New Jersey legalized casino gambling. This was done in order to promote urban redevelopment in Atlantic City. The first casinos were opened two years later, and long-time Atlantic City residents soon noticed an upturn in business activity (Figure 7.A). By the mid-1980s, the casinos of Atlantic City were the most visited in North America, exceeding even those of Las Vegas. Today, more than 35 million people visit Atlantic City each year. By the mid-1990s, nearly 50,000 persons were employed in Atlantic County's casinos, with more than 11,000 of these employees in Atlantic City itself. Yet some residents expressed concern that casino gambling was taking its toll on the local business community. Many small retail outlets and other local business went broke as they were unable to keep pace with larger competitors. Many of the casino jobs were low-paying, with little future, and land prices put a financial squeeze on elderly residents living on fixed incomes.

▲ **Figure 7.A** Atlantic City casinos and hotels

Of course, many tourists also visit the region's major cities to view their many cultural and historic attractions.

In recent years, truck farmers and other agricultural producers have faced increasing pressure to sell their acreages to developers. Generally, the value of land is determined by its highest and best use, or the land use associated with the maximum amount of money that a buyer would be willing to pay for it. Because developers can convert farmland to suburban developments and reap enormous profits, the assessed value of farmland can rise rapidly, forcing farmers or their heirs who cannot afford to pay the taxes on this valuable land to sell their lands. This financial pressure also encourages farmers who do remain in production to specialize in products that generate high-value outputs. In response to this problem, some states and counties have enacted laws specifically exempting farmland from assessment on the basis of its value to developers or other urban interests. For example, the Maryland Farmland Assessment Law requires that the assessed value of productive farmland be based on the agricultural productivity of the land alone.

The Future of Megalopolis

As this chapter has described, Megalopolis at the dawn of the 21st century represents the greatest concentration of wealth and power in one small area throughout the course of history. The influence of Megalopolis on the contemporary world is evident from even cursory examination of the political power of the United States, the economic clout of Wall Street, the international influence of U.S.-based global corporations, and the wealth of American cultural icons and symbols. Washington is the seat of government of the world's most powerful country; Wall Street is the world's most important financial center; Broadway is one of the world's leading cultural centers; and Madison Avenue represents the advertising industry, which has created and exported dozens of instantly recognizable corporate logos, slogans, and symbols throughout the world.

Like every region in North America and elsewhere, Megalopolis has its share of problems. The cities of Megalopolis are crowded. Commuters often face massive traffic jams, and the cost of living, especially in large cities, is very high. Large numbers of people crowded into a relatively small area inevitably results in increasing levels of air and water pollution. A potential long-term impact is flooding from rising sea levels due to global warming.

Numerous racial and ethnic groups inhabit the region, and tensions between these groups occasionally result in violence, usually based on socio-economic issues. Such problems are not unique to Megalopolis. The concentration of wealth and power in Megalopolis is accompanied by a concentration of creativity, and it is to be hoped that solutions to these and other problems of the region will be generated by the combined efforts of the large numbers of scientists, artists, economists, educators, and public officials who live and work in the region. In recent years, to be sure, economic and political power has become decentralized to other areas of North America and the world. In the United States in particular, the growth of Megalopolis has been eclipsed for several decades by even more impressive levels of economic and population growth in the Sun Belt, especially in California, Texas, and Florida. Nevertheless, Megalopolis has been the core of the North American political economy for over 200 years. It is unlikely to relinquish this position at any time in the foreseeable future.

Suggested Readings and Web Sites

Books and Articles

Borchert, John R. 1992. *Megalopolis: Washington, D.C. to Boston*. New Brunswick, NJ: Rutgers University Press. Overview of Megalopolis, with focus on its cities in the early 1990s.

Denevan, William M. 1992. "The Pristine Myth: The Landscape of the American in 1492," *Annals of the Association of American Geographers* 82: 369–385. Argues that Native Americans made substantial landscape modifications in eastern North America well before Columbus.

Domosh, Mona. 1996. *Invented Cities: The Creation of Landscape in Nineteenth-Century New York and Boston*. New Haven: Yale University Press. A history of urban development and the creation of the built environment in New York and Boston.

Eisenstadt, Peter, Ed. 2006. *The Encyclopedia of New York State*. Syracuse: Syracuse University Press. One of the most comprehensive reference books ever written on New York, with nearly 4500 entries and lavish photographs.

Gottmann, Jean. 1987. *Megalopolis Revisited: 25 Years Later*. College Park, MD: University of Maryland, Institute for Urban Studies, Monograph Series No. 6.

Gottmann, Jean. 1961. *Megalopolis: the Urbanized Northeastern Seaboard of the United States*. New York: The Twentieth Century Fund. The original conceptualization of the term megalopolis is discussed in this book with a focus on the Boston to Washington, D.C. metropolitan areas.

Hayes-Conroy, Allison. 2005. *South Jersey Under the Stars: Essays on Culture, Agriculture, and Place.*

Danvers, MA: Rosemont Publishing and Printing. A reflective book on ecological thought that weaves together the themes of farms, suburbs, capitals, and celebrations.

Jackson, K. T. 1985. *Crabgrass Frontier: The Suburbanization of the United States*. New York: Oxford University Press. A history of suburbanization in the United States, with special reference to the cities and suburbs of Megalopolis.

Lewis, Peirce F. 1972. "Small Town in Pennsylvania," *Annals of the Association of American Geographers* 62: 323–351.

Wacker, Peter O. and Paul G.E. Clemans. 1995. *Land Use in Early New Jersey: A Historical Geography*. Newark: New Jersey Historical Society.

Ward, David. 1971. *Cities and Immigrants: A Geography of Change in Nineteenth Century America*. New York: Oxford University Press. A classic study of the impact of waves of immigration on U.S. cities.

Wood, Joseph. 1997. "Vietnamese American Place Making in Northern Virginia," *The Geographical Review*, Vol. 87 (1): 58–72.

Web Sites

Digital Atlas of Megalopolis
http://www.umbc.edu/ges/student_projects/digital_atlas/credits.htm

Boswash: The Metropolitan Area from Boston to Washington
http://geography.about.com/cs/urbansprawl/a/megalopolis.htm

Ellis Island in New York City (searching for your family roots?)
http://www.ellisisland.org/

An Outline of American Geography: Megalopolis
http://usinfo. state.gov/products/pubs/geography/geog04.htm

Climate in Megalopolis: Insights from Northeast Snowstorms
http://www.infoplease.com/ce6/world/A0832515.html

8
The Great Lakes and Corn Belt

PREVIEW

Environmental Setting

- Most of the region consists of flat or rolling topography. Much of it was affected significantly by glaciation during past ice ages.
- The Great Lakes and the Mississippi, Ohio, Saint Lawrence and other rivers form the basis of an excellent transportation network, which humans reinforced early by the construction of railroads and highways.
- The region's climate is humid continental, with warm summers, cool to cold winters, and ample precipitation throughout the year.

Historical Settlement

- Much of the Great Lakes and Corn Belt region was contested among Britain, France, and local Native Americans prior to the American Revolution. After the United States became independent, the "Old Northwest" quickly filled up with settlers.
- The Erie Canal unlocked settlement in the region. By the late 1830s more than 200,000 people had used it to travel from New England and upstate New York to the Great Lakes.
- The population of the region has been influenced heavily by immigration. By the time the Census of Population of 1910 had been tabulated, almost one-third of the Great Lakes region of the United States was foreign-born.
- Between 1916 and 1920, some 500,000 African Americans left the rural South for the industrial centers of Chicago, New York, Detroit, Cleveland, Philadelphia, St. Louis, and Kansas City, fleeing racism and economic problems to seek work in northern and midwestern factories.

Political Economy

- Airline passengers crossing over rural portions of the Corn Belt today can see the impacts of the township and range system by observing the rectangular pattern of fields, highways, and secondary roads.
- Even though the Great Lakes and Corn Belt is the world's most profitable agricultural region, the number of farms in the region has declined steadily over the past century and the percentage of persons employed in farming also continues to drop. Those who remain in agriculture face continuing financial pressure, and it is often a struggle for these farmers to remain in business.
- For many years, the region has been the most productive manufacturing region in North America and one of the major industrial areas of the world. However, industry, like agriculture, has begun to decline in importance in many parts of the Great Lakes and Corn Belt.
- In recent years, there has been renewed growth and prosperity in diversified large cities, government centers, university communities, and other places.

Culture, People, and Places

- Located in the Great Lakes and Corn Belt region, Toronto, on Lake Ontario, is Canada's largest and most diverse metropolitan area.
- Chicago is the largest city on the Great Lakes. Other Great Lakes cities in both the United States and Canada are part of the old industrial "Rust Belt."
- Cities in the Corn Belt are older and more prosperous, and in many cases have better weathered transition to a post-industrial economy.

Introduction

The Great Lakes and Corn Belt region has often been called the "heartland" of North America. Car tags issued by the state of Ohio, whose shape is vaguely reminiscent of a heart, proclaim the Buckeye State to be "the heart of it all." Why is the term "heartland" often used to refer to the Great Lakes and the Corn Belt? How did the Great Lakes and Corn Belt region emerge as the American heartland? And, what are future prospects in a region whose economy has long been oriented to farming and manufacturing in a world increasingly dominated by the tertiary and quaternary sectors?

The Heartland label is appropriate for the Great Lakes and Corn Belt region for several reasons. Most obviously, the area is located near the geographic center of North America. The population centroid of the United States has crossed the region over the past 200 years, although it continues to move southward and westward away from the region—a metaphor perhaps for the increasing importance of the Sun Belt relative to the Heartland. The region's central status is reinforced by excellent transportation. The Great Lakes and the Mississippi River and its tributaries form a natural transportation network that was reinforced by

◀ Navy Pier and the Chicago skyline

19th century, water transportation along the rivers and in the Great Lakes was enhanced by a series of improvements, including the construction of locks and dams on the rivers, the building of the Erie Canal in 1825, and later the dredging of the Chicago River, providing a navigable connection between the Mississippi drainage basin and the Great Lakes. Later in the 19th century, railroads were built and these too provided for easy movement of people and goods.

The millions of persons who settled the Great Lakes and Corn Belt region came from many different origins. Natives of the Appalachians, such as Kentucky-born Abraham Lincoln, moved north and west across the Ohio River. These "Butternut" settlers were soon outnumbered by natives of New England, New York, and Pennsylvania who crossed the Allegheny Mountains in search of cheap, fertile farmland. They were joined by immigrants from England, Scotland, Ireland, Wales, Germany, the Netherlands, and Scandinavia. Many Canadians who a generation or two earlier had moved to Upper Canada returned to the United States, and were especially prevalent in Michigan.

By the end of the Civil War, however, most of the arable land in the region had been claimed and settled. In the late 19th and early 20th centuries, many migrants from southern and eastern Europe moved to the region's cities, where they provided the labor force for the region's large numbers of growing factories. Chicago, Detroit, Cleveland, and other cities south of the Great Lakes featured large numbers of Poles, Greeks, Italians, Hungarians, and Czechs. At the same time, many eastern and southern Europeans moved to Canada. These immigrants to Toronto, Hamilton, and Ottawa established Little Italy and Chinatown neighborhoods and ethnic enclaves of almost every nationality, many of which still exist today (Table 8.1). Other immigrants such as the Dutch, Swedes, Norwegians, Finns, and Danes joined ever-increasing numbers of Germans in farm and fishing communities in Michigan, Wisconsin, Minnesota, and southern Ontario.

TABLE 8.1 Historical Settlement of Ethnic Groups in Canada

Group	Dates	Major Occupations
Native Americans	pre–1600	Self-contained societies
French	1609–1755	Fishing, farming, fur trading, supporting occupations
Loyalists from the United States	1776–mid 1780s	Farming
English and Scots	mid 1600s on	Farming, skilled crafts
Germans and Scandinavians	1830–1850, c.1900, 1950s	Farming, mining, city jobs
Irish	1840s	Farming, logging, construction
African Americans	1850s–1870s	Farming
Mennonites and Hutterites	1870s–1880s	Farming
Chinese	1855, 1880s	Gold panning, railway work, mining
Jews	1890–1914	Factory work, skilled trades, small business
Japanese	1890–1914	Logging, service jobs, mining, fishing
East Indians	1890–1914, 1970s	Logging, service jobs, mining, skilled trades, professions
Ukrainians	1890–1914, 1940s, 1950s	Farming, variety of jobs
Italians	1890–1914, 1950s, 1960s	Railway work, construction, small business, construction, skilled trades
Poles	1945–1950	Skilled trades, factory jobs, mining
Portuguese	1950s–1970s	Factory work, construction, service jobs, farming
Greeks	1955–1975	Factory work, small business, skilled trades
Hungarians	1956–1957	Professions
West Indians	1950s, 1967–	Factory work, skilled trades, professions, service jobs
Latin Americas	1970s	Professions, factory and service jobs
Vietnamese	late 1970s–early 1980s	Variety of occupations

Source: From E. Herberg, *Ethnic Groups in Canada: Adaptations and Transitions.* © 1989 Nelson Canada, Scarborough, Ontario.

Most of the residents of the Great Lakes and Corn Belt in 1910 were of European ancestry. However, diversity in the region soon increased as African Americans began to move to the region in large numbers after World War I. The movement of African Americans from the South became a critical part of a new migration stream that dramatically changed the demographics and cultures of cities throughout the region. Between 1916 and 1920, some 500,000 African Americans left the rural South for the industrial centers of Chicago, New York, Detroit, Cleveland, Philadelphia, St. Louis, and Kansas City in what was known as the "Great Migration." Thousands more moved northward between the 1920s and the 1950s. These migrants were driven by both **push factors** and **pull factors**. Push factors are issues encouraging people to migrate away from their areas of origin. For southern African Americans, these push factors included institutionalized racism and segregation, lynchings, beatings, crop failures, and limited educational and economic opportunities. Pull factors are issues encouraging migrants to select particular destinations. Pull factors in the industrial cities of the Great Lakes and Corn Belt included jobs at much higher wages than were prevalent in the South, along with a call to freedom and a much sought after opportunity for equality.

The impacts of the **Great Migration** were massive and long-lasting on African Americans and on the cities where they made their new homes. Whole families or communities often made the trip together. Dreams of renting a big house for family and friends to share and a better lifestyle pushed them on despite reports by relatives of bad weather, hard work, and racist treatment by new neighbors. Despite these challenges, the populations in these industrial cities north of the Mason-Dixon Line and in the Midwest were expanded significantly from this mass migration. The cities of the Great Lakes and Corn Belt, notably Detroit, Chicago, St. Louis, and Kansas City, soon became well-known as African-American cultural centers. The African-American heritage of this region is commemorated today in Kansas City, which is the home of the National Jazz Museum and an adjacent museum commemorating Negro League baseball (Box 8.2).

Southern Ontario also has a lasting legacy of African-American migration, although from an earlier time period. Fascinating evidence of Canada's assistance in hiding and sustaining runaway slaves in the pre–Civil War era remains in unlikely places just north of the Great Lakes. Today, some parts of the Great Lakes and Corn Belt region continue to attract large numbers of immigrants. Chicago and Toronto in particular have attracted large numbers of migrants from Latin America, Asia, and elsewhere in recent years. Other metropolitan areas have also attracted substantial immigrant populations. For example, the Detroit area has North America's largest concentration of Arab Americans, especially around the city of Dearborn, where many businesses place signs in both English and Arabic. In part because of the transition from an industrial to a post-industrial economy, however, most of the region's smaller metropolitan areas have seen little immigration relative to cities of comparable size in the Sun Belt.

Political Economy

Long before 1900, the economy of the Great Lakes and Corn Belt was highly integrated, combining agriculture, industry, and services into a unified and highly cohesive regional economy. This cohesion has remained characteristic of the region ever since.

Agriculture and Other Primary-Sector Activities

Historically, agriculture has been the foundation of the economy of the Great Lakes and Corn Belt region. For nearly two centuries, the Great Lakes and Corn Belt region has been one of the most productive farming regions in the entire world. The agricultural productivity of the region is the product of several factors in combination: fertile soils, adequate rainfall, excellent transportation, technology, government policy, and sheer experience.

The large majority of farms in the Great Lakes and Corn Belt are family farms. A family farm is one run by an individual family, as opposed to a corporation. Family farms in the contemporary Great Lakes and Corn Belt are highly mechanized, capital-intensive operations. Most commercial farmers invest heavily in machinery, fertilizers, pesticides, high-yield crops, and other technological innovations. These farms are usually large. A typical midwestern grain farm, for example, includes several hundred acres (1 acre equals approximately 4050 square meters), yet this large tract of land is typically operated by a single farmer and members of his or her family. Families usually live in a farmhouse surrounded by barns, machine sheds, and other outbuildings (Figure 8.2).

▲ **Figure 8.2** Midwestern family farm

BOX 8.2 Goin' to Kansas City

During the 1950s, Wilbert Harrison sang "I'll be standing on the corner of Twelfth Street and Vine, with my Kansas City baby and a bottle of Kansas City wine." Even today, the corner of Twelfth and Vine is at the heart of Kansas City, Missouri's African-American community. Six blocks away, at Eighteenth and Vine, is a museum complex that houses the American Jazz Museum and the Negro Leagues Baseball Museum (Figure 8.A). Thus, two important developments in 20th-century African-American culture are memorialized in Kansas City, which was a major center of innovation for both institutions.

Kansas City, like other cities in the Great Lakes and Corn Belt, was a major migration destination during the Great Migration of African Americans out of the South during and after World War I. Many rural African Americans moved from the rural South to Kansas City and found work in meatpacking plants, railroad yards, steel mills, and factories. Although African Americans in Kansas City were not subject to the rigid segregation characteristic of the Deep South at the time, many worked at low-paying, dangerous jobs and many suffered from discrimination in education, employment, housing, and other spheres of life. Music and sports were important recreational outlets to Kansas City's African-American community. Legendary jazz musicians such as Jimmy Rushing, Count Basie, Coleman Hawkins, and Charlie "Bird" Parker began their careers playing in clubs near Twelfth and Vine.

During the early 20th century, baseball was the most popular spectator sport in the United States. Until Jackie Robinson broke the color line in 1947, however, African Americans were forbidden to play major league baseball. Separate professional leagues for African-American players were established and thrived from the early 1920s to the late 1950s. The Negro National League was founded in 1920. Its teams included the Kansas City Monarchs, Chicago American Giants, Chicago Giants, Dayton Marcos, Detroit Stars, Indianapolis ABCs, St. Louis Giants, and the Cuban Stars (who were based in New York). Thus, the cities represented by the League included many of the cities whose African-American populations were swelling in the early 1920s. Other leagues included teams in Pittsburgh, Philadelphia, and Newark as well as in industrializing Southern cities such as Atlanta, Birmingham, and Jacksonville. The rosters of these teams included many players who would go on to stardom once baseball was integrated, including Robinson, Willie Mays, Ernie Banks, Larry Doby, and Elston Howard.

The Kansas City Monarchs were the most successful team on the field and at the box office. They were known as the Yankees of the Negro Leagues: in their 37 years on the field, they won 12 league championships. Perhaps more importantly, the Monarchs provided an opportunity for economic and educational advancement for disadvantaged African-American youths. Attendance at Monarchs' games was integrated. Roughly half the fans at the average home game were white. The Monarchs' success brought a considerable infusion of money into Kansas City's African-American community. One sportswriter wrote, "From a sociological point of view, the Monarchs have done more than any other single agent to break the damnable outrage of prejudice that exists in this city."

▲ **Figure 8.A** A street party at the grand opening of the Eighteenth and Vine Jazz District in September 1997 in Kansas City. The district includes the refurbished Gem Theater on the right and the New Jazz Musicians Museum on the left.

Traditionally, agriculture in the Great Lakes and Corn Belt region has been associated with a set of core values that have long underlain the region's culture, as well as its economy. The well-known geographer John Fraser Hart identified the following characteristics as core values underlying the culture of the Great Lakes and Corn Belt:

- Each farmer should own the land he farms.
- Each farm should be large enough to provide a decent living.
- Hard work is a virtue, and a family is responsible for its own economic security.
- The farmer and family should do most of the work.
- Anyone who fails to make good is lazy.
- Farmers should receive a fair price for their products.
- The best government is the least government.

Of course, these statements are associated with a society whose values reflect Thomas Jefferson's conception of a commonwealth of independent family farmers (see Chapter 4). Hart's last two statements, however, are fundamentally contradictory. The Jeffersonian ethic is based on the concept of limited government; however, government intervention has often proved necessary to ensure that farmers receive fair prices and are otherwise able to achieve adequate rewards for their hard work. As we saw in Chapter 4, the **technological treadmill** process has driven those farmers who are less competitive or who own relatively isolated and lower-quality lands out of production.

As the name of the region implies, corn is one of the major crops of the Great Lakes and Corn Belt region. Not surprisingly, corn production is concentrated in the Corn Belt itself, which stretches from western Ohio to eastern Nebraska. The United States leads the world in corn production, and Iowa and Illinois are the country's two leading corn-producing states. Many corn farmers also plant soybeans. Not only do these legumes help maintain soil fertility, but they represent an increasingly valuable source of income. Although soybeans are native to Asia, the United States now produces over three-quarters of the world's soybean crop and is responsible for about 90 percent of the world's soybean exports.

The mixed farms that are prevalent in the Corn Belt specialize particularly in the production of cattle and hogs. A large majority of the region's grain crops are fed to cattle and hogs, which are then marketed. Hog production in the United States is especially concentrated in the Corn Belt—there are about five hogs in Iowa for every person—although an increasing number of hogs are being produced elsewhere.

Other types of farming are practiced in the northern and southern parts of the Great Lakes and Corn Belt region, where climatic and soil conditions are not so favorable for corn and soybean production. Dairy farming is prevalent in the northern part of the region, notably in Wisconsin, whose cooler climate and short growing season preclude competitive corn, cattle, and hog production. Because milk is bulky and highly perishable, milk sold to consumers directly is generally produced on dairy farms near large cities. Those dairy farmers living at greater distances from their markets typically sell milk to producers of butter, cheese, and other less perishable dairy products, as is the case in Wisconsin. Fruit is produced in western Michigan, northern upstate New York, and southwestern Ontario. These areas are adjacent to the Great Lakes, and the presence of the relatively warm water of the lakes retards winter freezing, making production of grapes, tomatoes, and other fruits profitable in these areas.

Even though the Great Lakes and Corn Belt is the world's most profitable agricultural region, the number of farms in the region has declined steadily over the past century and the percentage of persons employed in farming also continues to drop. Those who remain in agriculture face continuing financial pressure, and it is often a struggle for these farmers to remain in business. Farmers who cannot keep up with the technological treadmill are driven out of production. Thus the number of farms decreases, while the size of the average farm increases.

The technological treadmill has caused declines in acreage planted to many crops throughout the United States. In many areas of North America, crop production has largely disappeared in recent years. Even in the Great Lakes and Corn Belt, less profitable lands are no longer used as intensively as was once the case, with land once planted to corn, soybeans, and other crops now used for livestock ranching or converted to non-agricultural use.

Industry in the Great Lakes and Corn Belt

Historically, the highly profitable and productive agriculture of the Great Lakes and Corn Belt has been complemented by an equally profitable industrial sector. Food produced on the farms of the Corn Belt is consumed by industrial workers in the cities of the Great Lakes. For many years, the region has been the most productive manufacturing region in North America and one of the major industrial areas of the world. However, industry, like agriculture, has begun to decline in importance in many parts of the Great Lakes and Corn Belt.

The Great Lakes and Corn Belt region has many advantages that made the area into a major manufacturing center. The region is well equipped with nearby natural resources. For example, both coal and iron, the key components of steel, are produced on the outskirts of the Great Lakes and Corn Belt region, with iron being mined in northern Minnesota and coal produced in Pennsylvania, West Virginia, and Kentucky. As we have already seen, the area took advantage of natural and human-made transportation, providing ready access for its products to national and international markets. Labor was abundant, reinforced especially after the Civil War with the arrival of millions of immigrants from Europe and later, after the imposition of immigration quotas, by native-born Americans from Appalachia and the Deep South.

By the late 19th century, the Great Lakes and Corn Belt had become the leading heavy industry region of North America. Automobiles, trucks, aircraft, and other heavy machinery were produced in large quantities. Manufacturing activity quickly caused dramatic population increases. By 1900 Chicago was the second largest city in the United States, and Cleveland and St. Louis were among the country's ten largest. Many of these cities specialized in the production of a fairly small number of products. The primary industrial products produced in such cities soon contributed to their communities' image and self-identification. For example, everyone associates Detroit with automobiles, Pittsburgh with steel, and Milwaukee with beer. Although production of the principal products of such communities has generally declined, the images remain and are commemorated in many ways, including the names of sports teams such as the Detroit Pistons, Pittsburgh Steelers, and Milwaukee

▶ **Figure 8.3** The former headquarters of Motown Records in Detroit makes way for 21st-century development

Brewers, and in such icons of popular culture as Motown Records in the "Motor City" of Detroit (Figure 8.3).

By no means are these images limited to very large cities in the Great Lakes and Corn Belt region. Many smaller cities are also associated with specific products that were long manufactured there: for example, Battle Creek, Michigan (breakfast cereals), Akron, Ohio (tires), and Elkhart, Indiana (musical instruments). In most cases, the association of these cities with specific products is the result of local entrepreneurship, with one or more factories associated with production. In the Corn Belt portion of the region, many manufacturers specialize in food processing or the manufacture of agricultural implements. Communities such as Peoria and Moline, Illinois, and Waterloo and Dubuque, Iowa, are examples.

Industrial output and production in the Great Lakes and Corn Belt region has declined substantially in recent years. Many corporations have moved their manufacturing operations out of the Great Lakes and Corn Belt to other areas where production costs, especially costs associated with labor, are lower. Indeed, many of these manufacturing operations once concentrated in the Great Lakes and Corn Belt are now conducted in less developed countries. For example, dozens of steel mills once were located in the Pittsburgh area, but none remain there now; the world's largest steel mills are now located in Mexico, Korea, Brazil, and other countries. The United States, once the leading steel-producing country in the world, is now a major steel importer.

Despite overall declines, in some communities manufacturing activity has increased since the late 1980s. There are several factors underlying these increases. First, relative to many other parts of North America, the economy of the Great Lakes and Corn Belt is not dependent on military contracts. Most of the industrial production of the region is geared to civilian markets, and as a result the health of industries is not tied to changes in foreign policy or military appropriations. As well, there has been a substantial amount of foreign direct investment in manufacturing in the Great Lakes and Corn Belt. Foreign-based corporations such as Honda and Toyota have invested huge amounts of capital in the region; in fact, more than half of all Japanese direct investment in the United States is concentrated in Ohio, Michigan, Indiana, and Illinois (Figure 8.4). The factors that made the Great Lakes and the Corn Belt attractive to domestic investment in the past appear equally attractive to global investors today. These include a strong skilled labor force, economies of scale, transportation, and ready access to local, national, and international markets.

Generally speaking, those communities with diversified industrial bases have tended to weather regionwide declines in production and output more effectively than those whose industrial bases are highly specialized, or than those that specialize in production of goods for which demand has declined. For example, the continued decline in the number of farmers has reduced demand for agricultural implements. Those communities that specialize in producing agricultural implements, such as Peoria, Illinois, and Waterloo, Iowa, have struggled economically in recent years.

The Tertiary and Quaternary Sectors

Another factor that has been critical to determining the economic health of communities in the Great Lakes and Corn Belt region has been the extent to which these communities have been able to successfully replace their traditional dependence on agriculture and manufacturing with tertiary and quaternary sector activities. Those that have been able to do so—in particular, those that already had widely diversified economies to begin with—have weathered the recent economic downturns and indeed have prospered; those that were less successful have

Figure 8.4 Honda factory in Marysville, Ohio

been on the decline in recent years. Thus we have seen recent growth and prosperity in diversified large cities such as Chicago and the Twin Cities; in government centers such as Indianapolis and Columbus; in Pittsburgh, with an increasingly important health-care and high-tech economy; in university communities such as Ann Arbor, Champaign-Urbana, and Iowa City; in areas associated with computer technology such as Sioux Falls, and in medical centers such as Rochester, Minnesota. On the other hand, more traditional manufacturing-oriented centers such as Detroit, St. Louis, Hamilton, Waterloo, and Peoria have remained more dependent on agriculture and/or heavy industry and these areas have suffered population loss and economic declines.

Interestingly, tertiary and quaternary sector activities in the Great Lakes and Corn Belt region have often located and prospered at particular locales as a result of individual entrepreneurship and/or historical decision processes. Government centers, including state capitals, have generally experienced healthy growth in recent years. Cities such as Columbus, Ohio; Des Moines, Iowa; and Indianapolis, Indiana, are not only state capitals, but they have also attracted various white-collar activities such as insurance companies. University communities are also growth centers within their states. University communities often generate additional high-tech or white-collar employment, which is often associated with activities of the university and its faculty. For example, Iowa City is one of the two leading centers in the world for educational testing; University of Iowa faculty developed standardized testing methods for elementary and secondary schools in the 1930s and 1940s, and today the city's two largest employers, in addition to the university itself, are ACT Assessment and Westinghouse's division of educational research.

Medical services are important to some local economies. The Mayo Clinic, for example, is the linchpin of the economy of Rochester in southeastern Minnesota. In 1864, Dr. William Mayo, a native of England, decided to establish a medical practice in the then-new community of Rochester. His two sons, Will and Charles, also became physicians and joined their father in his practice. Will and Charles established a hospital in 1889, and invited other physicians to join the practice (Figure 8.5). In contrast to most medical clinics, the Mayo brothers paid their associates salaries rather than encouraging them to treat patients for their own profit. The profits earned by the associates in the practice were pooled, invested, and used to improve the quality of patient care and to promote medical research. The Mayo brothers took an active interest in promoting integration of medical research, teaching, and patient care. By the time the Mayo brothers died, both in 1939, the Mayo Clinic had a worldwide reputation. Medical research has spun off various other economic activities in the local area, further adding to the region's prosperity.

In contrast to other parts of the United States and Canada, neither tourism nor military activity plays much of a role in the economy of the Great Lakes and Corn Belt as a region. A few areas have become tourist centers. Large cities such as Chicago draw substantial numbers of tourists, and various resort areas such as the Door Peninsula and Lake Geneva in Wisconsin and Cedar Point on Lake Erie in Ohio also attract numerous vacationers. For the most part, however, the Great Lakes and Corn Belt is a net exporter of tourists. Many residents of the region own vacation property or take regular vacations outside the region.

As mentioned earlier, many farmers spend the winter months in the Sun Belt, particularly in Florida, south Texas, or Arizona. Many residents of Chicago, Detroit, Milwaukee, Minneapolis–St. Paul, and other places own vacation property or second homes in areas to the north such as northern Michigan, Wisconsin, and Minnesota. Similarly, the boreal forest and lake country of northern Ontario attracts thousands of vacationers from the Toronto metropolitan area. Frequently, persons who have vacationed in these communities for many years end up retiring there, and as a result population growth rates in various counties in the "North Woods" have been substantial. Many have attracted increasing populations of year-round residents, especially retirees. This increase in the number of year-round residents has resulted in an increase in business and service activities.

▶ **Figure 8.5** The Mayo Clinic, Rochester, Minnesota

Culture, People, and Places

From the air, perhaps no other region in North America looks as homogeneous as the Corn Belt and Great Lakes. Except for scattered rolling hills north of the lakes, and a few places where streams have cut into the land leaving small canyons in their wake, the physical landscape here looks much the same wherever you travel. Small lakes, rivers, and meandering streams are common, as are croplands alternating with forested stands of leafy oaks, maples, and other hardwoods.

In the U.S. portion of this region, land patterns appear regular as well as the patterns of its physical geography (thanks to the impacts of the township and range system). The regularity and rhythm of rectangular fields planted in row crops and seemingly endless linear highway and rail connections make this region look like a gigantic checkerboard from the air. Houses and an extensive series of outbuildings dot the landscape. Small towns appear at fairly regular intervals throughout the region as centers serving local populations and places for agricultural processing. Larger cities appear less regularly, but show consistent internal patterns featuring tall buildings located in the center of town and in nearby suburbs and more residential and commercial districts horizontally sprawling in rings located farther and farther out from the inner city.

The appearance of homogeneity, however, belies the substantial cultural, economic, religious, and ethnic diversity characteristic of this important and dynamic portion of North America. Historically, the mindset of much of the region was rural, with perceptions centered on quiet townscapes surrounded by the rustle of cornfields. Since the population of the rural parts of this region have been in decline for the past century and few new immigrants have moved in to replace earlier settlers, the original German-Scandinavian-eastern European cultural foundation persists. Ethnic identity here is often low-key and down-to-earth, as are other cultural and political traits. As National Public Radio's show "Prairie Home Companion" is fond of pointing out in its Saturday night broadcasts, Norwegians, Swedes, and others hold onto their identities and resist change or glorification of their ethnic heritage.

Places throughout the Great Lakes and Corn Belt are dealing with continued transition from an agricultural/industrial to a post-industrial economy. Here, we examine places in the region, dividing them into three groups: places in Canada, American places on and near the Great Lakes region, and places in the Corn Belt itself.

Canadian Places on the Great Lakes

The Canadian portion of the Great Lakes and Corn Belt region, including much of the province of Ontario, is Canada's economic core. It includes Canada's largest city, its major industrial centers, and prime agricultural land. Seven of the twenty largest metropolitan areas in Canada are located in southern Ontario. Four of them (Oshawa, Toronto, Hamilton, and St. Catherines–Niagara) form what is known as the "Golden Horseshoe" around the western end of Lake Ontario.

Toronto is located on the site of an old Huron community whose name meant "meeting place." Until the late 20th century, Toronto's demographics were dominated by two groups of Europeans in its earliest years—the British and the French. The Hurons and later the French fur traders saw it as merely a link in a land chain connecting Lake Ontario with Lake Huron. Even the British, who took over the city following the Seven Years War, were slow to recognize the supremacy of its location or its potential for growth.

Ultimately, for better or worse, political and economic development in Toronto were in the hands of the wealthy British elite over the course of the next century. When its growth expanded during the 19th century, Scots fleeing the problems in their Highlands and Irish fleeing the potato famine at home helped fill in the gaps in population and culture. However, the city did not blossom until the World War II era, when a tidal wave of immigrants arrived from all over the world to enlarge and enliven the city. After the war, the city began to grow rapidly. By 1970, Toronto had surpassed Montreal and become Canada's largest city. Today Toronto has been transformed into a vibrantly sophisticated place with a dynamism that is palpable to insiders and visitors alike. Although Toronto did not become a city of great cultural diversity until after World War II, the cultural diversity of Toronto today is parallel to that of Chicago and other major U.S. cities.

Between 1976 and 1991, the total population of the Toronto metropolitan area increased by over a million people, and it expanded to the north, west, and east onto previously productive farmland. It has been estimated that 100 acres (259 hectares) of new land are needed to support each increase of 1000 new people (Figure 8.6). Thousands of acres of cherry orchards and other fruit trees have been cut down and converted to suburban developments. As Toronto has expanded to the west and north, it has merged into a large conurbation reminiscent of Megalopolis (see Chapter 7). The city of Hamilton, a traditional center for the production of steel and other heavy industry, has become integrated into this metropolitan complex.

By no means is all of Ontario's population concentrated in Toronto, Hamilton, or their suburbs. The flat, fertile land between Lake Huron and Lake Erie has long been an important agricultural region. London, located halfway between Toronto and Detroit, and Windsor, to the south of Detroit, are both major industrial centers. East of Toronto are the cities of Kingston and the national capital of Canada at Ottawa. The north shore of Lake Superior is the site of several additional Ontario cities, including the port of Thunder Bay, the old nickel-mining community of Sudbury, and Sault Ste. Marie, which is an old steel-producing center as well as a port located between Lakes Superior and Huron.

▲ **Figure 8.6** Financial district in Toronto, Ontario

American Places on the Great Lakes

Given the Great Lakes' importance in transportation, it is not surprising that numerous cities have sprung up along all of the Lakes on both sides of the international boundary. The largest city on the Great Lakes—indeed, the third largest city in North America—is Chicago. Chicago's location on Lake Michigan, its favorable transportation connections, and its long-standing status as an industrial center have ensured it a lasting place of importance in North America's urban hierarchy (Figure 8.7). Popular movies like "The Blues Brothers," "Chicago," and the Al Capone dramas captured a few of the stories that have helped shape the city.

Chicago is not only the largest city in the Great Lakes and Corn Belt region, but it is also a global city with economic and transportation connections that extend well beyond the Great Lakes and Corn Belt to places throughout the world via rivers and waterways, railroads, airlines, and communication technologies. Chicago is a product of the second half of the 19th century. In 1850, only 29,000 persons lived in the city, but the population exceeded one million only 50 years later. The city's location on Lake Michigan at the mouth of the Chicago River gave the small town of Chicago its earliest growth impetus by being the head of navigation for all travelers and goods heading west by way of the Great Lakes. In 1848, the Illinois and Michigan Canal was completed, linking Chicago to the Mississippi River. Four years later, a railroad connecting Chicago with New York was completed. The city's role as the hub of the rail system of the nation was assured when Chicago was selected as the terminus of the transcontinental railroad during the Civil War years.

These unparalleled transportation connections soon made Chicago a leading industrial center. It became North America's leading meatpacking center, and many other industries, including furniture, clothing, and consumer durables, were produced there. Chicago and nearby Gary, Indiana, became the home of numerous steel mills. Labor for these and other industries was provided by a large immigrant workforce. Hundreds of

Environmental Setting

For purposes of this text, we divide the South into two major regions: the Inland South (discussed in this chapter) and the Coastal South (which is discussed in Chapter 10). The Inland South includes most of the old Confederacy. It includes parts of Virginia, West Virginia, the Carolinas, Georgia, Alabama, Tennessee, Kentucky, Mississippi, Arkansas, southern Missouri, northern Louisiana, eastern Oklahoma, and East Texas (Figure 9.1). All of Florida, along with the areas along the Atlantic and Gulf coasts from North Carolina to Texas, are discussed in Chapter 10 as part of the Coastal South.

Landforms

The Inland South includes portions of the Atlantic Coastal Plain, the Piedmont, the Appalachians, and the North American central lowland. Thus the topography is varied, from relatively flat and featureless terrain near the coast to rolling hills in the Piedmont to rugged mountains in the southern Appalachians.

The Atlantic Coastal Plain, which covers the eastern portion of Megalopolis (see Chapter 7), widens as it extends south and west into the Carolinas and Georgia. West of the peninsula of Florida it merges into the Gulf Coastal Plain, which has similar topographic characteristics. It includes most of southern Alabama, Mississippi, and Louisiana. The land is flat and soils are sandy. Rivers that originated in the Piedmont or the Appalachians meander lazily toward the Atlantic or the Gulf. Some parts of the Atlantic Coastal Plain are covered by large swamps, many of which are difficult to penetrate. The Great Dismal Swamp of northeastern North Carolina, the Okefenokee Swamp of Georgia, and the Big Thicket of Texas are well-known examples. In early 2005, ornithologists working in a poorly accessible, swampy area of southeastern Arkansas announced the rediscovery of the ivory-billed woodpecker, a bird that had been believed to have been extinct for more than 50 years (Box 9.1).

West and north of the Coastal Plain is the Piedmont region. As in Megalopolis, settlements sprang up throughout the Piedmont along the fall line, where rapids marked the head of inland navigation of many rivers. The Piedmont is the home of many of the Inland South's largest cities, including Atlanta, Charlotte, and Birmingham. The Appalachians, which extend as far north as Newfoundland, reach their maximum elevation in the southeastern United States. In contrast to

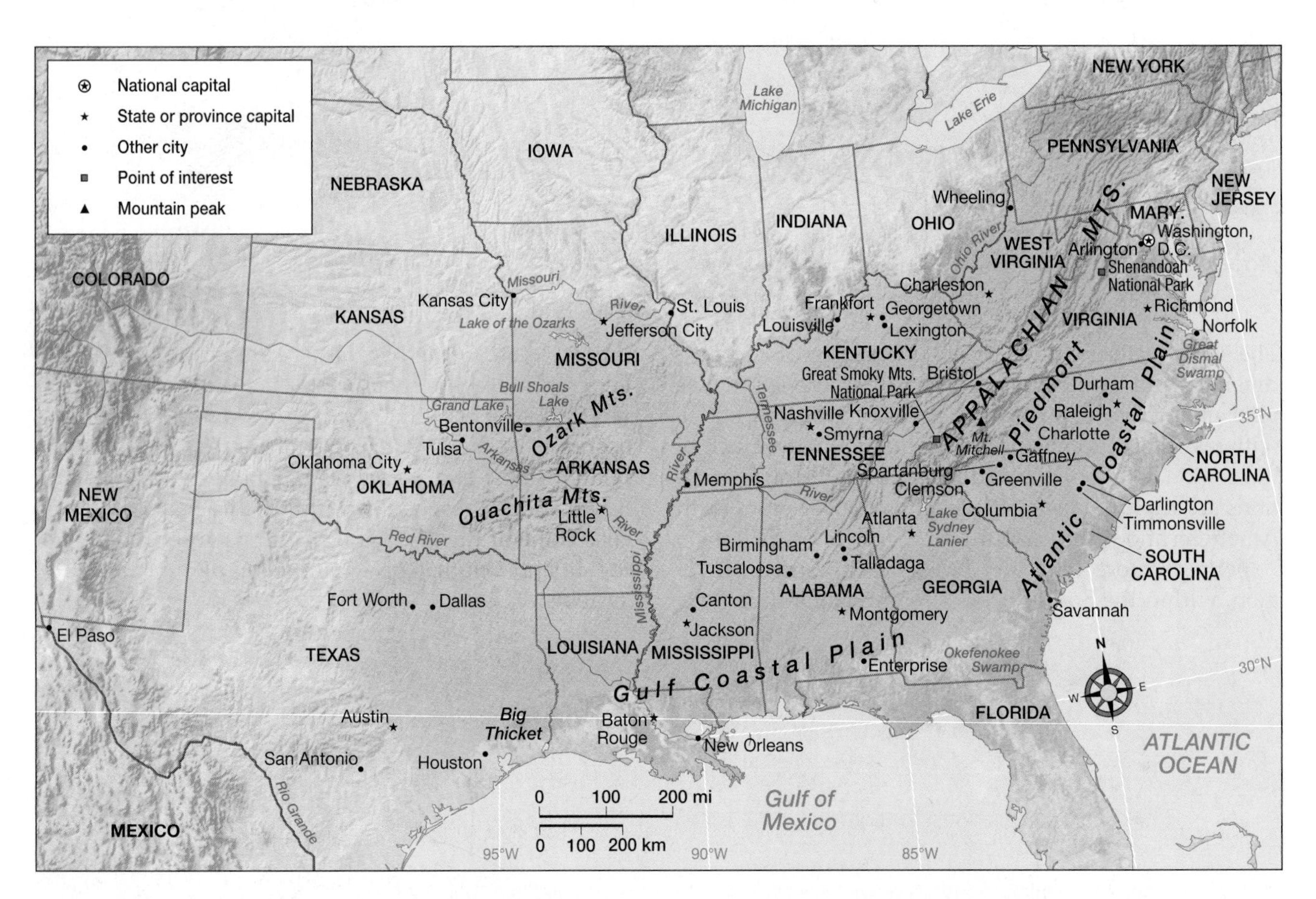

▲ **Figure 9.1** The Inland South region

BOX 9.1 The Rediscovery of the Ivory-Billed Woodpecker

The Ivory-billed woodpecker (*Campephilus principalis*) is the largest species of woodpecker in North America, and the third largest woodpecker in the world. The Ivory-billed woodpecker thrived in mature, swampy forests throughout the southeastern United States and in Cuba. The woodpecker's principal food was beetle larvae, which it extracted from dead and dying trees.

Over the past 300 years, most of the mature old-growth forests of the Southeast have been cut down, resulting in the destruction of the Ivory-billed woodpecker's habitat. Millions of acres of virgin forest were cut down between 1880 and the mid-1940s. Other Ivory-bills were shot by trophy hunters. By the 1930s, the bird was on the verge of extinction. No confirmed sightings were reported after 1944, when the last known Ivory-bill disappeared from an old-growth forest in Louisiana. An observer claimed to have heard the Ivory-bill's call in East Texas in 1967, and another was allegedly observed in Cuba in 1987, but neither of these observations were confirmed independently. A student claimed to have seen two Ivory-bills in southeastern Louisiana, but this sighting too was unconfirmed.

In recent years, efforts have been made to preserve old-growth forests. Amateur and professional ornithologists occasionally searched these areas for signs of the Ivory-billed woodpecker. In February 2004, an amateur naturalist, Gene Sparling, was paddling his kayak through the swampy "Big Woods" region of southeastern Arkansas. Sparling sighted a large woodpecker that appeared to be an Ivory-bill. Two weeks later, Sparling returned to the area with two professional ornithologists, Tim Gallagher and Bobby Harrison. All three men observed the bird. Over the next several weeks, other observers viewed and photographed it, and experts confirmed that the bird was indeed an Ivory-billed woodpecker, which had been thought extinct for more than 50 years. The discovery was announced in 2005.

The rediscovery of the Ivory-billed woodpecker excited the ornithology community around the world. Ornithologists are now using acoustic monitoring technologies as well as continued on-ground searches to determine how many breeding pairs may be living in the area. The rediscovery also reinforced continuing efforts to preserve the Big Woods. Currently an area of 18,000 acres is being preserved by the Nature Conservancy, which hopes eventually to preserve an additional 200,000 acres of land. Ornithologists are also making efforts to ensure that the area is not overrun by well-meaning birdwatchers who are intent on viewing the woodpecker, but who may be destroying the habitat upon which the bird depends.

New England, glaciers did not cover the southern Appalachians, reducing the impacts of erosion. The highest peak in the Appalachians, and indeed in all of eastern North America, is Mount Mitchell in North Carolina at 6684 feet (2037 meters) above sea level.

West of the Appalachians is the interior lowland of North America. Much of the land in this physiographic region is located in the Great Lakes and Corn Belt region to the north of the Inland South. However, the interior lowlands penetrate into the South in western Kentucky and Tennessee. The Ozark and Ouachita mountains form another physiographic province. Geologically, this area of Missouri, Arkansas, and Oklahoma is an extension of the Appalachians; many of the region's early settlers came from the highlands of Tennessee, Kentucky, and the Carolinas as well.

Climate and Hazards

The Inland South and Coastal South generally coincide with North America's region of humid subtropical climate. In this region, summers are hot and humid. Winters are generally mild, but the area is subject to occasional cold snaps and snowfalls when cold Arctic air spills southward. Precipitation is abundant throughout the year, but is heaviest in spring and summer when thunderstorms can dump large quantities of rainfall in short periods of time. In late summer and early fall, hurricanes often bring copious amounts of rainfall to the Inland South.

The southern Appalachians from northern Georgia to southern Virginia are high enough to impact local climatic conditions. In summer, the region is somewhat cooler than the lowlands to the east. Great Smoky Mountains National Park and other publicly and privately owned parks and resorts are very popular with tourists from Atlanta, Charlotte, Richmond, and other places who wish to escape the heat and humidity of southern summers. In the winter, the southern Appalachians often get heavy snowfalls, and average annual snowfall is high enough to sustain ski resorts in western North Carolina and eastern Tennessee.

Given the abundant rainfall and warm temperatures of the region, it is not surprising that the Inland South is heavily forested. Prior to human occupation, the entire region was covered with thick forests, whose presence impeded early Euro-American settlement. Most of the original forests were cut down, but today much of the area is once again forested. Commercial forestry is a significant contributor to local and regional economies in several parts of the region.

The abundant rainfall and frequent thunderstorms of the Inland South have meant that flooding is a common problem. Flash floods occur often, especially in the southern Appalachians, where a combination of especially high levels of annual rainfall and steep topography can result in very rapid discharge of water down the slopes. In the 19th and early 20th centuries, these factors in combination with the absence of trees that had been cut down generated several major floods, many of which resulted in extensive property damage and loss of life. Floods have also had a major impact on agriculture, washing away crops and soil. During the 20th century, the U.S. Army Corps of Engineers and other federal and state government authorities were active in trying to ameliorate flooding. During the 1930s, the **Tennessee Valley Authority** (TVA) was created, in part to control flooding on the Tennessee River and its tributaries as well as to generate hydroelectric power. Artificial lakes and reservoirs are numerous throughout the Inland South, and many, including Bull Shoals Lake in Missouri, Grand Lake in Oklahoma, and Lake Sidney Lanier in Georgia, are popular vacation spots.

Many of the floods of the Inland South are associated with storm systems that can be very destructive. Tornadoes occur regularly, especially in the relatively flat regions of central Mississippi and Alabama. Many tornadic storms generate very heavy rainfall, exacerbating the flooding problem. Hurricanes also generate heavy rains. The Inland South does not experience the very high winds and pounding surf associated with hurricanes making landfall along the coast (see Chapter 10), but very heavy downpours can occur as the moisture-laden winds associated with these tropical systems move inland and encounter the more rugged topography of the Piedmont and the southern Appalachians. Slow-moving hurricanes such as Hurricane Agnes in 1972 can drop two feet (0.6 meters) or more of rain in a few days, causing extensive flooding.

Historical Settlement

Given the region's mild climate and abundant natural resources, it is not surprising that the Inland South sustained a large and varied Native American population prior to the arrival of the Europeans. Most of the Native Americans of the Southeast, like those living in the Middle Atlantic states and New England, belonged to the Eastern Woodland culture complex and spoke languages that belonged to the Iroquoian, Sioux, or Algonquin linguistic families. They made their livings through a combination of agriculture and hunting and gathering. Many of the tribes inhabiting the present-day Southeast had established complex trade networks with other societies. Goods were traded over hundreds of miles; excavated sites in Oklahoma, for example, contain artifacts produced as far away as Central America. The Cherokee, Choctaws, and Creeks were numerically the most dominant groups. The Cherokee lived in the southern Appalachians in what is now western North Carolina, eastern Tennessee, northern Georgia, and northeastern Alabama.

English settlers began to colonize the Inland South in the 17th century, and by the time of the American Revolution thousands had crossed the Cumberland Gap into present-day Kentucky and Tennessee. After American independence, more and more settlers moved westward. The post-revolutionary expansion of American settlement east of the Mississippi River increasingly pushed the tribes to the west. In 1830, President Andrew Jackson signed the Indian Removal Act, which authorized the large-scale removal and resettlement to "Indian Territory" in what is now eastern Oklahoma. During the 1830s, thousands of Cherokees, Choctaws, Creeks, Chickasaws, and Seminoles were forcibly relocated to Oklahoma along what came to be known as the **Trail of Tears** (Figure 9.2). Many perished along the way, but the

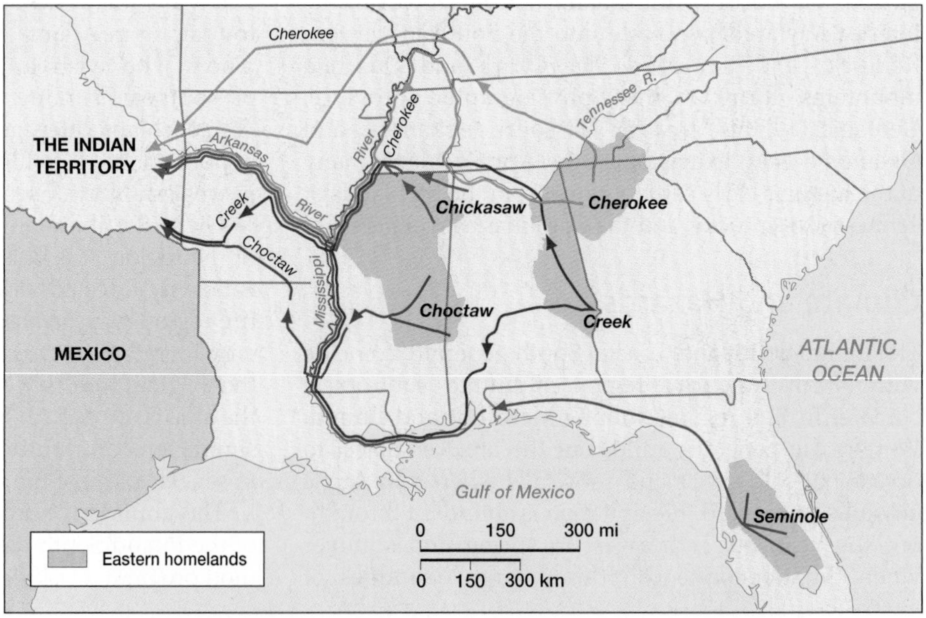

Figure 9.2 Route of the Trail of Tears

survivors and their descendants remain numerous in Oklahoma today. However, not all Native Americans complied with the government's orders to move westward. Many Seminoles moved into peninsular Florida, which did not see extensive Euro-American settlement until the 20th century. Some Cherokee remained in North Carolina, where their descendants are known as Lumbees. The population of Robeson County, North Carolina, is roughly one-third Lumbee, one-third African American, and one-third Euro-American.

Settlers from England, Ireland, Germany, and other origins soon claimed land that they had taken over from Native Americans. Many settlers were members of the English aristocracy or upper classes. These settlers tended to move to the lowland areas of the South, settling near the coast or along river valleys. There they established plantations on which cotton, tobacco, rice, indigo, and other crops that require semitropical environments were produced for export to Europe. Plantation agriculture requires a large labor force, and many settlers soon began to import slaves from Africa. By the time the slave trade was abolished in 1808, several million Africans had been captured and sold into slavery on the North American mainland as well as in the Caribbean, Central America, and South America.

Slavery was abolished in the United States during the Civil War. Following the Civil War, many freed African Americans remained in the South. Seeking to fill the labor shortages created by the emancipation of the slaves, southern planters and laborers began to enter into tenancy agreements. Under the tenancy system that eventually evolved from such arrangements, farmers were classified into two broad categories: tenants and **sharecroppers**. Tenants generally owned their own farming equipment and only rented land on which to farm, whereas sharecroppers often owned no equipment and paid no rent, but after working the land were obligated to divide the final harvest with the landowner. Under these circumstances, tenants and sharecroppers usually had little say in matters of rental fees and the distribution of profits from harvest. The tenancy system perpetuated the poverty in which many poor southerners already lived and became widespread throughout the Inland South, peaking in most areas during the 1920s and 1930s. Throughout most of the Inland South, poverty was combined with "Jim Crow" laws that relegated African Americans to second-class citizenship. These laws remained in force until they were eliminated during the civil rights movement of the 1960s.

The harsh economic conditions of the tenancy system, combined with rancorous race relations in the South, led many African Americans to migrate to northern industrial states after the turn of the century, especially during and after World War I. This mass movement, in which hundreds of thousands of African Americans relocated to urban centers in the north, has become known as "The Great Migration" (see Chapter 8). As we saw in earlier chapters, in the 1920s the United States reversed its long-standing policy of open immigration and established limits on the number of immigrants admitted from Europe and other parts of the world. Accordingly, northern industrialists turned to the South for labor. Millions of Southerners, both white and African American, moved to work in northern factories. Generally, these southern migrants followed migration paths trodden by friends and relatives from nearby communities. Thus natives of the Carolinas moved to New York, Philadelphia, and Baltimore; natives of Kentucky, Tennessee, Alabama, and Mississippi moved to Cleveland, Chicago, and Detroit. Many from further west, including East Texas and northern Louisiana, relocated to California. Baseball Hall of Famers Joe Morgan and Frank Robinson, born in Texas, and Willie Stargell, born in Oklahoma, all grew up in Oakland, California. Near the University of Southern California campus in Los Angeles is an African-American neighborhood known as "Little New Orleans", settled originally by natives of Louisiana. Such migration, in which destination choices are based on the migration decisions of friends and relatives, is known as **chain migration**.

Especially during and after World War II, out-migration of rural African Americans was hastened by the mechanization of agriculture, which reduced the need for manual labor in planting and harvesting cotton and other crops. The decline in farm employment due to agricultural mechanization and diversification compelled many former tenants and sharecroppers to move to urban areas in the South as well as in the North. Although some areas of the Inland South began to urbanize after the Civil War, many of the region's largest cities experienced most population growth from after World War II to the present.

Despite popular perceptions, a majority of the Inland South's Euro-American population did not consist of slave-owning plantation owners. The hilly lands of the Piedmont and the Appalachians were not conducive to this system of plantation agriculture. Rather, such lands were settled by small, independent yeoman farmers who worked their own fields, grew crops primarily for local consumption, and owned few if any slaves. Some of these highlanders were former indentured servants who were able to scrape together enough money to buy small farms; others moved directly from the British Isles or from other parts of the Middle Atlantic region to the Inland South. Prominent among these migrants were Scotch-Irish settlers, who were descendants of persons who had moved from Scotland to Ireland during the mid-1600s, when the Puritan government of Oliver Cromwell controlled the United Kingdom, and later moved to the United States.

By the time of the American Revolution, most states in the South were characterized by economic and political tensions between the competing interests of the lowland plantation owners and the highland small

Kentucky is known as the "Bluegrass State" in honor of the "blue" grass growing in the central part of the state between Louisville and Lexington. This area is perhaps best known as a center for thoroughbred horse racing. The Kentucky Derby, held on the first Saturday of May each year, attracts thousands of tourists annually. Eastern Kentucky, like neighboring West Virginia, is coal country and has long been the site of management-union tension and poverty.

Residents and visitors to the Inland South enjoy the diversity of the burgeoning music industry of Tennessee. Memphis is often called "the home of the blues" and "the birthplace of rock and roll" (a fact disputed by Cleveland, Ohio, location of the Rock and Roll Hall of Fame). Many of the nation's great blues singers and bands of the 1920s and 1930s got their start in clubs on Beale Street, and Elvis Presley recorded his first hit songs at Sun Records, near downtown Memphis. This studio was declared a national historic landmark in 2003. Not to be outdone, Nashville, Tennessee, is known as "the music city" and "country music capital of the world" (Figure 9.8). The city is home to the Grand Ole Opry and the Country Music Hall of Fame. Many of country music's biggest stars got their starts in the clubs and recording studios along 16th Avenue in Nashville.

The southernmost extension of the Appalachians is found in Alabama. Here the mountains extend south and west as far as Birmingham, the state's largest city. Located near iron ore deposits, Birmingham is well known as a steel-producing center. A statue of Vulcan, the Roman god of fire, symbolizes the importance of steel in Birmingham's history. Birmingham has designated a six-block area as a civil rights district. The district houses the Birmingham Civil Rights Institute and the Alabama Jazz Hall of Fame and is adjacent to Kelly Ingram Park, the site of numerous clashes between police and protestors during the civil rights movement.

South of Birmingham is the state capital of Montgomery. Located there are the Rosa Parks Library and Museum and the Civil Rights Memorial, which lists the names of those who died in the civil rights struggle (Figure 9.9). Also located in Montgomery are the Dexter Avenue Baptist Church, which was headed by Reverend Martin Luther King, Jr. in 1955, when Parks refused to step to the back of a bus. Montgomery is also the site of the Southern Poverty Law Center (SPLC). The SPLC, founded in 1971, argues civil rights and other discrimination cases and tracks the activities of hate groups in the United States.

West of Alabama is Mississippi. Like Alabama, Mississippi was the site of violent clashes during the civil rights movement. The state remains impoverished, but incomes in Mississippi are rising relative to the national average. One of the poorest places in

▲ **Figure 9.8** Nashville's Grand Old Opry district

▲ **Figure 9.9** The Civil Rights Memorial, Montgomery, Alabama

Mississippi is the Delta region between the Yazoo and Mississippi rivers. The flat, fertile land was long a stronghold for cotton production, but today is attempting to diversify its economy. Catfish raising in artificial ponds has become lucrative, with nearly half of the U.S. acreage devoted to catfish production found in Mississippi. The town of Belzoni bills itself as the "Catfish Capital of the World." Each year the town hosts the World Catfish Festival. Nearby Tunica has prospered from the introduction of casino gambling, which has generated hundreds of new jobs.

Places in the Western Inland South

West of the Mississippi River, the Inland South begins to merge into the Great Plains. Arkansas, like Mississippi, is a state of historic poverty and considerable potential. Eastern Arkansas is low and flat; western Arkansas is hilly. The latter area is known as a retirement center—so many midwestern retirees live in communities such as Mountain Home, in fact, that the *Chicago Tribune* is delivered door to door. Northwestern Arkansas also contains Bentonville, the home of Wal-Mart, as well as the home offices of J.B. Hunt trucking and Tyson poultry production. The state capital of Little Rock, like many state capitals across the Inland South, is located in the center of the state and dominates the state's political and cultural life.

The northern part of Louisiana is located in the Inland South (the southern half of the state is covered in Chapter 11). The contrast between North and South Louisiana is considerable. North Louisiana is historically agricultural and primarily Protestant, with settlers of English ancestry; South Louisiana is primarily Catholic and dominated by Cajuns of French ancestry. The largest city in northern Louisiana is Shreveport, an old river port on the Red River which has today become a center for distribution and, like many other southern communities, the home of a major military installation. The largest city in southern Louisiana is New Orleans, which is discussed in the next chapter.

The eastern part of Oklahoma was "Indian Territory" throughout much of the 19th century. Today, the Sooner State contains the headquarters for the Five Civilized Tribes and many other Native American groups including the Pottawotamie, Osage, Sac and Fox, and Kiowa-Comanche. Television advertisements for Texas describe the Lone Star State as "a whole other country." Only California rivals Texas for its diversity of people, places, and natural environments. The state is divided among several of the regions considered in this book. Southeastern Texas is treated as part of the Coastal South, while the western half of the state is part of the Great Plains. East Texas, in contrast, has the look and feel of the Inland South, and visitors to small East Texas towns might be excused if they mistakenly think that they are in Alabama or Mississippi.

The Future of the Inland South

Historically, the Inland South has been a backwater within North America. As recently as the 1950s, per-capita incomes in many Inland South states were barely half the U.S. average. Over the last half-century, however, the region has become one of significant growth. The last few decades have witnessed a steady integration of the once-isolated, insular Inland South into the mainstream of the national and global economies. This integration has been enhanced by improved transportation and communications. Aspects of Inland South culture, especially popular music and religious traditions, have had an increasing influence on popular culture across the North American continent and throughout the world. The integration of the region into the mainstream economy was hastened by the civil rights movement of the 1960s, which laid a foundation for eliminating many overt forms of bigotry and prejudice and helped expand the productive potential of the regional economy.

What does the future hold for the Inland South? There is little reason to doubt that the Inland South will continue to grow in prosperity and importance in the years ahead. North Americans are increasingly aware of the Inland South's geographic and economic advantages. This awareness is enhanced by the Inland South's significance as a communications center, as such communities as Atlanta, Nashville, and nearby Orlando have emerged as important centers for the communications media. People from throughout North America and other parts of the world have been attracted to the Inland South by the region's temperate climate, relatively low cost of living, and its relative lack of crowding, pollution, and crime.

As in other parts of North America, future prospects in the Inland South vary considerably within the region. Historically, the Inland South was a predominantly rural region, with relatively few major cities. This has changed over the past several decades and cities such as Atlanta, Charlotte, Birmingham, Nashville, and Memphis continue to grow in national and international importance. Much of the population and economic growth of the Inland South is taking place in these and other metropolitan areas, whose population is increasingly cosmopolitan. The region's vitality is being enhanced by communities of immigrants from Latin America, Asia, and many other areas of origin. Other places such as the Great Smoky Mountains and the Ozarks of Missouri and Arkansas are attractive to tourists and retirees and are also growing in prosperity. Unfortunately, not every place within the Inland South shares these advantages, and the future may be less bright for those places oriented primarily to agriculture or manufacturing. Overall, however, there is every reason to believe that population and income growth will continue to characterize the Inland South in the years ahead.

Suggested Readings and Web Sites

Books and Articles

Abramson, Rudy and Jean Haskell, eds. 2006. *Encyclopedia of Appalachia.* Johnson City TN: The Center for Appalachian Studies and Services. New book covering some of the well known and little known (and quite fascinating) facts about the Appalachian region.

Allen, James Lane. 1886. "The Blue Grass Region of Kentucky," *Harpers New Monthly Magazine* 72: 365–382. A 19th century writer discusses the cultural landscapes and mysteries of the Kentucky bluegrass region as a haven for horse lovers and a respite for urban dwellers.

Haley, Alex, Jr. 1974. *Roots*. New York: Bantam Doubleday Dell Publishing. The author traces his family history from the story of Kunta Kinte, who was captured into slavery from Africa in 1767, through the mid-20th century in Virginia, North Carolina, and Tennessee.

Anglin, Mary K. 1992. "A Question of Loyalty: National and Regional Identity in Narratives of Appalachia," *Anthropological Quarterly* 65: 105–116. Appalachian residents' stories about their comparative attachment to the lands and cultures of the region and their national vs. their regional identities.

Berthoff, Rowland. 1986. "Celtic Mist over the South," *Journal of Southern History* 52: 523–550. Tells the story of the influence of Irish immigrants on the cultures and landscapes of the Inland South.

Buckley, Geoffrey L. 2004. *Extracting Appalachia: Images of the Consolidation Coal Company 1910–1945.* Athens, OH: Ohio University Press.

Dorgan, Howard. 1987. *Giving Glory to God in Appalachia: Worship Practices of Six Baptist Subdenominations.* Knoxville: University of Tennessee Press. Analysis of the workings, patterns, and impacts of comparative congregations in the Bible Belt.

Finger, John R. 1984. *The Eastern Band of Cherokees, 1819–1900*. Knoxville: University of Tennessee Press. Study of the earliest residents of the eastern part of the Inland South region before and during their preparation for the devastation and death brought on by their long walk on the Trail of Tears.

Fowler, Damon Lee. 1995. *Classical Southern Cooking*. New York: Crown. Compilation of recipes associated with traditional southern cooking from grits to cornbread.

Hart, John Fraser. 1998. *The Rural Landscape*. Baltimore: The Johns Hopkins University Press. A geographic interpretation and analysis of the evolution of rural landscapes of the United States, with a focus on the Inland South and Midwest.

Hofstra, Warren R. 2004. *Planting New Virginia: Settlement and Landscape in the Shenandoah Valley*. Baltimore: The Johns Hopkins University Press. An exploration of some of the ways that the back country frontier culture of the Shenandoah Valley played a major role in the struggle between Native Americans and European nations.

Jordan, Terry G. 1966. *German Seed in Texas Soil: Immigrant Farmers in Nineteenth Century Texas*. Austin: University of Texas Press. A classic migration study from a geographic perspective focusing on the largest group of European immigrants to settle in Texas after the early 1840s.

Jordan, Terry G. 1981. "The 1887 Census of Texas Hispanic Population," *Aztlan International Journal of Chicano Studies Research* V: 271–278. A useful article that helps clarify how best to use historical census records to trace the migration and settlement patterns of various immigrant groups.

Jordan-Buchov, Terry. 2003. *The Upland South: The Making of an American Folk Region and Landscape.* Harrisonburg, VA: University of Virginia Press.

Otto, J.S. and N.E. Anderson. 1982. "The Diffusion of Upland South Folk Culture," *Southeastern Geographer* 22: 89–98. An "origin and diffusion" study of various folk beliefs and cultural expressions in the Inland South region.

Rehder, John B. 2004. *Appalachian Folkways.* Baltimore: The Johns Hopkins University Press. An engaging account of the food, folklore, music, architecture, cultures, customs, dialects, and other characteristics of one of the most stereotyped and mythologized regions in North America.

Williams, Jack. 2006. *E40°: An Interpretive Atlas.* Charlottesville, VA: University of Virginia Press.

Wilson, Bobby M. 2000. *America's Johannesburg.* Lanham, Maryland: Rowman and Littlefield. A historical geography of Birmingham, Alabama, focusing on the city's industrialization and the impacts of this industrialization on race relations.

Wilson, Charles Reagan. 1995. *Judgment and Grace in Dixie.* Athens: University of Georgia Press. Analysis of how Southern religious values have impacted North American culture.

Web Sites

Inland South Defined
http://encarta.msn.com/encyclopedia_1741500822_6/United_States_(Geography).html

Kentucky Derby
http://www.kentuckyderby.com/2006/

Appalachian Trail Conservancy
http://www.appalachiantrail.org

Peanut Production Policy
http://www.agtrade.org/digest.cfm?digestID=96

Cotton Seed Production
http://www.cotton.org/

Atlanta Tourism
http://www.atlanta.net/index.html

10
The Coastal South

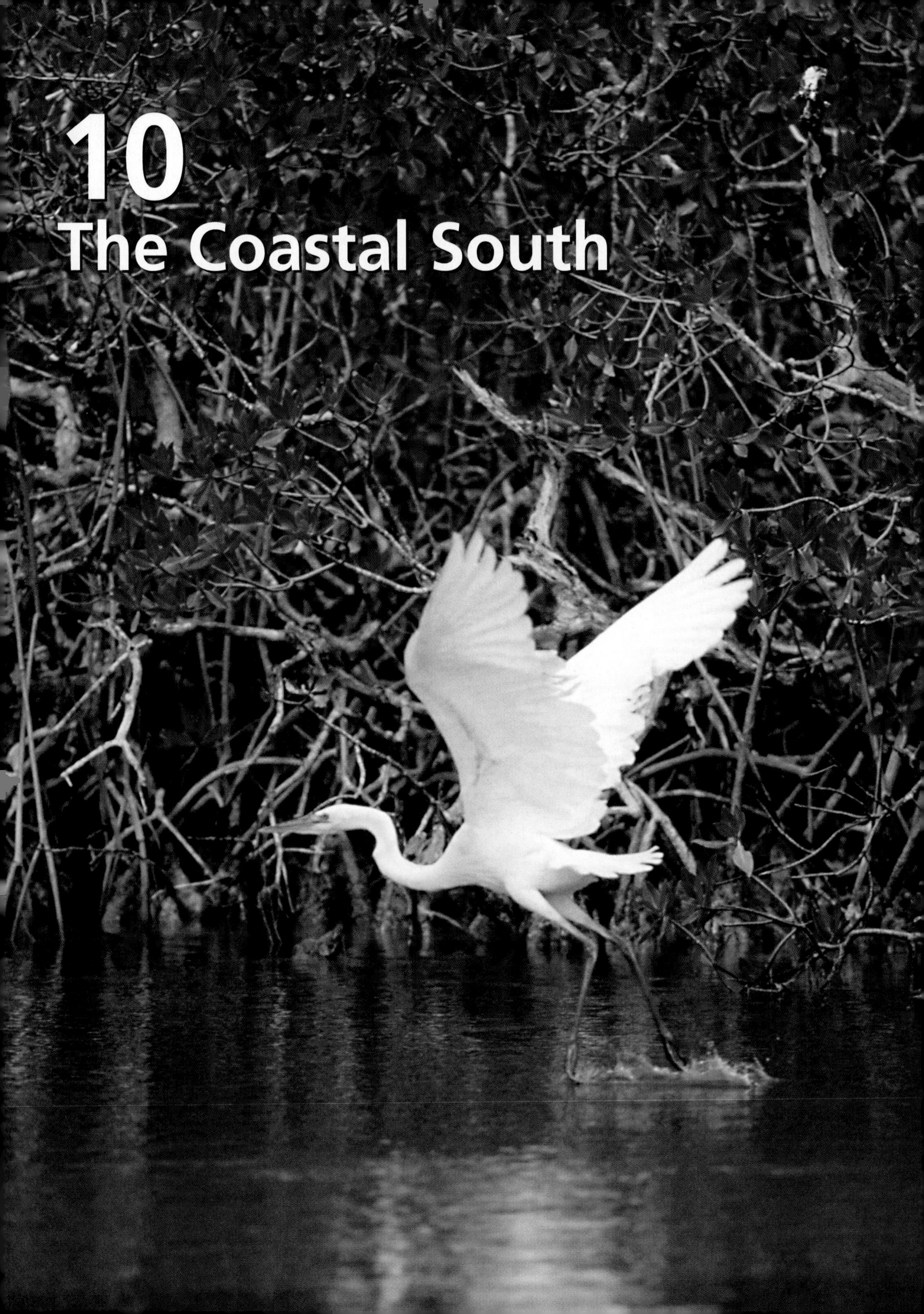

PREVIEW

Environmental Setting

- The Coastal South extends along the Atlantic and Gulf coasts from Virginia to Texas.
- The entire region is part of the Atlantic Coastal Plain, and the topography is low and flat.
- The climate of the Coastal South is warm and humid, with hot summers, mild winters, and abundant rain throughout the year.
- The region is subject annually to destructive hurricanes. The deadliest and costliest natural disasters in North American history have been hurricanes that struck the region.

Historical Settlement

- The original Native American inhabitants of the Coastal South were part of the Southeastern cultural complex.
- Prior to the American Revolution, much of the region was under Spanish or French control. French is still spoken in rural South Louisiana today, and Spanish and French place names and architecture can be seen throughout the region.
- During the 19th and 20th centuries large numbers of immigrants of European, African, Asian, and Latin American origin have moved into the region.

Political Economy

- The Coastal South has been oriented to commerce and trade. Port cities such as Charleston, Mobile, New Orleans, and Galveston were established near the few sites along the coast that offered good harbors.
- Agriculture in the Coastal South has historically been oriented to production of tropical and subtropical crops that can tolerate little or no frost. Rice, citrus fruit, sugarcane, and various vegetables are grown throughout the region.
- The region owes much of its current prosperity to military bases, space exploration, recreation and tourism, and retirement.

Introduction

Many people imagine the South as one big homogeneous region. In this book, however, we distinguish the Coastal South from the Inland South. Why do many geographers distinguish between the Coastal South and the Inland South? On what bases do they differ? The Coastal South includes the land along the Atlantic and Gulf Coasts from southeastern Virginia to the Mexican border. Relative to the Inland South (Chapter 9), its economic base is oriented far less to agriculture and much more to commerce and trade. Given its coastal location, it looks outward, particularly to the countries of the Caribbean and Latin America.

Culturally, the Coastal South is a pulsating and diverse *mixing zone* of peoples from Europe, Africa, the Caribbean and other parts of Latin America as well as others parts of the world. The Coastal South contains a much more heterogeneous population than does the Inland South. This heterogeneity reflects the fact that much of the region was at one time in the possession of Spain, France, or Mexico. This cultural and ethnic mixing is also evident in the region's past and current cosmopolitan outlook. The U.S. armed forces represent a very significant contributor to the regional economy. In addition, the region's resorts, beaches, and historic sites draw large numbers of tourists each year.

The region is one of the fastest-growing portions of North America, but it also suffers from devastating hurricanes. In August 2005, Hurricane Katrina struck New Orleans and the Mississippi Gulf Coast. The storm caused more than 1300 deaths and caused untold billions of dollars worth of property damage, making it the costliest natural disaster in U.S. history. Along with several other major hurricanes that struck the region in 2004 and 2005, Katrina has caused many local residents and government officials to rethink the desirability of continued, unchecked population growth in the Coastal South.

◄ A mangrove forest in Biscayne National Park, Florida

Environmental Setting

The Coastal South includes the land and offshore islands along the coast of the Atlantic Ocean and the Gulf of Mexico from Virginia southward and westward to south Texas. All of it is part of the Atlantic Coastal Plain, which continues northward as far as New York City (Figure 10.1).

Landforms

The topography of the region is low and flat. The highest point in the entire state of Florida, in fact, is a mere 345 feet above sea level. The soils tend to be sandy and infertile, making agriculture difficult in many localities. As a result of abundant rainfall, flat topography, and poor drainage there are many swamps, lakes, and coastal marshes in the Coastal South. Among the best-known of these features are Lake Okeechobee in Florida, the Great Dismal Swamp of Virginia and North Carolina, and Georgia's Okefenokee Swamp. Numerous additional swamps and marshlands are found throughout southern Louisiana. The lack of drainage along with frequent heavy rainfall make flooding a common problem in the Coastal South.

The shorelines of the Coastal South contain numerous sandy beaches. Many of these, including Myrtle Beach, South Carolina; Daytona Beach and Destin, Florida; and the "Grand Strand" of Mississippi, are popular resort areas that attract hundreds of thousands of tourists and vacation travelers each year. The coast of the region is also dotted with many bays, estuaries, and inlets. Some of the region's major cities are ports that were founded on bays that provided fine harbors for oceangoing ships. Examples of these bays include Tampa Bay in Florida, Mobile Bay in Alabama, and Galveston Bay in Texas.

Off the immediate coast, wave action has created numerous sandy **barrier islands**. Many of these barrier islands are extensively developed, including Hatteras Island in North Carolina, Palm Beach in Florida, and Galveston and South Padre Islands in Texas. Unfortunately, on some barrier islands extensive development has caused considerable erosion, imperiling the islands' already fragile ecosystems. In particular, development sometimes impedes the islands' natural tendency to replenish themselves with sand, and many barrier islands have consequently become reduced in size. National seashores have been created on Hatteras Island in North Carolina, Padre Island in Texas, and elsewhere in order to stem loss of land through beach erosion.

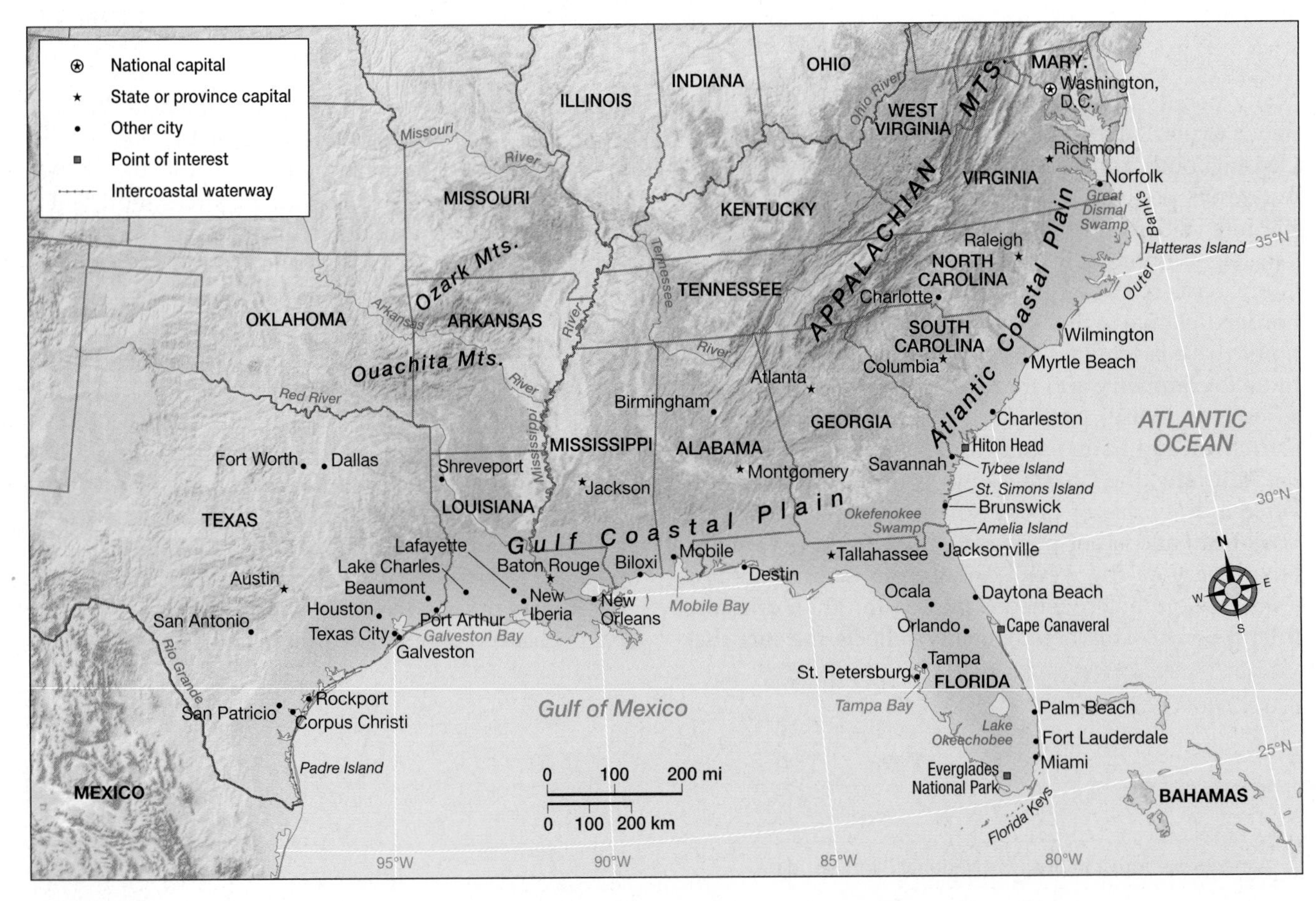

▲ **Figure 10.1** The Coastal South region

Climate and Hazards

Throughout the Coastal South, the climate is humid subtropical with long, hot, rainy summers and mild, wet winters. Snowfall and freezing temperatures are generally absent in southern Florida and rare elsewhere, allowing production of semitropical and tropical crops such as citrus fruit that can tolerate little or no frost. Summer days are hot and humid, with little day-to-day change in temperature for several months. However, areas near the immediate coast experience summer temperatures somewhat cooler than those outside the region located further inland. For example, the average July high temperature in Miami and New Orleans is 90°F, as compared to averages of 95° in Dallas and 93° in Oklahoma City and Shreveport, Louisiana.

The Coastal South is North America's most thunderstorm-prone region. Unfortunately, lightning kills more than 200 Americans each year. A majority of these fatalities take place in the Coastal South, notably in Florida. These thunderstorms generate sometimes copious quantities of rain. Hurricanes, which often strike the region between July and October, can bring torrential rainfalls, destructive winds, rip currents, and catastrophic beach erosion. Most of North America's best-known and most destructive hurricanes throughout recorded history have affected the Coastal South. Hurricane Andrew in 1992 caused extensive damage in both Miami and New Orleans, thus becoming one of the most expensive natural disasters in U.S. history. In 1900, a hurricane making landfall in Galveston, Texas, cost about 6000 lives, causing more fatalities than any other natural disaster in U.S. history. In 2004, the state of Florida alone was struck by four major hurricanes within a six-week period, and in 2005 New Orleans suffered catastrophic damage during Katrina. Climatologists are debating whether hurricane frequency is increasing in light of global warming and changes in global atmospheric circulation patterns.

Historical Settlement

The Coastal South has been settled by a large number of different cultures. A diverse population of Native Americans lived in the region prior to European settlement. Since that time, the region has been populated by Europeans, Africans, Asians, and Latin Americans.

Native Americans in the Coastal South

Aboriginal peoples lived in the Coastal South for thousands of years prior to the arrival of Europeans. The Native American residents of the Coastal South were part of the Southeast Culture Area. Most of these earliest residents of the South Coast and Inland South were farmers who also fished, hunted, and foraged for food. Despite these similar ways of life, groups were divided into many different language families at the time of European contact. The largest tribes included the Cherokee, Choctaw, Chickasaw, Creek, and Seminole (called the "Five Civilized Tribes" by Euro-Americans), as well as the Catawba, Caddo, Alabama, Natchez, and Timucua. One of the groups most often misidentified are the Florida Seminoles. They originated as a subgroup within the Creek people in Alabama in the late 1700s and ultimately migrated to Florida to escape colonial powers. Subsequently, the British and later the Americans called them Seminoles. All went well until the U.S. government decided to try to relocate them to Oklahoma. Then the United States and the Seminole nation fought a deadly guerrilla war in the 1830s with the goal of removing them to Oklahoma. This resulted in many of the native peoples seeking refuge and safety deep in the Everglades in south Florida.

The Natchez are the direct descendants of the ancient temple builders who lived along the Mississippi. The "Mound Builders" are perhaps the most studied aboriginal group in the South Coast region. During the 19th century, some anthropologists believed that lost European tribes had constructed the huge mounds. Scientific study finally established that these gigantic earthworks (some shaped like animals) were designed and built by aboriginal people. The Natchez were able to live in one place for a long period of time because they successfully grew corn, beans, squash, pumpkins, and tobacco on the fertile alluvial soil along the rivers of the Southeast. Other groups, such as the Seminoles who lived in the Everglades and the Karankawa who lived along the Texas Gulf Coast, survived by foraging for wild plants and fishing. The treatment of most of the Native American peoples and cultures in the Southeast by Euro-Americans was as life-threatening as in other parts of North America. By the late 18th century, many had been captured and forced to work as slaves. Others died of European diseases or were forced to relocate to other areas, notably to "Indian Territory" in present-day Oklahoma.

Early European Settlement

The Spanish were the first Europeans to attempt to found a settlement on the Atlantic Coast of Florida. In 1523, several hundred missionaries, colonists, and their slaves tried to colonize a site on the Savannah River. Their effort was unsuccessful, however, and the settlement was abandoned. Another group of Spanish colonists tried in vain to establish a supply center on the Gulf Coast in northern Florida. They were followed by French Protestants known as the Huguenots who were seeking a place to practice their religion in freedom. The Huguenots founded at least two settlements in Florida in the 1560s. These Protestant pioneers, who later successfully settled in South Carolina, were soon either dead or on their way home to France due to the damages done by hurricanes, attacks by the Spanish, and problems caused by warfare at home. Thereafter Spain assumed control of most of the region except for French settlements in Louisiana. After the British assumed control of eastern Canada in 1763, thousands of French-speaking settlers of present-day Nova Scotia were relocated to the swamps and marshes of south Louisiana (see Chapter 5). Today, thousands of "Cajuns" still call Louisiana home. Many continue to speak French at home.

Prior to the American Revolution, the coastal regions of Virginia, North Carolina, South Carolina, and Georgia were under British control. However, the rest of the Coastal South from Florida westward was under the control of France or Spain. The United States obtained sovereignty over this region in the 19th century. The Louisiana Purchase included New Orleans and much of present-day Louisiana. Florida, southern Alabama, southern Mississippi, and southeastern Louisiana were purchased from Spain in 1819. In 1845, American annexation of the Coastal South was completed when Texas joined the Union.

Because much of the Coastal South was at one time a colony of France or Spain, many inhabitants of the region at the time of annexation to the United States were of French or Spanish rather than English or northern European ancestry. This distinction was critical for two reasons. First, most of the French- and Spanish-speaking settlers of the Coastal South were Roman Catholics (although the Huguenots were Protestants). Moreover, France and Spain adhered to a legal tradition very different from the Anglo-American legal tradition that is the foundation of the legal and political system of the United States today.

France, Spain, and other countries that were part of the Roman Empire adhere to what is known as **Roman law**. Roman law differs in several important respects from **Anglo-American law**. Under Anglo-American law, disputes are settled with reference to precedent. The law is seen as an evolving body of legal principles augmented by previous cases in which specific laws are interpreted. Under Roman law, in contrast, disputes are to be resolved with specific reference to the written law alone. In theory at least, laws are to be written to cover every possible situation. As well, the principle that the accused is innocent until proven guilty is fundamental to the Anglo-American legal system; under Roman law, the accused must prove that he or she is innocent.

The differences between Anglo-American and Roman law had to be ironed out after the Louisiana Purchase of 1803. The transfer of sovereignty from France to the United States meant that more than 40,000 French citizens of present-day Louisiana including the city of New Orleans were to become Americans. For the first time, the United States obtained sovereignty over lands inhabited by "civilized" people of European ancestry. Many of these people were less than happy about becoming part of the United States without their consent. At the insistence of French ruler Napoleon Bonaparte, specific guarantees that these persons would be granted U.S. citizenship and would be allowed to keep their Roman legal traditions were included in the Louisiana Purchase agreement. The impact of French and Spanish settlement of the Coastal South remains evident in place names such as Corpus Christi, Beaumont, Mobile, and Saint Augustine and other landscape features like churches, house types, and land boundary patterns (Figure 10.2).

After the United States assumed control of the Coastal South, the region attracted many immigrants. In fact, the port of Galveston was second only to New York as the largest port of entry into the United States from Europe during the late 19th century. Immigrants

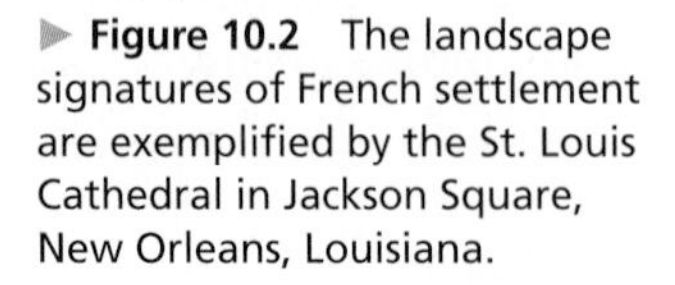

Figure 10.2 The landscape signatures of French settlement are exemplified by the St. Louis Cathedral in Jackson Square, New Orleans, Louisiana.

BOX 10.1 Galveston: Ellis Island of the Coastal South

Galveston, the largest city in Texas between 1850 and 1890, is on an island best known for its gentrified Strand shopping district and long stretches of sandy beaches. It was also the port of entry for tens of thousands of immigrants into Texas and the interior part of the United States from the early 1840s to the 1920s. Almost half of the population of this Coastal South "Ellis Island" in the mid-19th century, in fact, was made up of German immigrants. It was also the site of the worst natural disaster in the history of North America in terms of loss of life. The Great Storm of 1900 is a Galveston story filled with both optimism and despair in a place where Germans, Italians, Jews, Russians, and other immigrants joined African-American residents to start new lives in a strange and often hostile environment.

An amazing migration story known as the "Galveston Movement" is perhaps the least known aspect of Galveston's immigration history. This global effort to rescue Russian and Ukrainian Jews from the deadly pogroms of Eastern Europe in the early years of the 20th century resulted in more than 10,000 new émigrés finding their way safely to the shores of America's "Third Coast." Their arrival soon after the Great Storm wiped out much of the island in 1900 was badly timed—but filled with hope for better lives in cities located in interior Great Plains and Midwest such as Omaha, Denver, and Chicago. Here, after long train trips north from Galveston, Jewish communities welcomed the new arrivals and helped them find work and spiritual support. These Galveston Movement refugees were assisted throughout their journey to the United States by an extensive resettlement network that stretched from Galveston to New York City to London to Russia.

Russians, Germans, and immigrants from other parts of Europe joined new Galvestonians who were born in Latin America and other parts of the United States over the years. They came seeking employment opportunities at the city's port and in its service and construction sectors. The heritage of these diverse residents of one of Texas's most cosmopolitan cities remains today in the old but still beautiful German dance hall and the annual Italian festival that is still held on the island to commemorate the role of the city of Galveston as the Ellis Island of the Coastal South.

from Europe landed in Galveston and settled throughout much of the South and Midwest (Box 10.1). Other communities attracted various ethnic groups. Tampa became the home of thousands of Cubans, many of whom worked in cigar manufacturing. Baseball manager Lou Piniella may be the best-known descendant of this Cuban-American community. Numerous Irish-Americans settled the area around San Patricio (Spanish for Saint Patrick) in southeast Texas.

African Americans, Latin Americans, and Vietnamese Immigrants

Southern culture and lifeways have also been shaped by Africans who were forced to migrate into the region as slaves. The first small group was brought to Virginia from West Africa in 1619. African influences on music, food habits, speech patterns, and architectural styles are still evident to this day in the cadences of a southern drawl in southeastern Alabama; the smell of chicken being barbecued in a city park in Port Arthur, Texas; sounds of jazz and blues on a New Orleans street corner; and the ironwork often seen on an antebellum mansion in Georgia.

There were at least a half million African and African Americans living in the United States by 1790 when the first census was taken. Many of them lived in the Coastal South. It must have seemed more than a little ironic to African slaves owned by revolutionaries such as George Washington and Patrick Henry when they heard their masters crying out for an end to "British tyranny" and demanding liberty. Historian Edmund Morgan labeled the fact that American independence was, in many ways, purchased with slave labor and that many of the leaders of the American Revolution were slave owners the "American Paradox."

Of all the subregions in the Coastal South, perhaps no place remains as intensely "African" as the humid, marshy islands off the coast of South Carolina called "the heartland of Afro-America because it was . . . the port of Charleston that the largest number of African slaves were brought in the 18th century" (Hudson 2002, p. 154). Until recently, when resorts like Hilton Head were built along the coast and on islands just offshore, most of the residents of these "sea islands" were African Americans. Many spoke their own distinctive Gullah dialect.

In the decades after emancipation, African-American farmers faced the challenges of trying to survive on small farms, with limited access to credit, and racism. But despite racial bigotry, mistreatment, and poverty, more than 90 percent of all African Americans in the United States still lived in the South in the early 20th century. Racial tensions in both the South and the North often turned violent in these years of adjustment to so-called "equal treatment under the law." Extreme tensions, political underrepresentation, and poverty remain today as some of the many long-term impacts of the forced migration of Africans into the southern coastal region.

During the past few decades, many other ethnic groups have relocated to the Coastal South. In 1959, dictator Fidel Castro assumed control of Cuba. Over the next few years, thousands of Cuban refugees escaped to Miami and nearby communities. Even today, nearly half of all Cuban Americans in the United States call Dade County, Florida home. Growing communities of immigrants from other parts of Latin America and the Caribbean are also found in South Florida. For example, baseball star Alex Rodriguez was born in Miami to parents from the Dominican Republic. To the west, a substantial community of Vietnamese refugees has settled along the Gulf Coast between New Orleans and Corpus Christi. Many work in the fishing industry, while thousands of others work in urban jobs in Houston, New Orleans, and other cities. Dallas Cowboys professional football player Dat Nguyen, born in Vietnam and raised in Rockport, Texas, is representative of this community.

Political Economy

The distinctions between the Coastal South and the Inland South can best be observed with reference to the economic structures of the two regions. Relative to the Inland South, the economy of the Coastal South is much more oriented to the tertiary and quaternary sectors of the economy. Yet by no means are the primary and secondary sectors neglected.

▲ **Figure 10.3** Rice crop in Louisiana

▲ **Figure 10.4** Florida orange production

Primary and Secondary Economic Activities

As we have seen, soils in the Coastal South tend to be infertile and the drainage is often poor. Yet the region has a warm and humid climate, and it has a long growing season with little cold weather in winter. These factors help the Coastal South to produce large volumes of crops, particularly tropical fruits and vegetables that can tolerate little or no frost. Three of the most significant are citrus fruit, sugarcane, and rice (Figure 10.3).

Florida is the country's leading producer of citrus fruit, including oranges, grapefruit, tangerines, limes, and lemons. Much of the United States' total citrus output is produced in central and southern Florida, with a secondary center of production in the Lower Rio Grande Valley of Texas (Figure 10.4). Florida produces about 80 percent of the nation's grapefruit and oranges, with about 8 percent of grapefruit and 4 percent of oranges in Texas. In recent years, production in Mexico, Brazil, and other countries has increased, and this international competition has caused a reduction in citrus acreage in Florida. Much of the state's citrus crop is processed and sold as frozen concentrated juice. Citrus trees cannot tolerate frost, and in winter growers must keep a careful eye on the weather, burning smudge pots and running water through citrus groves in an effort to keep local temperatures above freezing. Despite these

◀ **Figure 10.5** Sugarcane harvesting in the south coast region

efforts, severe frosts have disrupted the citrus industries of Florida and Texas on many occasions.

The Coastal South also supplies much of the U.S. production of sugarcane. Historically, Louisiana has been the leading producer of cane sugar, but in recent years Florida has surpassed Louisiana in sugarcane production. Florida sugarcane production began in earnest in the early 1960s, after the United States cut off trade with Cuba, which had previously supplied large quantities of sugar to the United States. Although some of the sugarcane production process is highly mechanized, sugarcane grows best in wet, muddy soils that are difficult to access by machine, and cane must be cut by hand (Figure 10.5). Labor is imported on a seasonal basis from Jamaica, Haiti, and other Caribbean islands.

Rice is another major cash crop in this region. Rice production in the Coastal South is concentrated in southwestern Louisiana and southeastern Texas between Houston and Beaumont. The Coastal South also produces various other frost-intolerant vegetables such as tomatoes, carrots, sweet peppers, and radishes, especially in the winter months when cold weather precludes the growing and harvesting of these crops in the northern United States.

Not surprisingly, commercial fishing plays a significant role in the economies of various Coastal South communities. Shrimp are especially important, particularly in coastal Texas and in Louisiana, which ranks second to Alaska among the 50 states in total volume of seafood caught annually (Figure 10.6). Many shrimpers in coastal Louisiana and Texas are

◀ **Figure 10.6** Louisiana shrimper on the Gulf Coast

had a well-developed network of freeways and interstate highways as well as a former air force base in the process of being converted into a civilian airport. Second, its inland location, as opposed to a location along the coast, would allow for expansion in all four cardinal directions. Operating under a cloak of secrecy—land values would have skyrocketed had the public known Disney was planning to build an amusement park in the area—the corporation purchased 43 square miles (17.2 sq kilometers) of mostly undeveloped land south of Orlando.

Construction began in 1965 and Walt Disney World opened in 1970 (four years after Walt Disney himself, a lifelong smoker, died of lung cancer at the age of 65). Tourist traffic mushroomed. In 1969, a year before the park opened, the Orlando area hosted 3 million tourists. The number of tourists increased to 10 million in 1971, the first full year of the park's operation, and to 30 million by 1989. In 2001, more than 55 million tourists visited the Orlando area. These tourists visited not only Walt Disney World itself, but other Disney-owned parks such as Epcot Center and nearby non-Disney enterprises such as Sea World, Universal Studios Florida, and Cypress Gardens. Walt Disney World alone had over 55,000 employees (Box 10.2). The city of Orlando and surrounding Orange County have

BOX 10.2 EPCOT: A Mythical Global Geography?

Many of the families and other visitors who go to the Orlando area to enjoy a few days at Disney World also include a visit to EPCOT Center in their travels (Figure 10.A). EPCOT stands for the "Experimental Prototype Community of Tomorrow" and it was Walt Disney's vision for the future. This sprawling park consists of two areas of exploration for tourists: Future World and World Showcase. The first is a place where a blend of science and technology topics are featured in people-friendly exhibits, rides, and interactive displays. The second, World Showcase, involves a stroll through remarkably detailed displays of the world's peoples and cultures in a setting that could have been inspired by the creative ideas of students enrolled in a world geography class (but expressed through the lens of a Hollywood-inspired Disney-esque perspective). Each nation featured in this globally inspired part of the park is represented in colorful detail, even down to the cast members (all from the country they represent) who work on location throughout the exhibits. At night, an explosion of fireworks are displayed at the World Showcase lagoon paid for by eight different countries.

▲ **Figure 10.A** Epcot Center

Along with these two large sections of the park, EPCOT also houses the Living Seas Aquarium, the largest aquarium in the world, home to a variety of ocean creatures like sharks, turtles, and other sea life (as well as frequent appearances from a swimming Mickey Mouse). Another popular part of the park is Spaceship Earth, a journey that starts at the bottom of the huge EPCOT ball, and then travels up to the top of the ball as a walk through a history of communication, from prehistoric times to the future.

Filled with excitement, international food, colorful exhibits, and rides—and a plethora of stereotypes about world cultures and peoples—EPCOT is but one of a host of other entertainment centers in the Orlando area that have helped it evolve into one of the best known tourist destinations in North America today.

Figure 10.9 Miami cruise ship terminal

grown equally rapidly, with the four-county region surrounding Orange County increasing from less than 200,000 people in 1960 to more than 1.5 million by 2000. Within North America, only Las Vegas has more hotel rooms than the 100,000 rooms available in the Orlando area.

The Coastal South is also the headquarters for the United States' rapidly growing cruise ship industry (Figure 10.9). Over the past 15 years, the number of persons taking cruise ship vacations has increased more than tenfold. Miami is the leading center for cruise ship activity. The Miami area serves as the headquarters for several major cruise lines, which employ over 15,000 executives, planners, office managers, and other employees on land in addition to the thousands of persons who work on the ships themselves. Many cruises to the Caribbean and other destinations leave out of Miami and nearby Fort Lauderdale. Tampa, New Orleans, and Galveston are other major cruise ports. Each large cruise vessel employs well over a thousand cooks, waiters, stewards, casino dealers, shopkeepers, and other employees. Most of the crew members come from less developed countries, especially Asia, Latin America, and Eastern Europe. Much of the money earned by members of the ships' crews is remitted to family members and relatives in their home countries.

Cruises bring enormous revenues to ports as well. For example, the city of New Orleans reported an increase from 40,000 cruise passengers each year embarking in 1993 to over 450,000 annually by 2002. While central Florida is the major locus of tourist activity, other communities in the Coastal South also have very large tourist industries. Popular beach resorts, including Myrtle Beach, Amelia Island, Fort Lauderdale, Destin, Gulf Shores, Galveston, and South Padre Island, are found along both the Atlantic and Gulf coasts. Many other tourists visit popular cultural, historical, and recreational sites such as Charleston, Savannah, Miami, and New Orleans. The tourist trade has been augmented recently by the legalization of gambling in casinos in Louisiana and Mississippi. In 2005, several of the large casinos on the Gulf Coast of Mississippi were destroyed by Hurricane Katrina.

Seasonal activities draw large numbers of tourists. For example, millions of tourists visit Mardi Gras in New Orleans and other cities in February or early March each year. Others visit Florida in the spring to watch major league baseball's spring training. Spring break each March is also a major source of tourists. Beach resorts in the Coastal South, including Daytona Beach, Destin, and South Padre Island, continue to draw hundreds of thousands of students every spring, as do Caribbean destinations such as Cancun and Jamaica.

Tourism and recreation in the Coastal South has also spurred the growth of retirement migration into the region. In the 1940s through the 1970s, coastal Florida was a favored retirement destination of New Yorkers and other natives of the Northeast. As southern Florida became more and more crowded, land values rose and retirees, dependent on fixed income, began to identify less expensive destinations. Some parts of the Coastal South, notably eastern North Carolina, the Florida Panhandle, and the Gulf Coast from Alabama to Texas, continue to attract retirees, but for the most part the Coastal South has now developed to the point that it is out of the price range of many middle-class retirees.

Culture, People, and Places

Heterogeneity, Diversity, and Cosmopolitanism

The heterogeneity, diversity, and cosmopolitan nature of the Coastal South's past and present population can be symbolized by the celebration of Mardi Gras. Mardi Gras (French for "Fat Tuesday") is the climax of the carnival season, celebrated by many Catholics, especially from the Mediterranean region, prior to Lent. Mardi Gras is associated primarily with New Orleans, where it is a major tourist attraction; but many other communities in the region celebrate Mardi Gras with parades and other activities, notably Mobile, Biloxi, and Galveston. In the "Cajun" country of rural Louisiana, Mardi Gras is also celebrated but with a more family-oriented, less tourist-based atmosphere.

Not surprisingly, the diversity and heterogeneity of the Coastal South has created a variety of distinctive places. The four most significant metropolitan areas in the Coastal South are Miami, the Tampa Bay region, New Orleans, and Houston. We examine each of these areas and then briefly discuss other places in the region from North Carolina to Texas.

Miami, like Los Angeles and Phoenix, is a creation of the 20th century. In the 19th century, peninsular Florida was a hot, mosquito-infested, isolated, and sparsely populated region. In the early 20th century, however, population in the Miami area began to grow rapidly. Railroads and later commercial aviation made the region far more accessible, and the beaches of southeastern Florida became attractive to winter tourists and retirees. The development of aviation made Miami, the closest U.S. city to the Caribbean Islands and South America, an important gateway to the Caribbean region and to Latin America. The city profited not only from accessibility but also from the stability of the U.S. economy. Numerous banks doing business in the Caribbean and Central America are located along Brickell Street in Miami, the main business thoroughfare.

The population of Miami and southeastern Florida is very diverse, heterogeneous, and cosmopolitan. The area contains not only numerous Cuban Americans, but immigrants from Haiti, Jamaica, the Dominican Republic, and many other places in the Caribbean and Central America. Tourism and recreation continue to play major roles in the economy of Miami and central Florida. Miami is the headquarters for many popular cruise lines that take thousands of tourists to various Caribbean islands and Central American ports every week. Like other cities in the Coastal South, Miami is particularly vulnerable to natural hazards. For example, there were 41 tropical storms or hurricanes reported as causing damage in Florida from January 2000 to the end of 2006.

The conurbation surrounding Tampa Bay is an often overlooked but important urban center as well. The region contains the cities of Tampa and St. Petersburg on either side of the bay, as well as dozens of other communities along the coast to the north and south. In contrast to Miami, the Tampa Bay region has historically been less oriented to tourism and more to industry. Traditionally, Tampa was known for cigar manufacturing; although this industry is now far less important, Tampa remains a major port city and hosts a variety of manufacturing activities as well as retirement communities.

Before the devastation of Hurricane Katrina in 2005, New Orleans was one of the most colorful, unique cities in the United States. The city's location near the mouth of the Mississippi River allowed it to dominate trade throughout the North American interior, prompting the Louisiana Purchase of 1803. During the 19th and early 20th century, New Orleans was the major trade gateway between the continental United States and the Caribbean, although it has been eclipsed in recent decades by Miami and Houston as aviation has come to replace rail transportation. Many parts of the city are below sea level, so a series of faulty levees were built to prevent the Mississippi from flooding the city and its suburbs. As we saw in the late summer of 2005, New Orleans remains very vulnerable to flooding; a major hurricane or sustained, flooding rains over several days easily strained the city's levees and other defenses beyond the breaking point and resulted in catastrophic damage, as was seen in the aftermath of Katrina.

Four hundred miles west of New Orleans is the Coastal South's largest urban region—the Houston metropolitan area. New Orleans and Houston share a coastal location, emphasis on port facilities, and a hot, humid climate, but there the similarity ends. New Orleans is known for its laid-back lifestyle and its attractiveness to tourists; Houston is a working community. New Orleans has a long and colorful history; Houston is a 20th-century city (see Figure 10.10).

Nearby Galveston, located on a barrier island along the coast, was the largest and most cosmopolitan city in Texas in the 19th century. In 1900, Galveston was

◀ **Figure 10.10** Growth of Houston as a Sun Belt city

heavily damaged by a hurricane that took more than 6000 lives—the deadliest natural disaster in North America's history. Business leaders subsequently decided to redevelop inland, in a location less vulnerable to hurricanes. To promote ship access, the Houston Ship Channel was constructed, making Houston a seaport. The development of the oil industry in Texas also spurred Houston's development, as did the location of NASA south of the city in the 1950s. By the 1960s, Houston had become the fifth largest city in the United States. It remains an important industrial center, especially in petrochemicals, as well as an important center for transportation, trade, and other industries. The rivalry between Houston and Galveston remains intense in the early 21st century, with Galvestonians who call themselves BOIs (*Born on the Islanders*) and who are still hostile about the perceived "uppity attitudes" of many of the "mainlanders" from Houston.

Ironically, thousands of New Orleans natives relocated to the Houston area after the devastation of Hurricane Katrina; as New Orleans is rebuilt, many are facing the difficult decision whether or not to return home. This has been an especially agonizing decision for New Orleans' African-American middle class. One of New Orleans' oldest social clubs is "The Bunch," which sponsors an exclusive black-tie ball and parade each Mardi Gras. Membership in "The Bunch" is limited to 50 African-American men. Most of the members are physicians, dentists, attorneys, executives, and professional men. In 2005, 38 of the 46 members of "The Bunch" were forced to abandon their homes after Katrina, and many had to relocate not only their homes but their businesses and professions. The doctors and dentists who planned to return found not only that their homes and professional buildings were damaged or destroyed but also that many of their patients and clients had left, leaving them with too few patients and clients to remain in business.

Other Places

Many other communities add color, variety, and distinctiveness to the landscape of the Coastal South. The armed services play an important role in the economy of coastal North and South Carolina, with Camp Lejeune as the Marine Corps' major training base. The Outer Banks area, including Cape Hatteras, is a well-known ocean recreation area, and Kitty Hawk on Cape Hatteras is also known as the site of the Wright Brothers' first flight in 1903. To the south, Wilmington is known as the birthplace of Michael Jordan. Myrtle Beach, South Carolina, is one of the most popular and frequently visited beach resorts in the Southeast, while nearby Darlington is an important stop on the NASCAR tour.

The cities of Charleston, South Carolina, and Savannah, Georgia, were the largest cities in the Coastal South at the time of U.S. independence. They remain important ports, contain numerous navy and air force bases, and are popular tourist destinations. Georgia has several popular beach resorts, including Tybee Island, Brunswick Island, and St. Simons Island.

Jacksonville is the major city of northeastern Florida. It is an important navy home port and a significant industrial center. To the south, Daytona Beach is a popular resort and the home of NASCAR's Daytona 500. Nearby Cape Canaveral is the launching site for many U.S. space missions. As we have already seen, the Orlando area is the leading tourist destination in the United States. Indeed, the area from Daytona Beach through Orlando to Tampa Bay is an increasingly developed conurbation, also well known as the spring training home of many major league baseball teams.

West and south of Florida's capital city of Tallahassee are many other communities that epitomize the region's emphasis on tourism and government. Beachgoers flock to Destin, Gulf Shores, and other beaches of western Florida and southern Alabama. This area is sometimes called the "Redneck Riviera" because so many of the region's visitors come from Atlanta and elsewhere in the Inland South. Pensacola is a major navy port, as is Mobile.

West of New Orleans, the "Cajun" country of southern and southwest Louisiana was settled by French-speaking descendants of refugees from Nova Scotia. These "Cajuns" sought unsettled land under French control and moved to southern Louisiana in large numbers in the late 18th century. French is still heard on the street and in radio stations in these areas today, and oil revenues have allowed many Cajuns to remain on the land and maintain their distinctive culture. The communities of Lafayette, New Iberia, and Lake Charles represent the center of Cajun culture in Louisiana. Across the Sabine River are the oil-dependent cities of Beaumont, Orange, and Port Arthur, Texas.

The Future of the Coastal South

For centuries, the Coastal South has attracted a diverse population of settlers from throughout the world. The region's orientation to trade and its cultural diversity make it especially attractive in today's mobile, cosmopolitan, globalized world. The region's beaches, resorts, and historical landmarks and monuments attract millions of tourists from throughout the world every year, and many of these visitors eventually become permanent residents.

The geographical advantages of the Coastal South bode well for this region's future. In a world increasingly oriented to international interaction and trade, its location along the coasts of the Atlantic Ocean and the Gulf of Mexico make this region a natural gateway for cross-regional and international trade. The region's warm climate makes it attractive to new residents as well as to tourists. The area has long had a reputation for cultural diversity and a cosmopolitan outlook on life.

At the same time, however, residents of the Coastal South must remain aware of the impacts of natural hazards. Intense development along the coastlines and on barrier islands is imperiling their fragile ecosystems and enhancing the region's vulnerability to land subsidence, hurricanes, and tropical storms. Over the past two decades, powerful hurricanes have caused intense destruction and numerous fatalities throughout the Coastal South, as was made evident throughout the world during and after Hurricane Katrina. As more people move into the Coastal South region, an increased amount of attention must be paid to the region's environmental quality.

Suggested Readings and Web Sites

Books and Articles

Boswell, Thomas D., ed. 1991. *South Florida: The Winds of Change*. Washington, D.C.: Association of American Geographers. Overview of the development of south Florida through time and a discussion of key issues facing the region up through the early 1990s.

Boswell, Thomas D. and James R. Curtis. 1984. *The Cuban-American Experience: Culture, Images, and Perspectives*. Totowa, NJ: Rowman and Littlefield. A study of the migration, settlement, and ethnic landscapes of Cuban immigrants in the United States with an emphasis on their post-Castro relocation to the Miami area.

Braus, Patricia. 1998. "Strokes and the South," *American Demographics* 20: 26–29. A disturbing study of some of the health issues facing the American South due to economic challenges and racial differences.

Breen, T. H. 1985. *The Mentality of the Great Tidewater Planter on the Eve of Revolution*. Princeton, NJ: Princeton University Press. A history of the Tidewater area in the 18th century with an emphasis on regional perceptions, attitudes, and attachments to land, culture, and politics.

Carney, Judith. 2001. *Black Rice: The African Origin of Rice Cultivation in the Americas*. Cambridge: Harvard University Press. Rice is one of the most commonly eaten staple foods in the world. This book traces its origin in Africa and its ultimate diffusion to the southern part of North America.

Colten, Craig E. 2005. *An Unnatural Metropolis: Wresting New Orleans from Nature*. Baton Rouge, LA: Louisiana State University Press. An environmental history of the city that probably never should have been built (published a few months prior to the devastation of Hurricane Katrina).

Hardwick, Susan Wiley. 2002. *Mythic Galveston: Reinventing America's Third Coast*. Baltimore: The John Hopkins University Press. A study of the impacts of immigrant groups on the evolution of this island city set within the content of the urban historical geography of Texas.

Lewis, Pierce F. 2003. *New Orleans: The Making of an Urban Landscape*. University of Virginia Press.

Littlefield, Daniel C. 1991. *Rice and Slaves; Ethnicity and the Slave Trade in Colonial South Carolina*. Urbana, IL: University of Illinois Press. Explores the trading routes established between West Africa and the Americas, and the perpetuation of slavery.

Web Sites

Florida Keys Tourism
http://www.fla-keys.com/

Florida Population Information
http://quickfacts.census.gov/qfd/states/12000.html

City of Galveston
http://www.cityofgalveston.org/

City of New Orleans
http://www.neworleansonline.com/m

Hurricane Information
http://www.nhc.noaa.gov/

11
The Great Plains

PREVIEW

Environmental Setting

- The landscape of the Great Plains includes vast areas of flat, gradually sloping surfaces, but also includes a great deal of topographic diversity in the form of glacial deposits, dune fields, dissected plateaus, and isolated mountain ranges.
- Low precipitation and weather extremes characterize the Great Plains, with most crops in this region requiring irrigation.
- The Great Plains region is one of the great grasslands of the world and has soils that are excellent for growing grain where sufficient moisture is available.

Historical Settlement

- The Plains have been settled by Native Americans for thousands of years. Prior to the introduction of the horse in the 16th century, these peoples largely lived along rivers where access to water was guaranteed.
- Euro-American settlement of the Plains prior to the 19th century was delayed by the resistance from Native Americans and the harsh climate. Settlement in the 1800s was triggered by the invention of a better plow for breaking up grassland soils, free land distribution by the U.S. and Canadian governments, mineral discoveries, and the arrival of the railroads.
- The modern population is dominated by people of Euro-American descent. However, the Latino presence is growing in the southern Great Plains and Native Americans maintain a major presence, especially in Oklahoma and South Dakota.
- The average age of Great Plains' residents is increasing due to low birthrates and out-migration of young adults. This places severe strains on businesses, government agencies managing social welfare, and medical services.

Political Economy

- Agriculture has been the foundation of the Great Plains economy for much of the 19th and 20th centuries, but has always been subject to a boom and bust cycle due to climate variations and dependence on external market forces.
- Mineral extraction, particularly of oil, gas, and coal, are important to the Great Plains economy.
- In recent decades, a combination of low labor costs, interstate highway construction and greater ease of transportation, better refrigeration technology, and growth in the livestock industry have made the southern Great Plains the leader in the meatpacking industry in the United States.
- In the Plains, as elsewhere, university communities and state capitals have seen significant high-technology growth and higher per capita incomes, while other areas have languished.

People, Culture, Places

- The residents of the Great Plains tend to orient themselves to major cities at the periphery or outside of the region. These cities have larger populations, a wider range of cultural activities, and more diverse economies.
- Within the Plains, there is a wide variety of communities. Those that have been most successful in making a transition to the tertiary and quaternary sectors of the economy are prospering; others continue to lose population.

Future of the Region

- Present agricultural practices in the Great Plains are not sustainable. Future scenarios for the Great Plains economy and communities range from slow decline and depopulation to intentionally allowing the land to revert to its native state to placing greater emphasis on tertiary and quaternary economic development.

Introduction

For many non-residents, the words "Great Plains" conjures up images of flatlands stretching far into the distance, of long car trips driving through seemingly endless wheat fields, of unbroken horizons. But the Great Plains are far more diverse and complex than these simple images. A landscape that at first glance can seem monotonous contains a wide array of features ranging from true mountains to continental glacier features to wide plateaus. The climate of the Plains is semiarid and fickle, with searing heat one day and cold rain or freezing temperatures the next day. While agriculture, wheat, and other grains dominate the landscape in terms of total land use, much of the economy is transitioning towards a more diverse mix of manufacturing, high technology, and service sector activities. And the classic image of a Great Plains residence—an isolated farmhouse like Dorothy's in *The Wizard of Oz*—is no longer the standard. Few locations in the Plains are far from farms or ranches,

◀ A green prairie somewhere in South Dakota reflects a traditional view of the Great Plains yet a majority of Plains residents now live in urban areas and pursue non-agricultural occupations.

The Great Plains is a region in transition. Declining energy and water reserves, changing climate, an influx of new Latino and Asian-American settlers, and new transportation technologies are pushing the Plains towards a future that is very different than its past. But the direction of the future is uncertain, with residents and scholars alike debating what course is most likely, most reasonable, and most productive. In this chapter we examine the way in which the natural resource base of this region and the global economy are affecting changes on the environmental, economic, and cultural landscape of the Great Plains.

Environmental Setting

The Plains region includes all or part of 12 U.S. states (Minnesota, Iowa, North Dakota, South Dakota, Nebraska, Kansas, Oklahoma, Texas, Montana, Wyoming, Colorado, and New Mexico), along with parts of three Canadian provinces (Manitoba, Saskatchewan, and Alberta). To the east, the Plains blend imperceptibly into Ontario, the Great Lakes/Corn Belt and Inland South regions. The contrast to the west is far more dramatic, with the Front Range of the Rockies rising several thousand feet above the nearby plains (Figure 11.1).

Most people stereotype the Great Plains as a flat and featureless region. Some parts of the Great Plains are very flat indeed, for example the Llano Estacado of western Texas and eastern New Mexico, and the Lake Agassiz Basin of Manitoba, Saskatchewan, and North Dakota (Box 11.1). However, other areas within the Plains contain extensive dune fields, rugged badlands, or mountain ranges. Developing an ability to see and understand these variations requires us to look at the origins of the landscape and its location relative to major mountain and climate zones of North America.

Landforms

Much of the vast landmass comprising the Great Plains is relatively flat, sloping imperceptibly as it rises from about 500 to 1000 feet (150 to 305 meters) elevation in the east to around 5000 feet elevation in the west. The basement rock that provides the continental platform for this region formed over one billion years ago as small continental plates collided and welded together. Over the past 500 million years, the Great Plains' central location has kept them far removed from tectonic activity associated with continental margins.

Overlying the basement rocks are thick deposits of sedimentary rocks. Limestone, formed when a vast, shallow inland sea covered the region during two periods between 200 and 65 million years ago, makes up a large portion of this rock. Additional sedimentary layers were laid down by a variety of terrestrial processes when the area was above sea level. The limestone and other sedimentary deposits are often buried, although they do crop out in areas that have been eroded by rivers or where later deposits did not cover them up, as with the Edwards Plateau of central Texas. Although they are often not visible at the surface, these rocks are critical to

BOX 11.1 Is Kansas as Flat as a Pancake?

Many travelers crossing the Great Plains have described the region as being "as flat as a pancake." Is this statement literally true? Geographers Mark Fonstad, William Pugatch (Flynn), and Brandon Vogt did research to find out its accuracy.

To compare the flatness of Kansas to that of a pancake directly, it would be necessary to create a Kansas-sized pancake, or a pancake-sized Kansas. Since these tasks are impossible, the geographers compared the flattening ratio of Kansas to that of the pancake. The flattening ratio is the ratio of the semiminor to the semimajor axis, using the formula

$$F = (a - b)/a$$

where a is the length of the semimajor axis and b is the length of the semiminor axis. A perfectly flat surface has a semiminor axis with zero length, and therefore the value of F is one. On the other hand, any axis going through a perfect sphere has the same length, and therefore $F = 0$. Because its rotation causes the earth to be slightly flattened at the poles, the flattening ratio for Earth itself is $F = 0.00335$.

Using this technique to calculate the flattening ratio for the pancake and for the state of Kansas, the geographers found that a freshly cooked pancake from the International House of Pancakes had a flatness ratio of $F = 0.957$. The state of Kansas, however, is approximately 400 miles long with a net elevation change of about 3000 feet from its lowest to its highest point. Even taking local topographic variation into account, they found a flatness ratio of $F = 0.997$ for the state. The conclusion? Kansas is indeed flatter than a pancake! Not only did this study demonstrate the truth in the cliche that Kansas is as flat as a pancake, but it illustrated the importance of geographical scale in the analysis of geographical information.

Source: For details, see Mark Fonstad, William Pugatch, and Brandon Vogt, "Kansas is Flatter than a Pancake," *Annals of Improbable Research* 9:16–17.

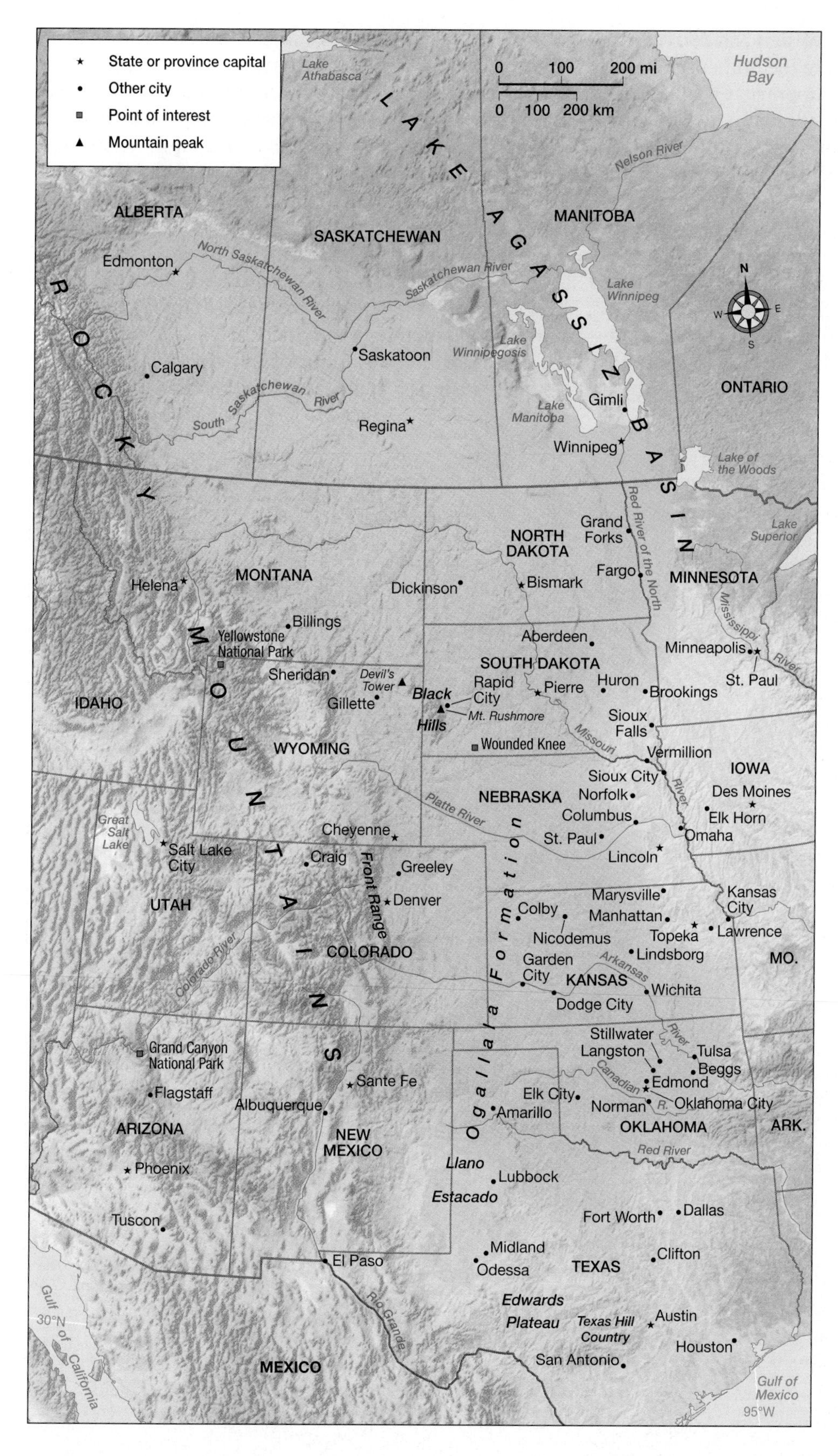

Figure 11.1 The Great Plains region

the vitality of the Great Plains. **Aquifers**, or water-holding rock formations, along with underground oil reservoirs, are located throughout the region.

The sedimentary deposits left behind by oceans contribute to the horizontal landscapes, but the primary reason for the gently sloping topography is the uplift of the Rocky Mountains to the west. As discussed in Chapter 13, the Rocky Mountain Front has uplifted about 0.9 to 1.2 miles in elevation over the last 10 million years. Eroded sediments from these mountains have been carried eastward and deposited over the older sedimentary layers, forming the immense, gently sloping surfaces of the contemporary Great Plains. Many of these gravel and sand alluvial deposits, such as the Ogallala Formation, are excellent aquifers. Water from these aquifers is essential to the success of agriculture in this semiarid setting.

There are many variations in the topography of the Great Plains. Glaciation of the northern plains has provided a mix of glacial landforms (Figure 11.2). These features include flat-lying glacial till plains, outwash plains, and lake bottom sediments. Loess, the windblown silt from glacial outwash, forms deposits up to 20 feet (6 meters) thick in some parts of the Great Plains. Interspersed with these glacial features are lumpy forms of moraine deposits that extend as far south as Kansas. The hummocky moraine deposits capture water in their depressions and therefore often form critical wetlands in the Great Plains. The prairie potholes of North Dakota, for example, were formed in moraines and are famed for their excellent bird habitat (Figure 11.3).

Aeolian processes are also important to the Great Plains. Windblown sand has formed extensive dune fields from Saskatchewan to Texas. Many of the dunes are covered with grass, which helps to hold them in place under the onslaught of blowing winds. But the grass also hides them from the perception of casual

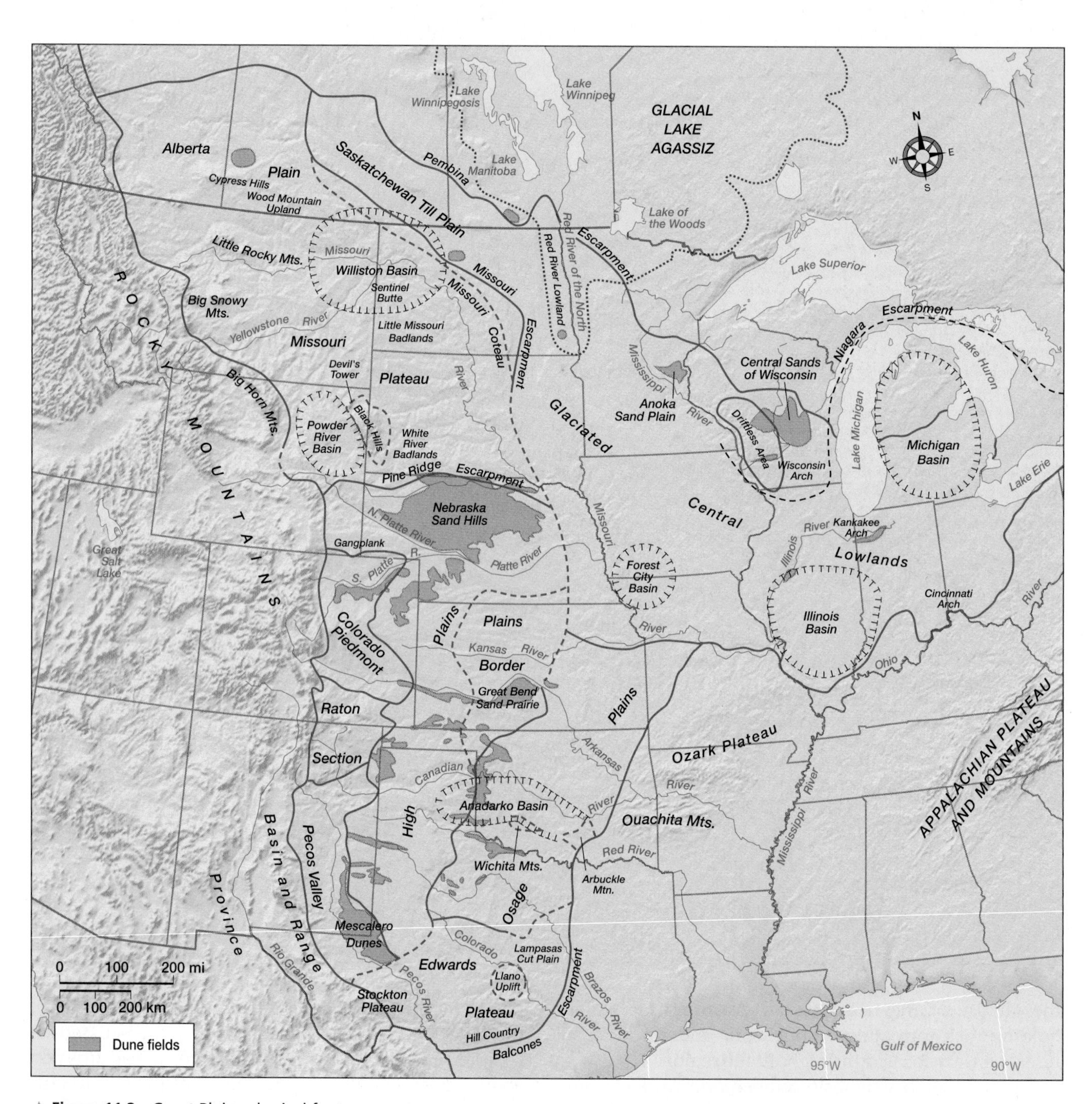

▲ **Figure 11.2** Great Plains physical features

▲ **Figure 11.3** Prairie potholes in North Dakota

passersby, who equate dune fields with large barren dunes as in the Sahara Desert. People are often surprised to learn that the dunes of the Nebraska Sand Hills cover over 19,300 square miles (50,000 square kilometers) and are the largest dune field in the Western Hemisphere.

Downcutting by rivers is also important to Great Plains topography. Steplike topography exists in large portions of the Great Plains where rivers have cut through resistant layers of rock, creating flat-topped mesas bordering broad alluvial valleys. In other areas, badlands have been created where erosion has cut thousands of gullies into low-lying, fine grained rocks.

Even in the Great Plains there are isolated pockets of high relief. Some of the greatest relief resulted from compression that occurred during the uplifting of the Rocky Mountains (see Chapter 13). This accounts for ranges like the Big Horn Mountains of Wyoming and the Black Hills of South Dakota. Other ranges were formed by igneous intrusions that were later exposed as the softer rock around them eroded away, as with Devil's Tower in Wyoming and the Crazy Mountains in Montana. Other uplands are all that remain of ancient sedimentary deposits, now whittled away to lone buttes and mesas. Some topographic breaks, such as the Balcones Escarpment that marks the edge of the Edwards Plateau, were formed by uplift and faulting. Finally, volcanic flows and cones, especially in the Raton Section of the Colorado–New Mexico border area, add drama to the landscape. The diversity of landforms and the variety of processes that formed them put to rest the idea that the Great Plains are an endless, uniform landscape of boring flatlands.

Climate and Natural Hazards

The climate of the Great Plains is characterized by unpredictability, variability, and extremes from day to day, month to month, and year to year. The tremendous north–south extent of the Great Plains, the significant changes in elevation from west to east, and the continental location lead to considerable climatic variability. Searing summer heat contrasts with bone-chilling cold. Months of drought give way to flooding rains and violent storms. At any given time, sharp differences in temperature and moisture may be observed even between nearby places. Weather phenomena are the primary hazards in the Great Plains.

Most places in the Great Plains are hundreds if not thousands of miles from the nearest ocean. Thus maritime influences on the climate of the Great Plains are minimal, and the climate of the Great Plains is continental. The region's climate is relatively dry with strong seasonal changes and dramatic variations in temperature and precipitation. The contrast between summer heat and winter cold is considerable, especially in the northern and central Plains. For example, the monthly mean temperature in Bismarck, North Dakota, is 70°F (38.9°C) in July, and 11°F (6.1°C) in January. Huron, South Dakota, is 73°F (40.6°C) in July, and 14°F

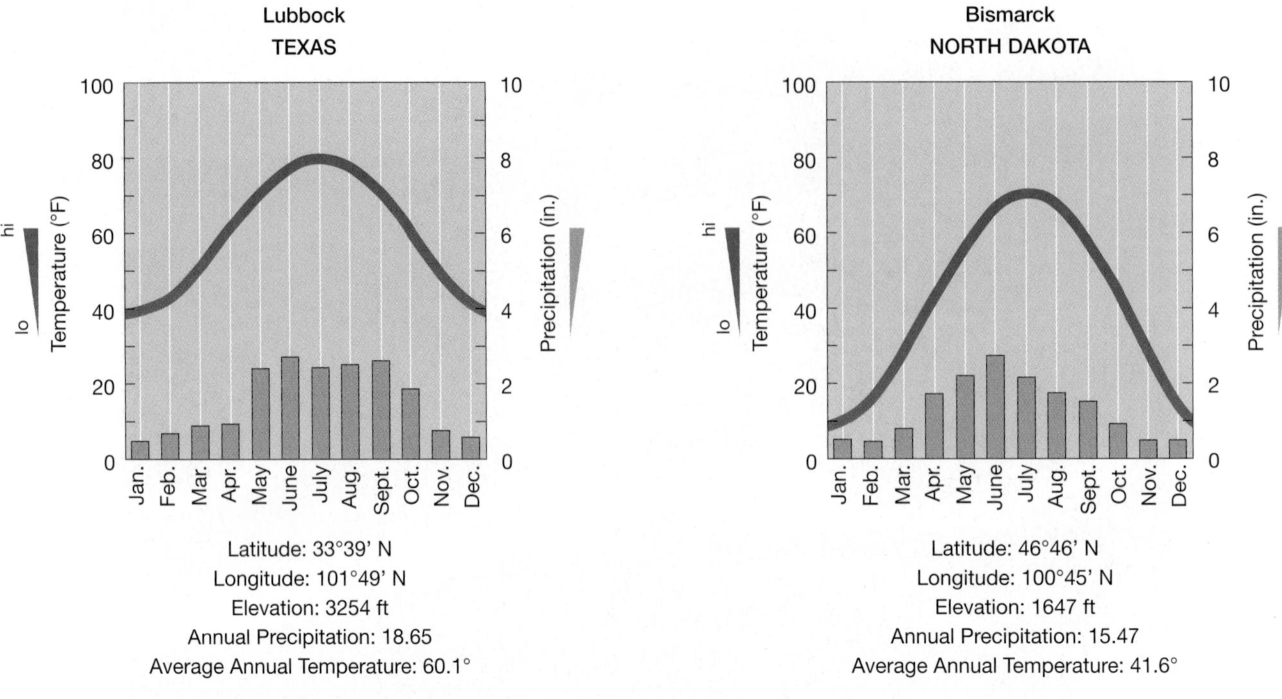

▲ **Figure 11.4** Climographs for Bismarck, North Dakota, and Lubbock, Texas

(7.8°C) in January. Further to the south, Oklahoma City has a July mean of 82°F (45.6°C) and a January mean of 37°F (20.6°C), while Lubbock, Texas, has a July mean of 80°F (44.5°C) and a January mean of 33°F (18.3°C).

Weather systems in the Great Plains, as is the case in most of the United States and southern Canada, generally move from west to east. The rain shadow effects of the Rockies and the great distance from the Pacific Ocean combine to prevent the westerly winds from bringing much moisture to the Great Plains. On average, annual precipitation is lowest along the western boundary of the Great Plains, where the rain shadow effect of the westerly winds is most pronounced.

Moisture affecting the Great Plains often comes from the Gulf of Mexico via two processes: cyclonic circulation and monsoon flow. In spring, strong low pressure cyclonic systems move from west to east across the Great Plains. The counterclockwise circulation of the low pressure pulls in warm, moist air from the Gulf of Mexico as well as cold air from the north. Squall lines form where the contrasting air masses come into contact. Violent thunderstorms with spectacular lightning, hail, and high winds often develop along these squall lines. Peak precipitation in the northern Great Plains generally occurs in May and June because of these cyclonic systems (Figure 11.4). This precipitation can result in extensive and sometimes severe flooding. In the northern Plains, flooding problems can be exacerbated by snowmelt, as is evidenced by the flooding of eastern North Dakota in 1997 (Box 11.2).

Especially in spring, such storms may develop into tornadoes. The southern plains are more susceptible to tornadoes than anywhere else in the world. Tornadoes are especially prevalent along "**Tornado Alley**," which extends from northern Texas across Oklahoma into Kansas (Figure 11.5). Winds of more than 200 miles (322 kilometers) an hour have been reported in large tornadoes, causing extensive property damage and occasional loss of life and injuries. Fortunately, in recent years the development and implementation of sophisticated tornado detection and warning systems have dramatically reduced the death and injury rate from tornadoes. In fact, the year 2005 was the first in recorded history in which no fatalities from tornadoes occurred in the United States during the peak tornado season in April, May, and June (although a November tornado resulted in 22 fatalities in and near Evansville, Indiana).

Another factor that generates rainfall in the Great Plains is the summer **monsoon**, which occurs when increasingly hot land surfaces cause the air above them to rise. The rising air forms a vast region of low pressure that disrupts the westerlies and sucks in moisture from the Gulf of Mexico to the south. Thunderstorms are generated as the moist air rises over the heated continental surface. The monsoon effect is most pronounced near the Gulf of Mexico and decreases with distance to the north. Peak precipitation in the southern Great Plains usually occurs over a period from late spring through late summer due to the combination of cyclonic and monsoon circulation.

The intense rainfall associated with tornadoes and thunderstorms in the Great Plains belies the fact that the region is characterized by aridity. Rain often falls so quickly that the ground cannot absorb the moisture, which runs off and contributes to flooding. Thus the

BOX 11.2 The Great Flood of 1997

Throughout North America, more people lose their lives and property to flooding than to more spectacular, more publicized natural disasters such as hurricanes, tornadoes, and earthquakes. Flooding is a significant hazard in many parts of the Great Plains given the region's level topography and unpredictable rainfall.

One of the most significant flood events in recent years took place on the Red River of the North in 1997. The Red River of the North rises in southwestern Minnesota and flows northward into Lake Winnipeg, Manitoba. It forms the boundary between Minnesota and North Dakota for much of its length. The topography on both sides of the river is flat, making the river susceptible to flooding when water levels get too high. At Grand Forks, dikes 28 feet high were built in order to protect the community against flooding.

The winter of 1996–97 was unusually snowy. By late March, 117 inches of snow had been recorded over the course of the winter in Fargo, North Dakota, with 98 inches upstream in Grand Forks. On April 4, the season's strongest blizzard pounded the region. The precipitation began as rain, turned to freezing rain, and eventually turned to heavy, wet snow. This new snow, along with snowmelt from previous storms throughout eastern North Dakota and western Minnesota, caused river levels to rise steadily. Local residents and volunteers attempted to protect the city using sandbags to restrain the raging waters, but on April 18 the dikes began to give way (Figure 11.A). Within the next two days, three-quarters of Grand Forks' 57,000 people along with virtually everyone in neighboring East Grand Forks, Minnesota (population 8000), were evacuated. A fire broke out in Grand Forks and firefighters who were unable to access the area watched helplessly as many downtown buildings were destroyed. Two-thirds of the buildings in Grand Forks and almost every building in East Grand Forks sustained damage. On April 21, the river crested at 54 feet, and it remained above flood stage for the next six days. The floodwaters moved northward into Canada, and on May 4 the river crested some 21 feet above flood stage at St. Agathe, Manitoba. Eight other towns south of Winnipeg also suffered extensive flood damage.

Since the flood, Grand Forks and East Grand Forks have undergone considerable transformation. Many houses and public buildings that had sustained extensive damage were torn down, and highly flood-prone areas were converted to parkland. New, higher dikes are scheduled for completion in 2007.

▲ **Figure 11.A** Flood damage in Grand Forks, North Dakota, 1997

amount of effective precipitation, or the precipitation that can be put to human use, is often significantly lower than the actual amount of precipitation.

Over the long run, average annual precipitation varies considerably from year to year. Periods of drought tend to alternate with periods of above-normal rainfall, generally in cycles of about 20 years. The 1890s, 1930s, 1950s, and 1970s were periods of unusually dry conditions in much of the Great Plains. When these droughts coincided with periods of economic distress, as during the Great Depression of the 1930s, considerable economic dislocation occurred. Although tornadoes are more terrifying in their damage at a local scale, it is drought that threatens the livelihood of more

meatpacking plants were located along railroad lines in major cities such as Chicago, Omaha, Sioux City, and Denver.

This shift in location of the meatpacking industry is the result of several factors. Per-capita meat consumption has declined in recent years, forcing producers to increase efficiency to remain profitable. One approach to increasing efficiency has been to move meatpacking plants closer to the animals, which reduces the high costs associated with transporting live animals. Cattle ranching has been a widespread activity throughout the Great Plains since the 1800s, and the number of cattle in the Plains has increased in recent decades because irrigation has provided more locally available animal feed. Large feedlots, with thousands of head of cattle, are found in various places in the central and southern Plains.

At the same time, improved refrigeration technology has reduced spoilage during transit, allowing meatpacking plants to be located at greater distances from markets. Ease of transport has been increased by the practice of processing beef and pork into boxed containers, which are cut into individual portions at the store. Interstate highways, which provide ready access to multiple destinations, have further reduced the need to be located near major railways or cities. Finally, the availability of low-cost Mexican-American and Asian-American labor has been a major incentive for meatpacking plants to move to the area. Some meatpacking firms pay bonuses to employees who recruit friends or relatives to work in meatpacking plants, reinforcing ethnic communities. The influx of immigrants has diversified the region but has also forced communities to deal with issues of bilingual education, cultural tension, and racial and ethnic discrimination.

Tertiary- and Quaternary-Sector Economic Activities

Agriculture and related manufacturing are more important to the economy of the Great Plains region than in any other region in North America, yet the percentage of Great Plains residents who earn their living directly in the primary and secondary sectors is still relatively low. Even in North and South Dakota, which have the highest percentage of agricultural employment among the 50 U.S. states, less than 10 percent of the people work in farming. Well over half of all employment in the Plains states is found in the tertiary and quaternary sectors.

Government is the largest non-primary employer in many parts of the Plains. Teachers, county agents, law enforcement officers, social service workers, and other local government employers outnumber farmers in many Plains communities. Although the tourist sector is generally not well developed, tourism and recreation can be locally important. Perhaps the leading tourist destination in the Great Plains is the Black Hills of South Dakota with Mount Rushmore, old West towns, the Homestead Gold Mine, and the annual Sturgis motorcycle festival. The nearby Theodore Roosevelt National Monument commemorates Roosevelt's time as a young rancher in North Dakota, as well as his many contributions to conservation. The Hill Country of central Texas is another major attraction for tourists interested in visiting quaint German towns and cattle ranches and enjoying river rafting.

At a more local level, the many flood control and irrigation reservoirs are important sources of tourism and recreation. On any summer weekend, thousands of Plains residents can be found fishing, swimming, and boating on the many artificial lakes that dot the prairie from Saskatchewan to Texas. Convenience stores, marinas, and boat repair businesses provide employment to many persons.

In recent years, many Plains states have attempted to attract high technology businesses, with varying degrees of success. The Great Plains has a highly educated labor force, a low cost of living, and low rates of crime and social disruption, all attractive to high tech businesses. The great distances and low population densities of the region also create a business niche for the Internet and communication industries. A notable high tech success story is Gateway Computers, which is headquartered in Sioux Falls and has become South Dakota's leading private employer. However, the aging of the Plains population and the perception by many that the Plains are dull and uninteresting may be impeding efforts to lure high-technology industry to the region.

The distribution of per-capita incomes across the Plains indicates the relative importance of the primary, secondary, tertiary, and quaternary sectors. Three types of places have the highest incomes. University communities, state capitals, and areas with urban and suburban development have benefited the most from high-technology employment. Other areas with relatively high incomes have high quality farmland but small populations, such as Cimarron and Texas Counties in the Oklahoma Panhandle, and Pembina County in the northeastern corner of North Dakota. The poorest regions of the Plains are those lacking any of these advantages.

A Typical Plains Economy: Marysville and Marshall County, Kansas

The economy of the contemporary Great Plains is typified by Marshall County, Kansas, located along the Nebraska border midway between Manhattan, Kansas, and Lincoln, Nebraska (Figure 11.10). In 2000, the population of Marshall County was 10,904. 3275 persons, or 30 percent, of Marshall County's people lived in the county seat of Marysville. More than 99 percent of the county's population is of white, non-Hispanic

◀ **Figure 11.10** Marysville, Kansas, a typical Great Plains town and economy

ancestry, with only a handful of Native Americans, African Americans, Asian Americans, and Hispanics. The population tends to be older; in 2000 more than 13 percent of the county's residents were at least 75 years of age.

Agriculture is a mainstay of the county's economy. In 1998, approximately 100,000 acres (259,000 hectares) of soybeans, 79,000 acres (204,610 hectares) of sorghum, and 29,600 acres (76,664 hectares) of corn were harvested. Together, these crops covered more than a third of the county's 903 square miles (365.4 square kilometers), or nearly 600,000 acres (1.6 million hectares) of land. The county's 1998 labor force of 7422 persons included 1100, or about 15 percent, working as farmers or farm employees. 979 (13 percent) were employed in manufacturing. The total primary and secondary sector employment in Marshall County was nearly 30 percent of the workforce, or almost twice the percentage in these sectors nationwide (see Chapter 4).

More than two-thirds of the county's work force nevertheless worked in the tertiary and quaternary sectors, including 1524 in services, 1090 in retail trade, 468 in transportation, and 418 in finance, insurance and related businesses. 931 persons were employed by government agencies. Of these, 69 worked for the federal government, 54 for the military, 43 for the State of Kansas, and the remaining 765 for local government agencies. These included police officers, teachers, and county government and county hospital employees.

Marshall County's largest employer is the Landoll Corporation, which manufactures agricultural tools, machinery, and transportation equipment. Landoll, like many locally owned Great Plains companies, had a modest beginning. Its founder, Don Landoll, began his career as a welder in a three-person shop in Marysville. Landoll began to experiment with the design of farm implements and subsequently launched his own business, branching into trailers, forklifts, crash rescue vehicles, deicing equipment for airlines, and other heavy machinery. Indeed, the Landoll Corporation now controls 85 percent of the U.S. market for towing trailers used to haul buses and cars (Figure 11.11).

◀ **Figure 11.11** Landoll Products, Kansas

eastward in search of bargains. South of Colby are Dodge City and Garden City, which are Old West communities that today have increasing numbers of Hispanic and Asian Americans working in the meat-packing industry.

The Southern U.S. Great Plains

The southern Plains, from the Kansas–Oklahoma border southwards, share the climatic variability and economic dependency of the Plains sections to the north, but are characterized by somewhat greater economic and cultural diversity than the rest of the Plains.

Eastern Oklahoma was set aside as "Indian Territory" by the U.S. government in the 1830s in order to house Native Americans displaced from more productive and valuable lands to the east. Large numbers of members of the Five Civilized Tribes—Cherokee, Choctaw, Chickasaw, Creek, and Seminole—were forcibly removed to what is now eastern Oklahoma from the southeastern United States on the Trail of Tears. Other Native American groups were relocated to eastern Oklahoma later in the 19th century. Many Native American nations retain their tribal headquarters in Oklahoma; for example, Tahlequah is the capital of the Cherokee Nation.

The major cities of Oklahoma are Tulsa and Oklahoma City. Both are strongly oriented to the oil industry, while the latter is the state capital. Today most of Oklahoma's growth is taking place in and near the Tulsa and Oklahoma City metropolitan areas, including the university communities of Norman, Stillwater, and Edmond. Roughly two-thirds of Oklahoma's population can be found in the greater Tulsa and Oklahoma City metropolitan areas.

Oil is also very important to the economy of the region. The cities of Midland and Odessa in West Texas owe much of their growth to the oil industry. Midland is a white-collar community, while Odessa—featured in the movie *Friday Night Lights*—is a more blue-collar city. Lubbock is a center for agricultural service and cotton production, as well as the home of Texas Tech University. To the north, Amarillo is a center for helium production. Both of these communities also serve nearby farming areas, which are imperiled by the continued depletion of the Ogallala Aquifer. The Texas Plains has seen a substantial growth in its Mexican-American population in recent years.

The Future of the Great Plains

Geographers Frank and Deborah Popper of Rutgers University have proposed that large areas of the Plains be taken out of agricultural production and be allowed to revert to their condition prior to the arrival of Europeans, with large herds of bison and perhaps a few Native Americans leading traditional lifestyles. The Poppers argued that such measures are called for because of the continuing depopulation of some parts of the Plains and the unsustainable irrigated agriculture characteristic of others. Stating that such lifestyles are not sustainable in the long run, the Poppers said the best use of the Plains is to turn them into a **buffalo commons**. These buffalo commons could be sources of ecotourism revenues for wildlife viewing, for hunting and fishing, and for people wanting to learn about native cultures. The Poppers' arguments have generated considerable discussion among scholars, journalists, politicians, and business leaders in the Plains concerning the future of the region. At the same time, continued depopulation in parts of the Plains may make their vision a reality, whether planned or not.

What can the Plains expect in the future? For more than a century, this region has been characterized by a boom–bust economy, dependent on agriculture and natural resources. This dependency is likely to continue, although leaders in the Plains are making concerted efforts to wean the region from this historical dependency and to promote the development of the tertiary and quaternary sectors. Universities in the region are playing an important role. In some cases, research institutes associated with universities have begun to spin off small but significant high-tech employment bases. For example, Norman, Oklahoma, is the home of the National Severe Storms Laboratory, and the community has become a center for small firms undertaking storm prediction and meteorological research. Such activities may be critical in promoting a higher degree of sustainability than is characteristic of the Plains in recent years.

By no means is the future of the Plains bleak. The region is characterized by dependency, but it has many features that are attractive to people across North America. The sense of local community is strong, crime rates and rates of social disruption are low, and congestion, pollution, and the cost of living are minimal. Family ties are strong, and some residents of the Plains see the region as a repository of traditional North American family values, regarding it as a safe, positive, and constructive environment for raising children. These factors can provide a basis for successful development in the Great Plains in the years ahead. By paying careful attention to sustainable development, promoting local control over local development, cherishing traditional Jeffersonian values, and thoughtfully incorporating the technologies of the information age, residents of the Plains may be able to prosper and integrate themselves more fully into the North American economy in the years ahead.

Suggested Readings and Web Sites

Books and Articles

Cather, Willa. 1913. *O Pioneers!* Boston: Houghton Mifflin. Classic novel about rural Nebraska settlement in the late 1800s and early 1900s.

Frazier, Ian. 1989. *Great Plains*. New York: Farrar, Straus, and Giroux. An account of travel through the Great Plains in the 1980s.

Johnsgard, Paul. 1999. *Earth, Water, and Sky: A Naturalist's Stories and Sketches.* Austin: University of Texas Press. Vignettes describing the natural history and ecology of the Great Plains, and how the Plains environment has been impacted by humans.

Heat-Moon, William Least. 1991. *Prairy Earth.* Boston: Houghton Mifflin. An account of life in a rural Kansas county in the late 1980s.

McMurtry, Larry. 1966. *The Last Picture Show.* New York: Dial Press. Classic novel about small-town life in West Texas in the 1950s.

Opie, John. 2000. *Ogallala: Water for a Dry Land,* 2nd ed. Lincoln: University of Nebraska Press. Describes the Ogallala Aquifer and federal, state, and local efforts to control the aquifer's depletion.

Riney-Kehrberg, Pamela. 1994. *Rooted in Dust: Surviving Drought and Depression in Southwestern Kansas.* Lawrence: University Press of Kansas. Describes how ordinary rural Kansans coped with Dust Bowl conditions during the Great Depression of the 1930s.

Rolvaag, Ole. 1927. *Giants in the Earth*. New York: Harper and Brothers. Dramatic novel about Norwegian settlement of the northern Great Plains in the 19th century.

Shortridge, James R. 2004. *Cities on the Plains: The Evolution of Urban Kansas*. Lawrence: University Press of Kansas. A chronicle of the founding and growth of Kansas cities from 1850 up to the present, showing how the state's hierarchy of cities emerged from a complex series of promotional strategies.

Webb, Walter Prescott. 1931. *The Great Plains*. Boston: Ginn. Classic historical geography of the Great Plains region, emphasizing impacts of environmental differences between the Plains and areas farther to the east on settlement, culture, and economy.

Wishart, David J., ed. 2004. *Encyclopedia of the Great Plains*. Lincoln: University of Nebraska Press. A groundbreaking reference book on the Great Plains with entries by 1316 authors on topics such as the region's images and icons, climate, politics, historical development, and folklore.

Worster, Donald. 1979. *Dust Bowl: The Southern Plains in the 1930s*. New York: Oxford University Press. A history of the Southern Plains region in the 1930s.

Web Sites

Geologic History of the Great Plains
http://www.lib.ndsu.nodak.edu/govdocs/text/greatplains/text.html

City and County of Denver, CO
http://www.denvergov.org/

Population and Demographics of Denver
http://denver.areaconnect.com/statistics.htm

Southeast Wyoming Tourism
http://www.wyomingtourism.org/

Early Novels about the Great Plains
http://www.willacather.org/

Agricultural Economy
http://www.iisd.org/agri/

Drought History
http://news.nationalgeographic.com/news/2005/06/0614_050614_drought.html

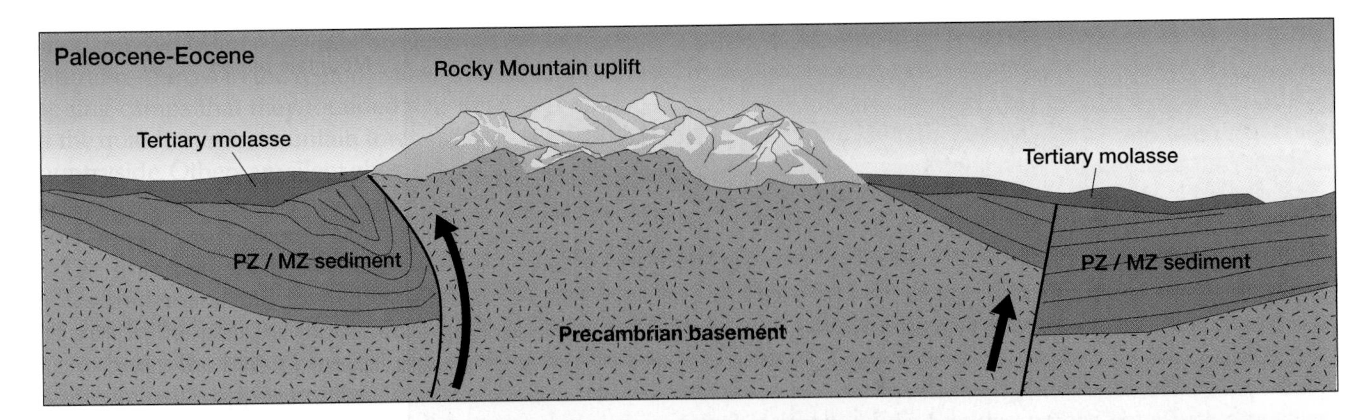

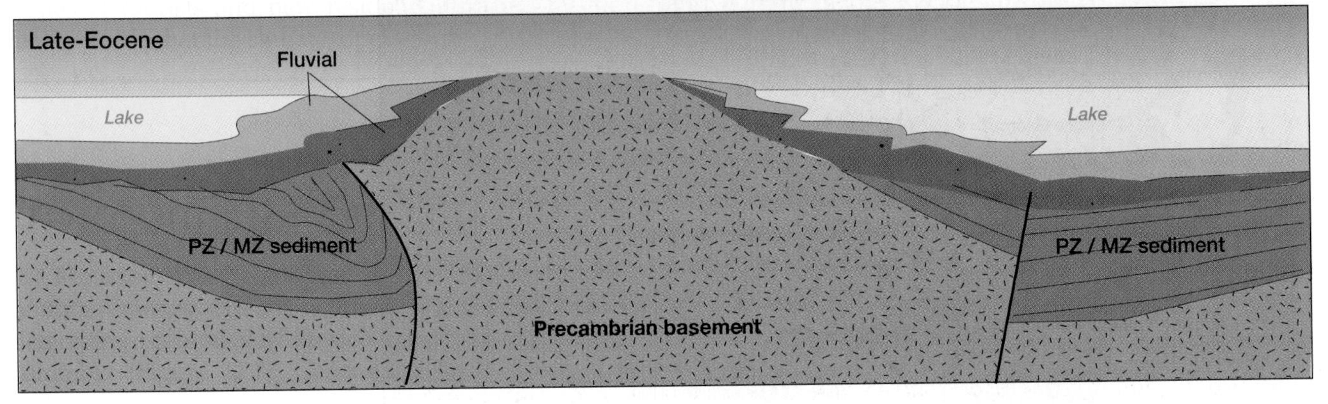

▲ **Figure 12.2** Landform processes and landform features in the Laramide period

unstratified sediment. When deposited at the edges of a glacier the till is referred to as an end **moraine**. The mountainous features remaining in the wake of glaciers can be seen throughout the Rocky Mountains. At the head of glacial valleys, bowl-shaped depressions are left behind to create **cirques**. When two or more sides of a mountain are worn down by glaciers, the resulting pyramid-shaped mountain is called a **horn**. One of the more famous horns in the Rockies is Mt. Assiniboine in British Columbia, Canada (Figure 12.3). Erosion caused by two parallel glaciers can create a ridge of mountains called **arêtes**, which are known for their sharp and steep mountains that often resemble the teeth of a saw, such as the aptly named Sawtooth Mountains of Idaho.

The most prominent feature spanning the entire Mountain West is the **Continental Divide**. From northwest Canada it follows the crest of the Rocky Mountains for over 2000 miles (3220 kilometers), where it continues in the Sierra Madre Occidental Mountains of Mexico. This north–south running divide dictates where precipitation falling in North America will flow. Snow and rain falling on the west side of the divide will flow to the Pacific Ocean, while that on the east side will flow to the Atlantic Ocean and the Gulf of Mexico (Box 12.1).

The Rockies can generally be divided into distinct geographic regions covering six states and two Canadian provinces. The southern Rockies of Colorado contain the highest mountains in the region, with 54 mountains above 14,000 feet (4267 meters). The highest peak in the Rockies is Colorado's Mount Elbert at 14,433 feet (4400 meters). To the west, the mountains are slightly lower and include the Uinta and Wasatch Mountains of Utah, and mountains in southeast Idaho and southwest Wyoming. To the north are the greater Yellowstone mountains, covering most of western Wyoming and parts of Montana and Idaho. North of these are the mountains of central Montana and Idaho, which reach their crescendo at Glacier National Park in northwest Montana. In the far north are the jagged peaks of the Canadian Rockies. With elevations usually in the range of 9000–13,000 feet (2740–3960 meters), they are known for the regal settings of Banff and Jasper National Parks of Alberta and Kootenay National Park in British Columbia.

Mountain Hot Spot: Yellowstone National Park

Moving westward millions of years ago, the North American Plate eventually passed over a volcanic **hot spot**, where volcanic activity is increased by magma consistently rising to Earth's surface. Three major eruptions in the last two million years provided the volcanic debris that eventually cooled and settled before an underground magma chamber collapsed, forming the 1300-square-mile (3366 square kilometers) Yellowstone Caldera. One of the largest active calderas in the world,

▲ **Figure 12.3** Mt. Assiniboine, British Columbia

it is home to the park's famous geysers and hot springs. It also created the horseshoe formation of volcanic mountains that line the Yellowstone Caldera.

The volcanic activity that created the park continues today. It fuels the geysers, hot springs, and mud pots that make Yellowstone famous. Other prominent features of the park are **fumaroles**, or hot springs so hot that their water boils away creating steam. The resulting steam, gas vents, and sulfur deposits create the brilliant yellow linings and smells for which they are famous. The sulfuric acid also helps create the barren land that surrounds fumaroles.

Climate

In the eyes of most people outside the region, the Rockies are a region of cold, snowy winters and sunny alpine meadows lined with wildflowers. Yet these

BOX 12.1 Hiking the Continental Divide Trail

For a really exciting way to experience the Rocky Mountain region, we suggest saving up your road trip funds and heading up to the 3100 mile long Continental Divide Trail. It provides spectacular backcountry views along the length of the Rocky Mountains from Mexico to Canada. The hike extends along the crest of the Rockies and includes magnificent national parks such as Glacier, Yellowstone, and the Great Basin Divide.

The trail is about 70 percent complete, with most of the unfinished segments found on Bureau of Land Management land in New Mexico and Wyoming. Its route follows the Continental Divide across the five western states of Montana, Idaho, Wyoming, Colorado, and New Mexico from the border of Mexico to Canada. The trail crosses five ecological zones that include all the different landforms, climate, vegetation, and wildlife ecosystems of the Rocky Mountain region.

The trail begins in the south at the Mexican border near Antelope Wells, New Mexico, and then travels north through the desert into the Gila National Forest and the Aldo Leopold Wilderness and up into the El Malpais National Monument. It then follows the 1000-year-old Zuni-Acoma trade route. Before reaching Colorado, the trail travels through the Cibola and the Carson and Santa Fe national forests. From Colorado, it follows the Continental Divide along the crest of the Rockies across Idaho's Caribou-Targhee National Forest, and then follows the boundary between Idaho and Montana. Finally, the path of the Continental Divide Trail heads north through Glacier National Park in northern Montana.

To learn more about how it feels to hike along this world-famous trail, follow the experiences of intrepid hikers Sarah Heidenreich, Adrianne Gass, Simon Dyer, and Darryl Riches at *http://gorp.away.com/gorp/resource/us_trail/continen.htm.*

BOX 12.2 Chief Joseph Speaks Out

Nez Perce leader Chief Joseph (1840–1904) was best known for his resistance to the American government's attempts to force his tribe onto reservations. The Nez Perce was a peaceful nation that lived from Idaho to northern Washington state. The tribe had maintained a good relationship with explorers and settlers from the United States after the Lewis and Clark expedition. Joseph, in fact, spent much of his early life at a Christian mission.

In 1855, Chief Joseph's father signed a treaty with the United States that allowed his people to retain most of their native lands. Later, in 1863, a new treaty was discussed that would severely reduce Nez Perce land and give much of it to the U.S. government. However, Chief Joseph's father maintained until his death that his people had never agreed to this second treaty. In 1877, when Chief Joseph became the new leader of his tribe, it was this treaty/nontreaty that caused the showdown that eventually led to his death. After months of fighting over who should control the land that had been the Nez Perce nation's for centuries, many of these brave resistance fighters were captured and sent to reservations as far away as Oklahoma. Many of those who survived the trip later died from malaria and starvation.

Meanwhile, Chief Joseph continued to try every type of possible appeal to the federal authorities to return the Nez Perce people to the lands of their ancestors. After losing this battle, he and his remaining family members and fellow fighters were sent to a reservation in Washington where, according to a doctor on the reservation, he later died of a broken heart.

Chief Joseph was an eloquent speaker as well as a brave fighter. Here are a few of his inspiring quotes:

- We were contented to let things remain as the Great Spirit made them.
- The earth and myself are of one mind.
- It does not require many words to speak the truth.
- Good words do not last long unless they amount to something. Words do not pay for my dead people. They do not pay for my country, now overrun by white men. They do not protect my father's grave. They do not pay for my horses and cattle.
- I am tired of fighting. . . . From where the sun now stands, I will fight no more forever.

For more information on Chief Joseph, see *http://www.powersource.com/gallery/people/joseph.html.*

explore the Rocky Mountains. After the Louisiana Purchase, the Lewis and Clark expedition was routed through northern portions of the Rocky Mountains in Montana and Idaho in 1805. A year later, Zebulon Pike was commissioned to explore the newly acquired territory around the Front Range of Colorado. Then in 1807 John Colter returned to the East with almost unbelievable descriptions of what is now Yellowstone National Park. These expeditions brought news of natural wonders, wide open territory, and vast prospects for economic development. The West would never be the same.

Stories and legend brought the West to people in the East, yet it was gold that brought the people in the East to the Mountain West. While nearly 1000 people had traveled on the Oregon Trail by 1843, it was the discovery of gold in California in 1849 that fueled settlement and the search for the precious minerals in other regions. Gold had been discovered in Colorado by 1858, bringing with it 50,000 people in one year. In 1860, gold and potato farming became Idaho's calling card. Soon Idaho's native peoples would be forced onto reservations and railroads would crisscross the state, delivering a 90,000 person settlement boom in 1890 alone.

The Montana Rockies suffered a similar fate. Gold brought the initial settlers, but it was the Northern Pacific Railroad and the epic battles with legendary Indians such as Chief Joseph that cleared the way for extensive settlement. In the 1880s, Montana grew by more than 100,000 people with the discovery of gold, silver, and eventually vast reserves of copper in the mountains outside Butte, which by 1900 was the largest city in the region. Wyoming's settlement was accelerated by virtue of being located along the Oregon Trail and later the Union Pacific's transcontinental railroad. However, settlers came primarily for the opportunities in ranching on the plains. The splendor of what would become Yellowstone National Park spared the Wyoming Rockies from the hordes of miners and other settlers who moved into other parts of the 19th century Mountain West in such large numbers.

Before the fur trappers arrived, the southern portion of the region had been under Spanish control. Spanish adventurers, lured by tales of gold, explored many parts of the region. The Sangre de Cristo Mountains in southern Colorado and northern New Mexico were named by Francisco Torres in the early 1700s. Spanish settlement of the Rio Arriba area in northern New Mexico began in the 1700s before Mexico became independent of Spain. Descendants of these settlers are still there in large numbers in this early Hispano Homeland. More recently, the Latino population has increased because of immigration directly from Mexico and from such places as Texas and New Mexico; highest recent Mexican-American populations are in urban areas such as Denver.

Settling the Canadian Rockies

Much of the Canadian Rockies was initially under the ownership of the Hudson's Bay Company and thus was explored by fur traders in the 18th century. Sold to the Dominion of Canada in 1870, growth in the region was spurred by the construction of the Canadian Pacific Railway (CPR) and, later, by the Klondike gold rush in the 1890s. Lumber quickly became one of the chief economic resources of the Canadian Rockies and remained so until the discovery of vast mineral deposits, especially oil, in the 20th century. European settlement in the Canadian Rockies, however, was slow due to poor economic conditions in the late 19th century. Canada was unable to compete with the United States in attracting European immigrants and much of the Canadian West actually experienced population losses prior to the 1890s.

However, the establishment of Fort Calgary in 1876, the completion of the Canadian Pacific Railroad in 1885, and the announced closing of the American frontier by the Census Bureau in 1890 all led to major surges in immigration to the Canadian West. The Canadian Minister of the Interior, Clifford Sifton, began an advertising campaign to attract people to western Canada using attractive adjectives and banning from publication temperature reports and words referencing the cold. Towards the turn of the century, hundreds of thousands of settlers began moving west seeking land, opportunity, and adventure. Many of them would settle in the Canadian Rockies, with a majority choosing the Front Range close to Fort Calgary, which eventually became the city of Calgary.

Twentieth-Century Settlement: Location, Location, Location

Economic struggles in the first half of the 20th century severely damaged the economic opportunities that originally brought settlers to the Rockies, especially in agriculture and mining. By the post–World War II era, urban growth was defining settlement in the region, as metropolitan areas such as Denver and Salt Lake City grew rapidly. In the later decades of the 20th century, a number of phenomena brought new residents to the mountains. Today young and old alike venture to the Rockies, seeking abundant and inexpensive land, a pleasant climate, less congested metropolitan areas, infinite recreational activities, scenic beauty, and lucrative jobs.

Between 1980 and 2000, the U.S. portion of the Rocky Mountain region grew by an astronomical 65 percent. It now boasts the fastest-growing labor force, the largest increase in new businesses, the most impressive growth in per capita income and gross domestic product, the highest rise in upper-income households, and the greatest number of new households. Even the Canadian Rockies are getting in on the trend. Calgary is expected to have one million residents by 2008, making it the largest Canadian city between Toronto and Vancouver and the fourth largest in the country.

Political Economy

As we have seen in earlier chapters, most of the eastern United States and Canada is or was at one time used for agriculture. Only a few places east of the Rockies, such as northern Maine, Michigan's Upper Peninsula, the swamps of south Louisiana, and Florida's Everglades, are more than five miles from the nearest road, railroad, or navigable waterway. Many of the cities and towns of eastern North America were established to serve nearby farms. As a result, eastern North America is characterized by continuous settlement, with relatively few large tracts of uninhabited, uncultivated territory.

West of the Front Range, in contrast, much of the land is too dry, too steep, or too high in elevation for agriculture. Farming is concentrated along river bottoms and in other areas with sufficient flat land. Most of the early settlers who traversed the Oregon Trail with the intention of farming bypassed the Rockies, preferring instead to settle in the low, fertile farmlands of California, Oregon, or Washington State. Many of those who did settle in the Rocky Mountains were drawn by other potential sources of wealth, notably minerals, lumber, and other natural resources. These factors have all influenced the developing political economy of the Rocky Mountain region.

The Primary Sector

Historically, the economy of this region has been dominated by the primary sector. Industrial activity is limited, but the importance of the primary sector is now dwarfed by continued growth in the tertiary and quaternary sectors. As we have seen, a defining characteristic of the Rocky Mountains is its relative unsuitability for agriculture. Most of the terrain in the region is rugged, steep, and arid. Only a few areas are suitable for commercial crop agriculture. Most of these are located along the valleys of major rivers, although the amount of commercially viable farmland is increased by the use of irrigation. Throughout the region, irrigators have been able to take advantage of the principle of appropriation discussed in more detail in the chapter on the Great Plains. The appropriation principle as presented in Chapter 11 allows for long-distance transfer of irrigation water.

Throughout most of the Mountain West, ranching is more prevalent than farming. Ranchers run cattle and sheep on extensive tracts of land, some of which are privately owned but many of which are leased from the federal government. Indeed, there are approximately six cattle for every person in the state of Montana (Figure 12.8).

many long-term residents nostalgic for the days when Denver was a quiet, rustic, dusty old cow town. Even though historic preservation has retained much of Denver's old character, this city is growing into a lively, cosmopolitan metropolis . . . and fast!

The Future of the Rocky Mountain Region

The 21st century will create both opportunities and challenges in the Rocky Mountain region. As this chapter has shown, the most serious problem facing the region is finding a way to deal with a rapidly expanding population. The natural landscape has always been the magnetic force drawing people to the mountains. The problem now is developing a means to continue the economic benefits gained from abundant natural resources, while preserving pristine natural settings that attract tourists, businesses, and new residents.

The gains of a growing economy and population are evident in an abundant and lucrative job market but are balanced by increases in pollution and the degradation of vulnerable ecosystems. Climate modeling in the region has demonstrated that minimum snow levels will rise in elevation and land-use changes and increased pollution can and will cause local level climate changes of considerable magnitude. Broad-based ecosystem management will be needed to insure that the residents of the Rocky Mountains continue to benefit from and enjoy the natural beauty they covet. Any longtime resident of the region will testify to the drastic landscape changes that have recently occurred. These changes often induce old and new residents into social and political battles over the future of the Mountain West and will continue to do so with more people on the way.

In addition, most of the Rockies are arid, meaning that the most pressing environmental issue here is water shortages. An increasing population requires water not only for residential use but also for commercial and agricultural use. This natural resource is already strained, and it is not just the Rocky Mountains that are dependent on the water that comes from its mountains. Reservoirs and aquifers in the region are being depleted, while continued droughts are incapable of replenishing water reserves and greatly increase the fire danger. The dry conditions that fuel vastly destructive wildfires are a huge concern for homeowners, businesses, agriculture, and the tourist industry. Management of this limited resource is a very contentious issue in the Rockies, and will only worsen as water sources continue to be strained.

Environmental concerns are but one side of the complicated picture painted by excessive growth. There is also a pressing need to develop effective transportation networks to battle congestion. New residents from inside and outside the United States will spark social, cultural, and political change. Sustained economic growth will be tested by an uncertain future for natural resource extraction and the effects of a rapidly globalizing national economy. All of this will come to dominate life as the Rocky Mountain region continues to seek to define and redefine itself in the 21st century.

Suggested Readings and Web Sites

Books and Articles

Abbott, John S.C. *Kit Carson: Pioneer of the West*. 1973. New York: Dodd and Mead. The life and times of frontiersman and mountain man Kit Carson is chronicled in this exciting story set in the American West with a focus on the Rocky Mountains.

Adams, Andy. 1903. *The Log of a Cowboy*. Boston: Houghton Mifflin Company. Written by a real cowboy who became a writer, this book is the first one to describe the life of a cowboy on trail drives.

Borne, Lawrence R. 1983. *Dude Ranching: A Complete History*. Albuquerque: University of New Mexico Press. Traces the evolution of dude ranching in the Rockies from its earliest development up through the present day.

Ferguson, Gary. 2004. *The Great Divide: The Rocky Mountains in the American Mind*. New York: W.W. Norton. The author of this book weaves a series of colorful tales about some of the characters and places in the Rockies. He then uses these stories to trace the ebbs and flows of attitudes about the open spaces and vast expanses that these mountains represent that relate to a sense of freedom in North America.

Jordan, Terry G., Jon Kilpinen, and Charles F. Gritzner. 1997. *The Mountain West: Interpreting the Folk Landscape*. Baltimore: The Johns Hopkins University Press. This book depends upon log folk buildings of the American West to provide evidence that artifacts on the landscape such as barns, fences, and dwellings can help determine regional boundaries and expand our understanding of the origin and diffusion of regional cultures.

Michener, James A. 1974. *Centennial*. New York: Random House. Novel about the North American West, especially Colorado, that spans the settlement and development of the region from prehistory to the early 1970s, including Native Americans, trappers, traders, homesteaders, farmers, hunters, and speculators.

Wislizenus, F.A. 1969. *A Journey to the Rocky Mountains in the Year 1939*. Rio Grande Press. A historical travel guide to the Rockies filled with memorable stories of mountain adventures.

Web Sites

National Parks of the Rocky Mountains

Rocky Mountain National Park
http://www.nps.gov/romo/

Waterton-Glacier International Peace Park
http://www.americanparknetwork.com/parkinfo/gl/

Canadian Rockies
http://www.canadianrockies.net/

Mountain and Ski Resorts
http://www.mountaindestination.com/

City of Missoula, Montana
http://www.ci.missoula.mt.us/

City of Prince George, British Columbia
http://www.city.pg.bc.ca/

▲ **Figure 13.3** Dramatic vistas frame the Colorado Plateau

▶ **Figure 13.4** Alluvial fan formation in southern Utah

glacial meltwater that covered this part of the Pacific Far West during the retreat of Pleistocene glaciers. In much of the far eastern edges of this plateau, only grazing, mining, and scattered small farm agriculture is practiced. But across large parts of eastern Washington and extending down into Oregon on the southern side of the Columbia River is a more fertile part of the Columbia Plateau known variously as the "Palouse" (in east-central Washington), the "Inland Empire," and the "Great Columbia Plain." More precipitation falls here and consequently it is a wheat-growing area known for its agricultural productivity, large landholdings, heavily mechanized equipment, and fascinating historical development.

Perhaps no one has written more eloquently about the Palouse than historical geographer Donald Meinig. In his classic, *The Great Columbian Plain*, Meinig opens his first chapter by describing the boundaries and physical features of this complex region:

> In the far Northwest of the United States lies an unusual land. So sharply is it set apart from its surroundings that it can be recognized immediately, at a mere glance. Approached from any direction the visible change at its borders is striking: the forest thins then abruptly ends, the mountains lower then merge into a much smoother surface, and a different kind of country, open and undulating, rolls out before the viewer like a great interior sea.
>
> Donald Meinig, 1968, p. 3

Weather and Climate

Local and regional patterns and processes of weather and climate are controlled primarily by elevational differences. Much of the area is arid or semiarid, and air moving inland from the Pacific is a dominant influence over most of the region. The concepts of **rain shadow** and **orographic lifting** are vividly expressed here. Moisture-bearing winds blow in from the Pacific and drop their precipitation on the **windward** (western) sides of the Sierra Nevada and the Cascades. These high mountains form dry zones in the rain shadows located east of these mountain barriers. The eastward-flowing air then picks up new moisture as it passes over other mountains and is uplifted yet again.

This is illustrated by rainfall totals in various parts of British Columbia. Along the coast, communities on the western slopes of the Coast Mountains receive as much as 150 inches (380 centimeters) of rainfall each year. On the **leeward** (eastern) side of the mountains in British Columbia's Okanagan Valley, in contrast, annual precipitation averages less than 10 inches (25 centimeters) (Figure 13.5). These same processes are duplicated again and again throughout the Intermontane West, where mountain barriers block moisture from reaching places such as the Columbia Plain, eastern Oregon, much of Idaho and Utah, the Colorado Plateau, and southeastern California. Indeed, Nevada is the driest state in the United States, and Death Valley in the rain shadow of the Sierra Nevada in southeastern California is the driest locality in North America, with less than 3 inches (8 centimeters) per year.

▲ **Figure 13.5** Vineyards near Osoyoos Lake and the town of Osoyoos in British Columbia

1847, the beginnings of this unique western society were planted in the Salt Lake valley just west of the Wasatch Ranges by members of the Church of Jesus Christ of Latter Day Saints. After Joseph Smith was inspired to found the Mormon faith in upstate New York in the 1830s, the church grew rapidly. However, many Mormon doctrines and practices, including the practice of polygamy, differed dramatically from prevailing Christian practices and many Mormons were persecuted. In 1844, Smith was killed by an angry mob in Nauvoo, Illinois. After Smith's death, the Mormons decided to move to an unsettled area outside the United States to avoid further persecution and interference. Led west by their new leader, Brigham Young, thousands of Mormons migrated to an isolated region near the Great Salt Lake, which at that time was still part of Mexico. There they eventually established their religious headquarters in Temple Square in downtown Salt Lake City, forming a well-organized and well-regulated community (Figure 13.7).

Young's intention was to create a vast Mormon-controlled region known as Deseret, with Salt Lake City at its center. As more and more Mormons moved westward, Young sent groups to establish new settlements to the north and south of Salt Lake City. The church defined how settlements would be laid out thoughout Mormon country, thus creating a somewhat uniform landscape that covered an increasingly large area from the Snake River in the north to the Colorado Plateau, and then eventually as far south as the California city of San Bernardino and into Mexico (Figure 13.8).

Distinctive Mormon landscape features still stand out in much of this part of the Intermontane region. A series of small villages were laid out with privately owned agricultural fields located all around them. Some communal use of resources was common. Houses were built in a uniform Greek Revival style, adding another layer of uniformity and uniqueness to the desert landscape. Irrigation ditches bordered streets and property lines. Rows of Lombardy poplar trees added another dimension of order and sameness to Mormon landscapes. As geographers Richard Francaviglia and Donald Meinig pointed out, these and other features make Mormon country a fascinating place to study the impacts of religion on the land.

Farmers and Ranchers

Irrigated farming was carried on successfully by aboriginal peoples and later by Spanish and Mexican settlers long before the Mormons settled the Salt Lake Valley. Later, other farmers from the East and Midwest came into the area to try their hand at setting up additional irrigated farmlands at three different scales of settlement. These included: (1) farming large acreages such as the Columbian River Basin, the Snake River Plain, and the Gila Valley; (2) creating farmland near widely dispersed trading center locations such as towns along the upper Colorado River and its tributaries; and (3) building local irrigation systems for fields and pastures on individual farms that helped support cattle and sheep-raising operations on family farms.

Farmers from other parts of the United States and Canada relocated to the Intermontane West on wagon trains and the railroad. Some left their homes in places like North Dakota and Montana to move across the Canadian border to begin life anew. Experienced ranchers from the Great Plains also relocated to the American and Canadian Intermontane West, preferring to use land for sheep and cattle grazing that was located in dry basins surrounded by foothills and mountains, or on top of plateaus where level land made optimal grazing possible. Basque sheepherders were also a part of this ongoing migration stream into

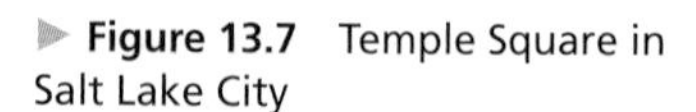
Figure 13.7 Temple Square in Salt Lake City

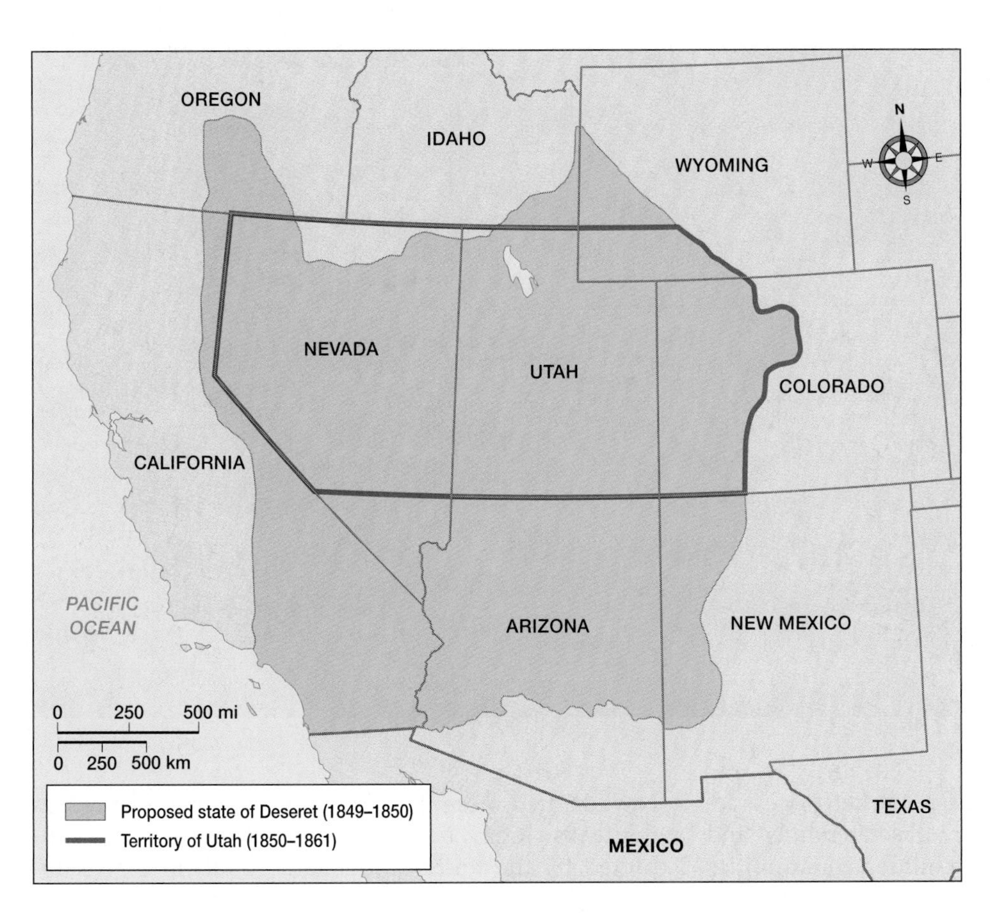

Figure 13.8 Extent of Brigham Young's planned territory of Deseret

the Intermontane West. Traveling to the American West from their homeland in the Pyrennees Mountains in western Europe, this unique group of migrants traveled by train to San Francisco and Sacramento on their way to find suitable grazing lands east of the Sierra Nevada Mountains and Cascade Ranges. The impact of Basque sheepherders on the cultural landscapes of places such as Reno, Nevada, and Boise, Idaho, remains today in Basque restaurants featuring delicious sheepherder bread, folk festivals, and distinctive architectural styles.

Lack of available water for urban development, intensive agriculture, and extensive settlement is a defining feature of this part of North America. Disputes about water ownership, including long-standing and contentious debates between Arizona and California over the right to use water from the Colorado River separating these two states, are characteristic throughout the Intermontane West. The greater Phoenix metropolitan region is also receiving attention as a center for retirees, seasonal migrants, and rapidly increasing numbers of year-round residents (Figure 13.9). In Quartzite, Arizona, not far from the California border, for example, tens of thousands of retirees and their recreational vehicles swell the local population from a small town of a few hundred fulltime residents in the summer to a densely populated city of RVs and campers arriving daily from northern climes during winter months (Figure 13.10). Places such as St. George, Utah, and Sedona, Arizona, are also attracting growing communities of retirees, many of whom reside there year-round.

Political Economy

Minerals and Other Natural Resources

The settlers who moved into the interior west were motivated by a variety of factors. Some sought agricultural land; some, like the Mormons, were searching for social isolation where they could practice their religion without persecution. Others were motivated by a search for mineral wealth. Early explorers brought back news of extensive deposits of gold, silver, copper, and other minerals, and news of these and other discoveries triggered large-scale in-migration as prospectors sought to strike it rich. Mining became the primary economic activity in many portions of the Interior West and adjacent regions. Mining companies and mine owners frequently dominated state and regional politics in Nevada, Colorado, and Montana.

As we saw in Chapter 11, mining is an economic activity subject to major **booms and busts**. The discovery of a new mineral deposit brings a significant influx of people, especially if prices for the mineral on the world market are high. Historically, mineral discoveries have attracted large numbers of relatively rootless persons intent on striking it rich quickly and moving away soon thereafter. A disproportionate percentage of these migrants were young men who had no families or left their families elsewhere with the intention of making quick profits and then returning home. Place ties to the mining communities were tenuous. As mineral

▶ **Figure 13.9** Urban sprawl in Phoenix

deposits played out and as world market prices declined, production decreased or ceased entirely and many people moved away. Many mining communities were abandoned entirely, becoming "ghost towns." For example, the silver-mining community of Virginia City, Nevada, had more than 20,000 residents at its peak in the 1880s. Today, barely a thousand residents are supported by the tourist attractions found in the reconstructed ghost town (Figure 13.11). Similar stories can be told about other Great Basin and mountain towns

▲ **Figure 13.10** On the road in Quartzite—RV heaven in summer (left) and winter (right)

▲ **Figure 13.11** Re-created touristy landscape in Virginia City, Nevada

such as Leadville, Colorado; Bisbee, Arizona; and Carson City, Nevada.

Today, the mining of precious minerals remains a significant although less important contributor to the economy of the interior west. In terms of total value, copper is the most valuable precious metal of the region, and nearby regions. Major copper deposits are found in Montana, southern Arizona, central Utah, and British Columbia. In recent years, production levels have dwindled because of foreign competition and lowered world market prices. Mines in the Butte region; in Ajo, Arizona, just north of the Mexican border; and elsewhere have been shut down. Idaho is the region's leading producer of silver and gold, which are also mined in Nevada, Montana, and Utah. Uranium is also an important mineral found in the Intermontane West.

Agriculture

While many early Euro-American settlers of the interior west moved to the region in search of mineral wealth, many others moved there in search of agricultural land. However, topography and lack of rainfall prevented or constrained agriculture. Much of the region is too rugged for farming, and many other places are very dry.

The principle of **appropriation** enabled farmers in the Intermontane West and other dry places to irrigate their fields and thereby to overcome the effects of aridity. Water law in the United States is derived from the water laws of England, which like the eastern United States has a humid climate with ample access to water. According to English water law, the person owning land adjacent to a watercourse has the right to use the water. Thus, the owner of lakefront or riverfront property can use the water in the lake or river to irrigate crops. This principle was used in the eastern United States, but early settlers soon realized that it was inappropriate for the western part of the country, where rainfall is scarce and available watercourses for irrigation are few. In the West, the principle of appropriation was developed. The appropriation principle grants the first person to put water to beneficial use the right to use it, regardless of location. Thus, a settler could build a ditch or canal from a distant lake or river to irrigate fields and claim appropriative water rights. The appropriation principle enabled settlers to build canals, irrigation ditches, and channels between permanent streams such as the Colorado, Snake, and Green rivers and their own lands.

of the region remains one of the poorest in North America. Thousands of residents of MexAmerica, especially Texans, the large majority of whom are of Mexican ancestry, live in **colonias**. Colonias are settlements located on the outskirts of cities or in rural areas in which many people live in trailers, wooden shacks, or other makeshift housing units. They are typically outside city limits and therefore not subject to minimum housing standards associated with nearby communities. Many colonias lack electricity, sewage treatment, and running water; and residents face poverty, disease, and inadequate transportation, education, and health care.

This rapid growth also has exacerbated the region's environmental problems. Much of MexAmerica is arid or semiarid, and the region's increasing population has meant an increased demand for water resources. Water is taken from the Rio Grande and other rivers faster than these streams can replenish themselves. At its mouth between Brownsville, Texas, and Matamoros, Mexico, the Rio Grande is often dry. Air pollution has become an increasingly serious problem made worse by smoke coming from the region's thousands of factories, waste dumps, and smelters. Some efforts are being made by the U.S. and Mexican governments to reduce pollution and to promote cross-border environmental protection.

The increased flow of trade and people across the boundary has also meant an increase in trafficking of drugs and other illegal goods. By 2000 it was estimated that two-thirds of the illegal drugs brought into the United States were brought across the U.S.–Mexican boundary. Illegal drugs are cached and concealed within shipments of legal goods and transported across the border. To combat the illegal drug trade and desperate illegal immigrants, the U.S. federal government employs more than 11,000 agents and spends nearly two billion dollars a year. Various other goods including firearms, weapons, and pharmaceuticals are also smuggled across the border in large quantities. Security at the international boundary has increased on both sides of the border since September 11, 2001.

And yet MexAmerica continues to grow in total population, economic development, and in its importance to the global economy and its nearest neighbor, the United States. With this growth has come an increasingly vocal, viable, and visible Latino identity. The importance of the distinct languages, cultures, and values of the region's diverse ethnic and racial groups are emerging in the first decade of the 21st century as one of the most observable measures of the growing impact of MexAmerica on North American culture. As pop music, movies, and television programs expand their content and marketing to appeal to Latino viewers, changes have come in the economic and political climate of the region as well. From mass acceptance of pop music icon Selena in the 1990s and other Mexicans in the public eye, the peoples and cultures of MexAmerica are becoming increasingly visible central players in the theater of North American cultural change. Indeed, as the media reminds us almost on a daily basis, the peoples, cultures, and places of MexAmerica provide a glimpse into the future of many other parts of North America.

Ample and often dramatic evidence exists that the region we have identified as MexAmerica in this text is truly a unique place. This evidence includes movies, music, and books about the region; Nostrand's experiences at the village of El Cerrito; and the work of other cultural-historical geographers such as Daniel Arreola (who continues to document the ever-changing Latino landscapes of San Antonio and South Texas) and Donald Meinig (who wrote the now classic book, *Southwest: Three Peoples in Geographic Change*). The region's distinctive transnational values and cultures, complex political economy, and dramatic environmental setting clearly set it apart from the rest of the continent.

Despite this uniqueness and isolation, however, this important region of North America is also "America becoming." As tens of thousands of people born in Mexico and many other places in Latin America continue to migrate to cities and towns far north of this borderland region in the coming years, these words written more than two decades ago seem prophetic:

> In a strange way, this strange space exists all over the American Southwest, for the Southwest is now what all of Anglo North America will soon be—a place where the largest minority will be Spanish-speaking. It's a place being inexorably redefined—in politics, advertising, custom, economics, television, music, food, and even the pace of life—by the ever-growing numbers of Hispanics in its midst. It is becoming MexAmerica
>
> Garreau, 1981, pp. 210–211

Suggested Readings and Web Sites

Books and Articles

Arreola, Daniel D. 2002. *Tejano South Texas: A Mexican American Cultural Province.* Austin: University of Texas Press. Award-winning book written by a cultural geographer who has spent his career documenting South Texas landscapes.

Arreola, Daniel D. 1995. "Urban Ethnic Landscape Identity," *Geographical Review* 85 (October): 518–534. A case study of the city of San Antonio, Texas, as a site of historical and present-day Mexican ethnic identity.

Ganster, Paul and David Lorey. 2007. *The U.S-Mexico Border in the Twentieth and Twenty-first Centuries.* Lanham, MD: Rowman and Littlefield.

Garreau, Joel. 1981. *The Nine Nations of North America.* Boston: Houghton Mifflin. Best-selling book written in a journalistic style that defines a set of new regions in North America that are based primarily on cultural, social, and economic criteria.

McWilliams, Carey. 1949. *North from Mexico: The Spanish-speaking People of the United States.* Philadelphia: J.B. Lippincott Company. Another book written by a journalist that tells the dramatic story of Mexican migration to the United States up to the late 1940s.

Meinig, Donald W. 1971. *Southwest: Three Peoples in Geographical Change, 1600–1970.* New York: Oxford University Press. A classic in regional geography that uses a "sequent occupance approach" to analyze the layered identities, cultures, and landscape impacts of Native American, Spanish, and Mormon peoples in the Southwest.

Murguia, Alejandro. 2002. *The Medicine of Memory: A Mexican Clan in California.* Austin: University of Texas Press. Powerful story of the lives and landscapes of one Mexican clan in California as an illustration of the lives of many other Latin American migrants in the United States.

Nostrand, Richard L. 2003. *El Cerrito, New Mexico: Eight Generations in a Spanish Village.* Norman: University of Oklahoma Press. Personalized geographic analysis of changes in people and place in one small village in the Southwest through the generations.

Nostrand, Richard L. 1992. *The Hispano Homeland.* Norman: University of Oklahoma Press. A richly detailed book on historical and present-day impacts of long-term Hispano families in the Southwest that laid the earliest foundation for the "Homeland Theory" debated by ethic and population geographers for the past decade or so.

Richardson, Chad. 2002. *Batos, Bolillos, Pochos, and Peldos: Class and Culture on the South Texas Border.* Austin: University of Texas Press. An anthropological and sociological analysis of peoples and places in deep south Texas told through a cultural and socioeconomic lens.

Wood, Andrew Grant, ed. 2004. *On the Border: Society and Culture Between the United States and Mexico.* Lanham, MD: Rowman and Littlefield.

Web Sites

All-American Canal
http://www.iid.com/water/works-allamerican.html

Kartchner Caverns State Park
http://www.desertusa.com/azkartchner/

Maquiladora—Mexican Manufacturing
http://www.mexconnect.com/business/mex2000maquiladora2.html

Festival of the West
http://www.festivalofthewest.com/

Mesilla, New Mexico
http://www.oldmesilla.org/

15 California

PREVIEW

Environmental Setting

- California consists of four major topographic regions: the Coastal Ranges, the interior valleys, the Sierra Nevada and Cascade ranges, and portions of the Great Basin.
- Most of California has a mediterranean climate characterized by warm, dry summers and mild and rainier winters. Generally speaking, coastal and mountain locations are cooler and wetter; inland and valley locations are warmer and drier.
- Much of California is subject to periodic earthquakes because of its location along the boundary of the North American and Pacific plates.
- California has long faced a scarcity of water because of its dry climate, large population, and high per-capita water demand.

Historical Settlement

- Prior to the arrival of the Spanish in the 18th century, California supported a wide diversity of Native American tribes.
- The Spanish and Mexican influence remains in many buildings, place names, and other features on the California landscape.
- Today California is home to about half the United States' immigrant populations, with large populations of Mexican Americans, Asian Americans, African Americans, and other minority groups.

Political Economy

- California was annexed by the United States in 1848, and the Gold Rush a year later brought California a large influx of Euro-American, Chinese, and other new settlers.
- California's natural resource base includes abundant agricultural land, large areas of frost-free climate, and petroleum and other minerals. California is the leading agricultural state in the United States, and is ranked first in the production of many fruits, vegetables, and other frost-intolerant crops.
- California's 20th-century economic growth can be attributed to its strong natural resource base, defense and government activity, technological innovation, entertainment and tourism, and the large influx and contributions of immigrants and domestic migrants.

Culture, People, and Places

- Due to early settlement patterns, most major California cities are located along the central and southern coast.
- Increasingly, California's population is concentrated in suburban areas, many of which are dominated by non-white ethnic and racial groups.

The Future of California

- California represents an important component of the Pacific Rim, and trade and cultural exchange with East Asia is increasingly important to the California economy.

Introduction

California! For some readers of this textbook, the word brings to mind postcard views of surfers, sandy beaches, and Hollywood signs on rocky hillsides. For others, images of California center on crime, traffic, and urban sprawl spilling across fertile farmlands. Our perceptions of California are often mixed because California is shaped by a remarkable diversity of physical environments, residents, economic systems, and culture. California is rich in its variety and extremes of people and place (Figures 15.1 and 15.2); indeed there are almost as many Californias as there are types of Californians. It is because of this unparalleled diversity and contrast that the California region offers so much opportunity and appeal to immigrants, businesses, and youth.

How, in the face of such diversity, can California be defined as a region? Four common geographic characteristics provide the answer. These include isolation relative to other North American regions, a high rate of suburbanization, a major role in cultural innovation, and links to the Pacific Rim. But while these traits provide a basis for geographers to define the region, they are not traits that necessarily help forge a common bond among residents of California—these traits are *not* like language or religion, which often provide a shared sense of identity. As a result, while California functions as a region, its many differences can often pull it apart. In this chapter, we discuss how California's four common characteristics help create and bring the California dream to life for many millions, but how the diversity of those dreams and aspirations pose major dilemmas for the future of the region.

◀ The Beverly Hills Hotel in Los Angeles, California: an architectural symbol for a region

▶ **Figure 15.1** Venice Beach in southern California

a)

b)

c)

▲ **Figure 15.2** California contrasts: (a) Mojave Desert, (b) High Sierra, and (c) Pacific coastline

Environmental Setting

Even a casual visitor to California cannot help but be aware of the region's great variety of landforms, climates, and vegetation. This is not surprising given that California spans a northsouth distance similar to the distance between Philadelphia and northern Florida. But there is more driving the diversity than California's large size—its location is also key. California's location at the boundaries of the North American and Pacific plates is responsible for the complex mosaic of mountains and valleys, as well as the earthquake and landslide hazards in the region. The location of the mountains relative to prevailing winds and pressure cells creates a diversity of climates ranging from hot dry deserts to areas receiving more than 300 inches (760 centimeters) of snow per year. The mosaic of landforms and climates in turn create a vast variety of ecological niches supporting vegetation types ranging from cactus to redwoods over 30 stories tall. Location of tectonic plates, resulting landforms, and climate zones therefore explains a large part of the California environment. A map of the region discussed in this chapter is shown in Figure 15.3.

Landforms

California can be divided into three large areas—mountains that parallel the coast, interior valleys, and the high mountains of the east (Figure 15.4). All of these landform areas are oriented to the boundary of the North America and Pacific plates, running northwest to southeast. The one major exception is the Transverse Ranges (also called the Tehachapi Mountains), which provide a boundary between the Los Angeles and the more rural and agricultural Central Valley.

Coastal Mountains and Valleys The northerly movement of the Pacific Plate relative to the North America Plate has created a series of lateral faults, also known as

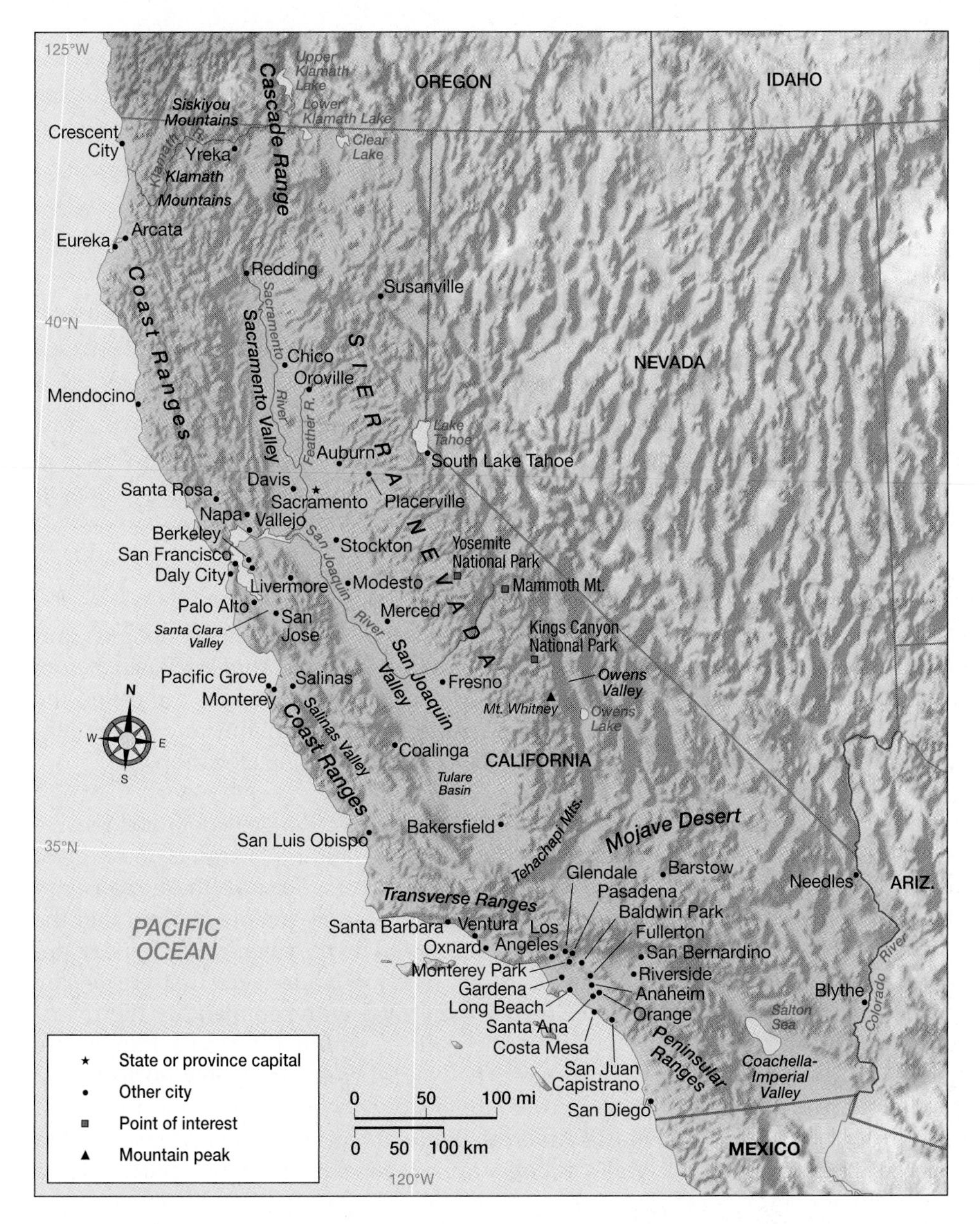

Figure 15.3 The California region

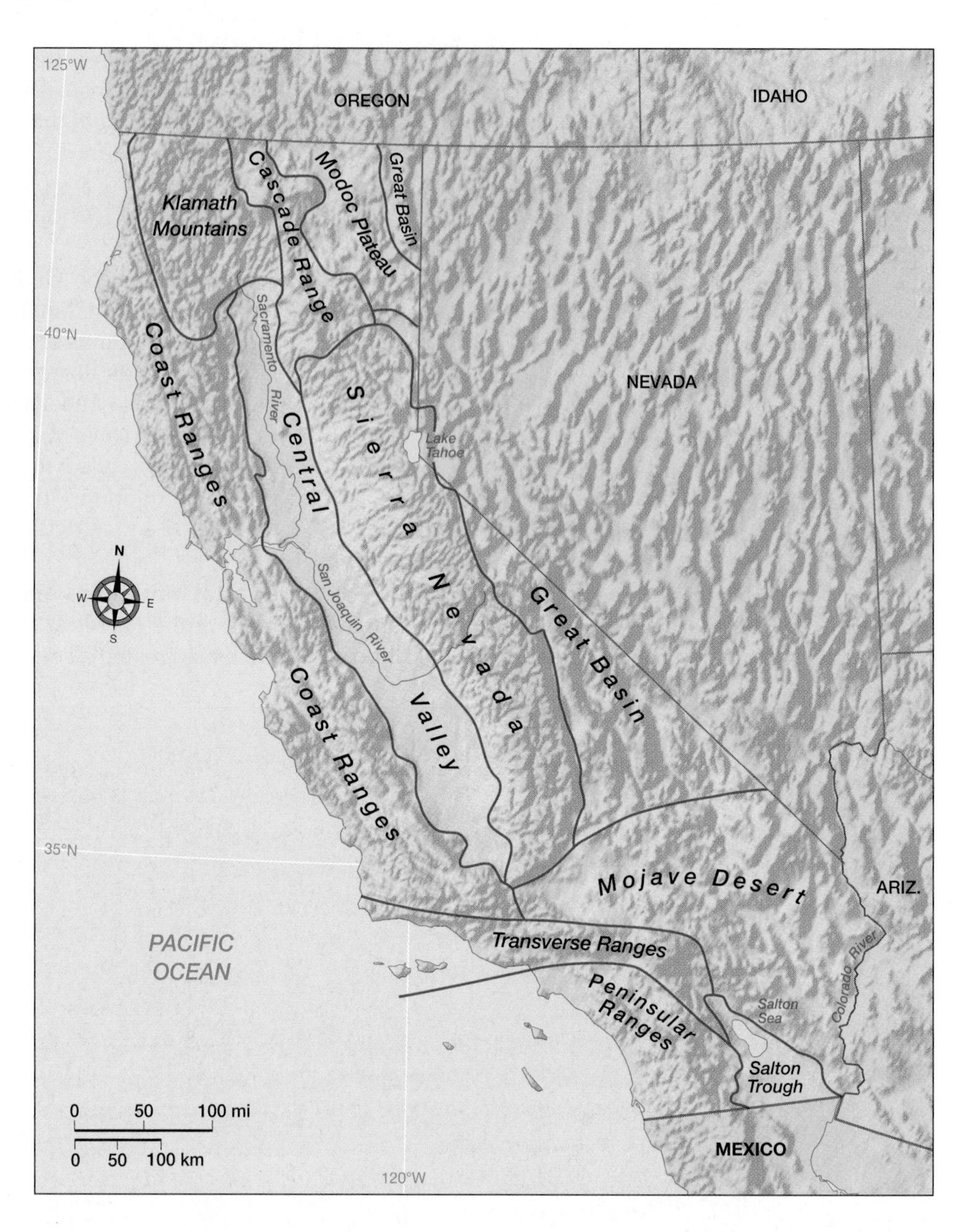

▶ **Figure 15.4** California's landform regions

strike slip faults along the California coast. The primary direction of movement along these faults is horizontal, with northwest movement on the Pacific Plate relative to southeast movement on the continental plate. The most famous of these faults is the San Andreas Fault, which bisects Los Angeles and San Francisco (and is discussed in more depth later), although there are many other lateral faults through the coastal region.

Although motion along these faults is mostly horizontal, there is enough vertical motion to build a series of mountain ranges and valleys from northern California to Mexico. These mountains are referred to as the "Coast Ranges," with portions of the ranges having other local names. Most of the Coast Ranges have elevations of 2000 to 4000 feet (610 to 1220 meters), although the Klamath Mountains in the northwest and the Transverse Ranges north of Los Angeles rise to nearly 9000 feet (2740 meters). South of the Transverse Ranges, the ranges move inland and traverse to the east of the Los Angeles Basin before extending into Mexico. Here the ranges are called the Santa Ana and San Jacinto Mountains and sometimes exceed feet 10,000 (3050 meters) in elevation.

Interior Valleys The Coast Ranges contain extensive valleys, and the center of California consists of one giant valley drained by two rivers. Because these rivers run through low-lying areas, some people assume that the valleys have been carved by the rivers. In fact, the opposite has occurred. The lowlands were first created by tectonic forces, then filled with sediment by the rivers.

Many of the valleys within the Coast Ranges are well known. The Napa-Livermore Valley north of San Francisco is the leading wine-producing region of North America. South of San Francisco, the Santa Clara Valley is better known as "Silicon Valley" because of its high-technology industry. The Salinas Valley southeast of San Jose is a major vegetable-producing area made famous in the stories of writer John Steinbeck. The landscapes of the San Fernando and San Gabriel valleys in the Los Angeles area form the backdrop for many television shows and movies.

All of these valleys are dwarfed, however, by the huge Central Valley. Approximately 430 miles (692 kilometers) in length and more than 50 miles (80 kilometers) across, this valley is the region's largest and most productive agricultural complex (Figure 15.5). Portions of the "Great Central Valley," as this area is often called, are referred to by the names of the rivers that flow through it. The Central Valley thus includes the San Joaquin Valley in the south and the Sacramento Valley in the north.

The High Sierra and Cascade Ranges To the east of the Central Valley are the Sierra Nevada, some 400 miles (645 kilometers) long and 60 miles (100 kilometers) wide. The east face of the Sierra rises abruptly, sometimes over 10,000 feet, to reach a crest with peaks that typically vary between 12,000 and 14,000 feet (3650 and 4270 meters) in height. These peaks include Mount Whitney, the highest peak in the lower 48 states. From this high crest, the western side of the range slopes gently downward until it merges into the foothills and flatlands of the Central Valley.

The rocks that make up the Sierra Nevada were created by an ocean plate that sank beneath the continental plate between 210 and 85 million years ago. Sediments carried downward by the ocean plate melted, then solidified as they rose to create granite bodies beneath Earth's surface, or to spew lava and ash at the surface. Most of the areas that now make up the Central Valley and Coast Ranges were underwater at this time. Erosion reduced the original Sierra to lowlands, before uplift in the last ten million years raised the mountains to their present heights. The reasons for the recent uplift are still widely debated. Glaciation of the High Sierra has since carved a dramatic landscape of scenic spires, deep canyons, and waterfalls that can be seen in national parks such as Yosemite and Kings Canyon.

Large, isolated volcanoes appear along the western flank of the Sierra near their northern end. These large peaks result from the subduction of the Juan de Fuca oceanic plate beneath the North American Plate and indicate the beginnings of the Cascade Range that continues further to the north (see Chapter 16). The highest and best known is Mount Shasta, more than 14,000 feet (4270 meters) in elevation.

Climate and Vegetation

Spatial variations in temperatures and precipitation across California are extreme. Regardless of location, however, the distribution of precipitation and temperature generally follow the same seasonal pattern. Most of California has a winter precipitation maximum with moderately cool temperatures and a long summer period with warm to hot temperatures and little or no precipitation. Summer days are generally clear. In winter, clear days alternate with longer periods of clouds and rain. This seasonal pattern is known as a mediterranean climate regime and is typical of all west coast locations around the world located at latitudes similar to that of California, such as Mediterranean Europe. Summer dryness is caused by the northerly movement of subtropical high pressure over the region. In winter, the jet stream and prevailing westerly winds migrate

◀ **Figure 15.5** Bales of cotton ready for ginning near Bakersfield in the San Joaquin Valley

southward, blowing ocean moisture and subsequent precipitation into the region.

Spatial variations in precipitation and temperature in California are driven by latitude, altitude, and distance from the ocean. In general, the climate is cooler and wetter in the north, along the coast, and in the mountains. The east side of the mountain experiences a rain shadow effect, so that the interior valleys become increasingly dry as one moves inland. Coastal California and the lowland valleys are generally frost-free year-round, while the mountains receive heavy snow. Indeed, some communities in the Sierra Nevada receive upwards of 300 inches (760 centimeters) of snow each winter. Snowmelt provides needed water to San Francisco, Los Angeles, and other cities and for agriculture.

Like climate, vegetation varies broadly with latitude, distance from the ocean, and elevation. Deserts occur at inland locations in southern California, including portions of the Central Valley, where warmer temperatures, rain shadow effects, and distance from ocean moisture create extremely dry conditions. Cacti and other drought-tolerant plants dominate the sparse vegetation. In areas receiving slightly more rain or cooler temperatures, grasses with extensive root systems and slender blades that minimize evaporation quickly sprout after fires or rainfall. These grasses are adapted to semiarid conditions.

Enough winter-spring rainfall to support shrubs and trees coupled with summer drought creates the distinctive mediterranean vegetation community. Adaptations of shrubs and trees to summer drought and fire include roots that are deep enough to reach water in the dry season, waxy coating on leaves to reduce transpiration, and thick bark to resist fires. Vegetation in California can also include a mix of dense shrubs (chaparral), cacti, and grasslands, all of which are adapted to periods of months without rainfall. This community occurs along the coast in southern California, on drier slopes throughout the Coastal Ranges, and in foothills along the Central Valley. The classic vision of California as a land of scattered dark green oaks with golden grasslands—the natural vegetation of Hollywood and Los Angeles—is one version of this mediterranean landscape, although one that is now threatened by the onslaught of a new virus called Sudden Oak Death, a fungal disease that attacks oaks and other large trees.

Greater rainfall, more fog, and cooler temperatures support a wide variety of coniferous forest types farther north along the coast, on more moist (generally windward) portions of the Coast Ranges, and throughout much of the Sierra Nevada. The needles on these trees allow them to rapidly initiate growth during the relatively brief period when moisture is available and there is also enough sunlight for photosynthesis. Gigantism, as in the case of the spectacular California redwoods, allows trees to sieve summer fogs that blow through the vast expanse of foliage, capturing as much as four inches of moisture in a single day!

Much of the natural vegetation that once covered California is now gone or disappearing. Extensive channelization has drained wetlands in the Central Valley, while irrigation and cultivation has altered vast areas of desert and grassland. Excessive salts added to soils and water via irrigation have converted once lush portions of the Central Valley into desert. In recent years, urbanization and exurbanization have caused the most significant loss of natural lands. Even undeveloped areas are changing due to climate change and the introduction of large numbers of exotic species that are displacing native plants.

Water Resources

California's large and growing population and its high-intensity agriculture depend on importing vast amounts of fresh water. "Imports" are necessary because more than 70 percent of the state's precipitation falls in its northern mountains, but more than 80 percent of California's water is used by urban residents, industries, and agriculture in the much drier south. Some of the state's most agriculturally productive farmland is located in places classified as deserts on vegetation maps, such as the San Joaquin Valley.

The discontinuity between water availability and water needs means that access to and control of water has been the single most important factor controlling the development of much of California. The result has been the construction of massive water projects that are among the largest and most extensive in the world. Although irrigation was used by Native Americans and later by early Spanish gardeners and vintners, large-scale irrigation became legal throughout the state with the passage of the Wright Act in 1887. Extensive dam building by the federal government in the mid-20th century made possible the storage and delivery of vast amounts of water. More than 90 percent of the state's crops are now grown using irrigation and almost the entire population of the Los Angeles metropolitan area uses water imported from far outside the basin.

The impacts of this dam building, channel construction, and water diversion are profound. For example, despite protests from John Muir and the newly organized Sierra Club, the Hetch-Hetchy Valley, a valley akin to Yosemite in splendor, was flooded to deliver water to San Francisco along a 186-mile- (300 kilometers) long aqueduct. Three other major water rearrangement systems include the California Water Plan, with its snakelike California Aqueduct extending from Lake Oroville in the north to Los Angeles; the Los Angeles Water Project, which siphons off water from the far-distant Owens Valley (and the dream of developer-politician William Mulhulland); and the federally funded Central Valley Project. All of these projects and others transfer huge quantities of water from the wetter north to the much drier south. Many northern Californians resent the large-scale transfer of "their" water to the southern part of the state, and accuse southern Californians of wasting the water for frivolous use in swimming pools, suburban lawns, and fountains. There may be some truth to this statement—the average resident of Los Angeles uses almost twice as much water as the average San Francisco resident.

There can be no doubt that careful protection and enhanced conservation of water resources will be essential to California's future. The region's population is projected to double or perhaps triple in the next 100 years. At the same time, water supplies might decline due to global warming, and political battles could lead to loss of water imported from other areas (Figure 15.6). Regardless of the specifics, it is clear that California's survival depends on finding new and hopefully more environmentally sensitive methods for reducing consumption and providing unpolluted water to its residents, farms, and factories.

Because of its location astride the boundary between the North American and Pacific plates, California is one of the most seismically active places in North America. San Francisco experienced a catastrophic earthquake in 1906, and the aftermath of the 1989 Loma Prieta earthquake that disrupted baseball's World Series was witnessed by millions on TV. The first written account of California's seismic activity was recorded in 1769 by Gaspar de Portola, who testified to "a horrifying earthquake which repeated four times during the day."

The earthquakes result from the slippage of the North American and Pacific plates as they grind past one another, splintering Earth into many slivers along a 700 mile contact zone. These many faults are known as the San Andreas Fault System, which has been active for the last 10 million years and extends from the Gulf of California to Point Arena far north of San Francisco. Along these faults, slivers of western California from

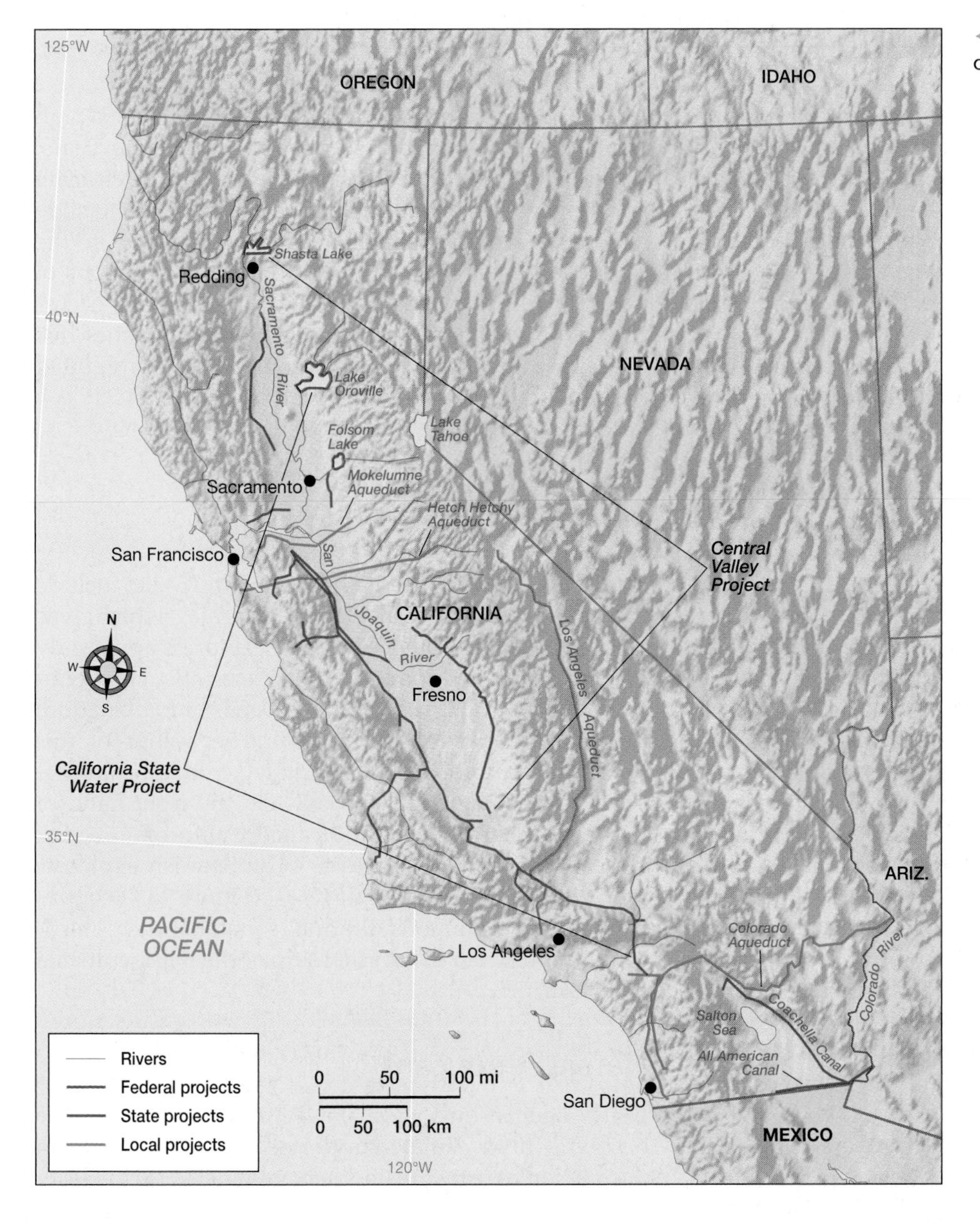

Figure 15.6 California's water conveyance systems

BOX 15.1 The Northridge Earthquake

On January 17, 1994, many southern Californians woke up at 4:30 A.M. to the jolting and jerking of a major earthquake. Although residents of coastal California have grown accustomed to the notion that earthquakes are a normal part of life, many were awakened that winter's morning by the thought that this 6.7 quake could be the "big one" they had been warned about for years.

This Northridge Earthquake (as it came to be known) was the first major quake to strike directly under an urban area of the United States since the Long Beach earthquake in 1933. It occurred along a blind thrust fault which produced the strongest motion ever recorded on instruments under an urban setting in North America. Major freeways collapsed, parking structures crashed apart, and apartment buildings pancaked to the ground. Wood-framed houses in Los Angeles, especially the San Fernando Valley and Santa Monica areas, were especially hard hit. The intense vertical and horizontal accelerations of movement took their toll on all kinds of buildings, but those with lower level parking garages and first floors that had been built directly on the ground suffered the most damage.

Newscasters outside the area continued to broadcast the latest updates of frightening stories. Many predicted that that this really had been the big one that geologists had been expecting for years. Surely, they postulated, southern California was history—a place damaged beyond repair that was destined never to be rebuilt again.

However, cities have a habit of emerging from disaster. Examples are: the tsunami that wiped out much of Crescent City, California, after the Alaska earthquake of 1964; the devastation wrought by the great fires in Chicago and San Francisco; or the Galveston, Texas, great storm that killed more than 6000 people in 1900. And, of course, most of us remember the incredible devastation and flooding caused by Hurricane Katrina along the Gulf Coast in 2005. Instead of relocating, the residents of most such places rebuild with caution, optimism and hope.

San Diego to Mendocino are moving northward at an average velocity of 5 cm per year. Unfortunately, these movements occur in abrupt jumps, with decades or centuries of quiet followed by large, damaging jolts when meters of movement occur at one time (Box 15.1). It is these jolts which have led to the totally false myth that all of California will someday fall into the Pacific Ocean.

Unfortunately, the state's densest populations and largest urban areas are located on top of or alongside this active fault zone, which increases the potential for loss of life, injury, and damage to property. The State of California created maps of these "Special Studies Zones" and now requires special construction techniques for new structures in them. People selling homes in California must inform potential buyers if the house is in a seismic hazard zone, but don't be misled if a seller hypes a house sale based on its location in a "safety zone." Moreover, not all earthquakes in California occur along the San Andreas system. Towns such as Coalinga located on the western edge of the San Joaquin Valley and Oroville on the eastern edge of the Sacramento Valley in northern California have experienced damaging earthquakes in recent years. Anyone living in California should be prepared for an earthquake.

In contrast to its high seismic hazard, California is only rarely affected by severe weather. In fact, the more populated areas of California have little day-to-day weather variation, which is part of the region's appeal for residents, tourists, and farmers. Hurricanes and tornadoes are rare in California, although moisture from Pacific hurricanes occasionally brings major rainfall and flooding to the region.

California is also known for its extreme landslide hazard in many areas. These landslides typically result from a combination of steep topography and high winter rainfalls, often associated with El Niño years. In addition, loss of vegetation due to fires or clearing by humans and seismic activity can contribute to slide occurrence. Annual damage from landslides in the state of California is estimated to be in the millions of dollars, with the great majority of this damage occurring in hilly or mountainous urban areas. Most of this damage could be avoided if zoning laws prohibited building in landslide-prone locations.

Historical Settlement

The name "California" was first used in 1542 to include the area we now call Baja California, which was thought to be an island. The early explorers apparently learned the word from a popular romantic novel entitled *Las Sergas de Esplandian* (The Deeds of Explanation) published in Seville about 1510. The author of this book, Garcia Ordonez de Montalvo, romantically described the imaginary land as: "On the right hand of the Indies there is an island called California . . . very close to the Earthly Paradise." The Spanish explorers who voyaged up the Pacific Coast (Figure 15.7) continued to use the name "California" for all of Spain's territorial claims from the southern tip of the Baja peninsula to Alaska.

Native Americans

It has been estimated that California's rich resource base and mild climate supported a Native American population density greater than anywhere else in North

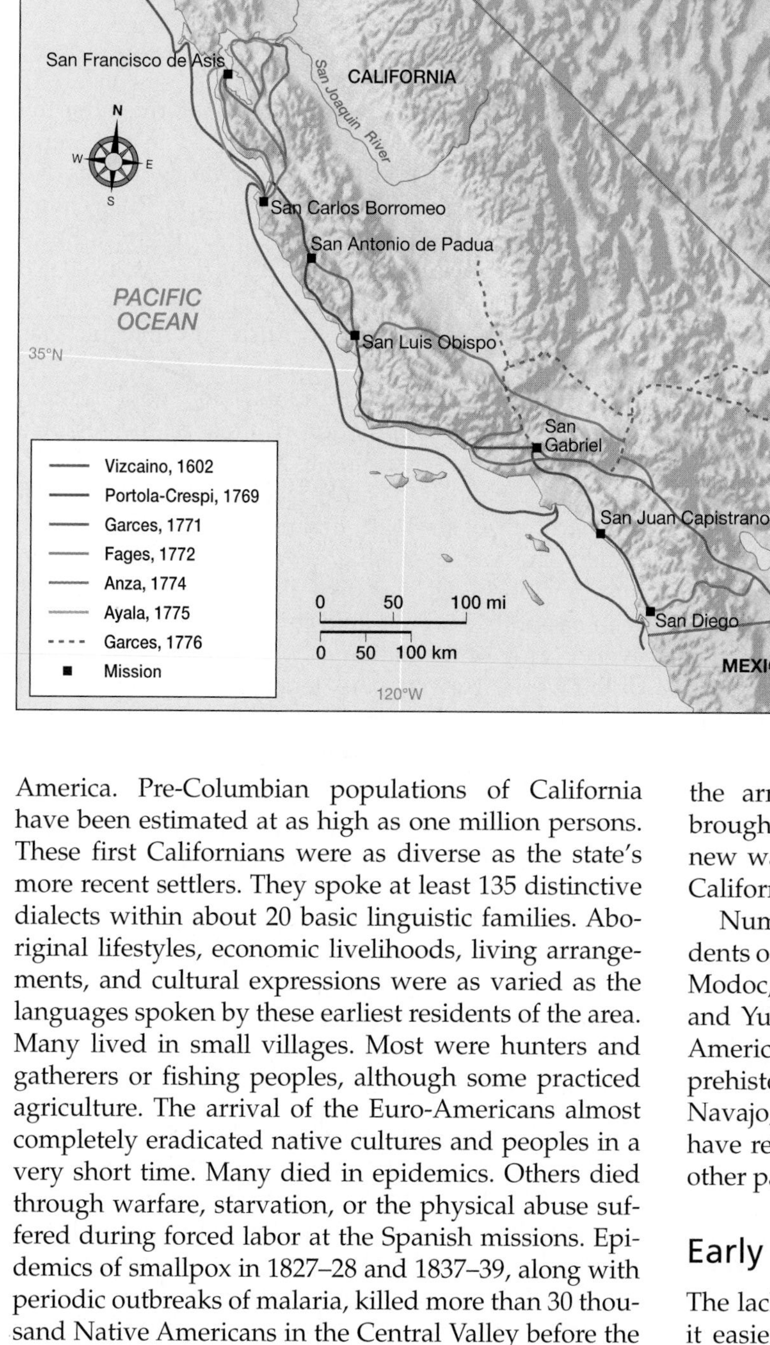

Figure 15.7 Early exploration routes in California

America. Pre-Columbian populations of California have been estimated at as high as one million persons. These first Californians were as diverse as the state's more recent settlers. They spoke at least 135 distinctive dialects within about 20 basic linguistic families. Aboriginal lifestyles, economic livelihoods, living arrangements, and cultural expressions were as varied as the languages spoken by these earliest residents of the area. Many lived in small villages. Most were hunters and gatherers or fishing peoples, although some practiced agriculture. The arrival of the Euro-Americans almost completely eradicated native cultures and peoples in a very short time. Many died in epidemics. Others died through warfare, starvation, or the physical abuse suffered during forced labor at the Spanish missions. Epidemics of smallpox in 1827–28 and 1837–39, along with periodic outbreaks of malaria, killed more than 30 thousand Native Americans in the Central Valley before the late 1830s. This demise was followed soon thereafter by the arrival of Gold Rush miners, many of whom brought new diseases, new technologies, and yet more new ways for exploiting both the natural resources of California and its native peoples.

Numerous legacies were left by these earliest residents of the state, including place names such as Colusa, Modoc, Mono, Napa, Shasta, Tehama, Tuolumne, Yolo, and Yuba counties. In 2000 there were 333,511 Native Americans in California. Many are descendants of local prehistoric cultures, whereas many others belong to the Navajo, Dakota, Cherokee, and other nations. Many have relocated to urban and suburban California from other parts of the United States.

Early Spanish and Mexican Settlement

The lack of a cohesive aboriginal tribal structure made it easier for incoming groups to take control of both land and people in early California. Juan Rodriquez

Cabrillo sailed into the harbor at San Diego in 1542, before Cartier arrived in Quebec and nearly 80 years before the Pilgrims landed at Plymouth Rock. Soon thereafter, the Spanish administrators divided the region into Baja California in the south and Alta California in the north. During the later Mexican control of the region from 1821 to 1846, Alta California was claimed to include all the land between the Continental Divide and the Pacific Ocean—the present day states of California, Nevada, Utah, and Arizona plus parts of Wyoming, Colorado, and New Mexico.

During the 16th through the 19th centuries, California was visited by Europeans from several different countries of origin. During the late 1500s, Sir Francis Drake and other English explorers arrived and attempted to secure bases for trade in California. Added to this threat to Spanish dominance in the region was the arrival of the Russians, who extended their Alaskan and Siberian fur trade to California's north coast in the late 1700s. This interest in the area by other strong European powers encouraged the Spanish government to tighten its grip on California through a combined mission and military system of control. Missions were designed with two main purposes in mind: to convert local populations to Roman Catholicism and to use new converts as agricultural laborers. A typical mission consisted of a church, workshops, storerooms, living quarters, and agricultural fields surrounding it. The first Franciscan mission was set up at San Diego in 1769, followed by the establishment of the first presidio (military center) at Monterey in 1770. The central coast location of Monterey and its mission at nearby Carmel made it ideal as a site for trade throughout the region and for the arrival of new immigrants. Monterey soon became the capital of the Spanish colony.

In the following decades, the fertile valleys near the coast became popular sites for new Spanish settlements. Other missions and their associated pueblos were soon established in places such as San Gabriel Archangel near today's city of Pasadena, San Luis Obispo, San Francisco, and San Juan Capistrano, followed by missions and pueblos in San Jose in 1777 and Los Angeles in 1781. Indeed, many of California's major coastal cities owe their origins to the Spanish missions.

Spanish settlements were carefully located based on natural resource availability and advantages of soils, slope, and water for irrigation. Irrigation ditches were dug; fields and vineyards planted; and horses, sheep, and cattle herds brought in to graze the pastoral vegetation. The first longhorn cattle were brought from Mexico into San Diego; and by the early 19th century, livestock-raising was a productive industry. An 1811 census of California mission livestock reported over 67,000 cattle and oxen, over 100,000 sheep, and nearly 20,000 horses. As time went on, however, the missions began to decline. An earthquake in 1812 destroyed the mission at San Juan Capistrano and damaged several others. Still others were damaged by floods or fires. In the end, however, it was politics that closed down the mission system. In 1821, Mexico gained its independence from Spain. Confusion and unrest were widespread during this time of transition, which ended in 1848 when California became a territory of the United States. The rapid influx of Mexican and American settlers, many of whom settled Spanish and Mexican land grants known as ranchos, also hastened the demise of the mission system. Moreover, the Mexican government halted the use of forced Native American labor by the missions, which undercut the mission's economic base. Today, many of the mission buildings of California still stand and some, including San Juan Capistrano and Santa Barbara, are well-known tourist attractions.

The Gold Rush

The United States gained control of California after the Mexican War, and gold was discovered in the Sierra Nevada foothills in 1848. Only a year after gold was discovered at Sutter's Mill on the American River, more than 40,000 newcomers had arrived. California's Euro-American population increased from less than 40,000 in 1840 to 100,000 in 1850, at which time it became a state. The impact of the Gold Rush on California was phenomenal.

The immediate effect of the news of "gold in them thar hills" was depopulation of most of the region's towns and cities. According to an account by John Fremont during a visit to San Francisco in June, 1848: "All or nearly all of its male inhabitants had gone to the mines. The town, which a few months before was so busy and thriving, was then almost deserted." In less than a month, news of gold in California had reached every corner of the nation as well as many other parts of the world. Newspapers, letters, and posters spread the word far and wide. The front page story in the Hartford, Connecticut *Daily Courant* on December 6, 1848 is typical. It reported that:

> The California Gold Fever is approaching its crisis. We are told that the new region that has just become a part of our possessions is El Dorado after all. Thither is now setting a tide that will not cease its flow until either untold wealth is amassed, or extended beggary is secured. . . . Soldiers are deserting their ranks, sailors their ships, and everybody their employment, to speed to the region of the gold mines.

Few of the many prospectors who sought gold in California found enough to get rich. Yet the Gold Rush, along with the completion of the transcontinental railroad, had several major long-term impacts on California's settlement. The first and most profound one was an increase in the state's population. Between 1850 and 1900, the population grew from 93,000 to 1.5 million. The cities of San Francisco and Sacramento emerged as major supply points and by 1900, San Francisco had

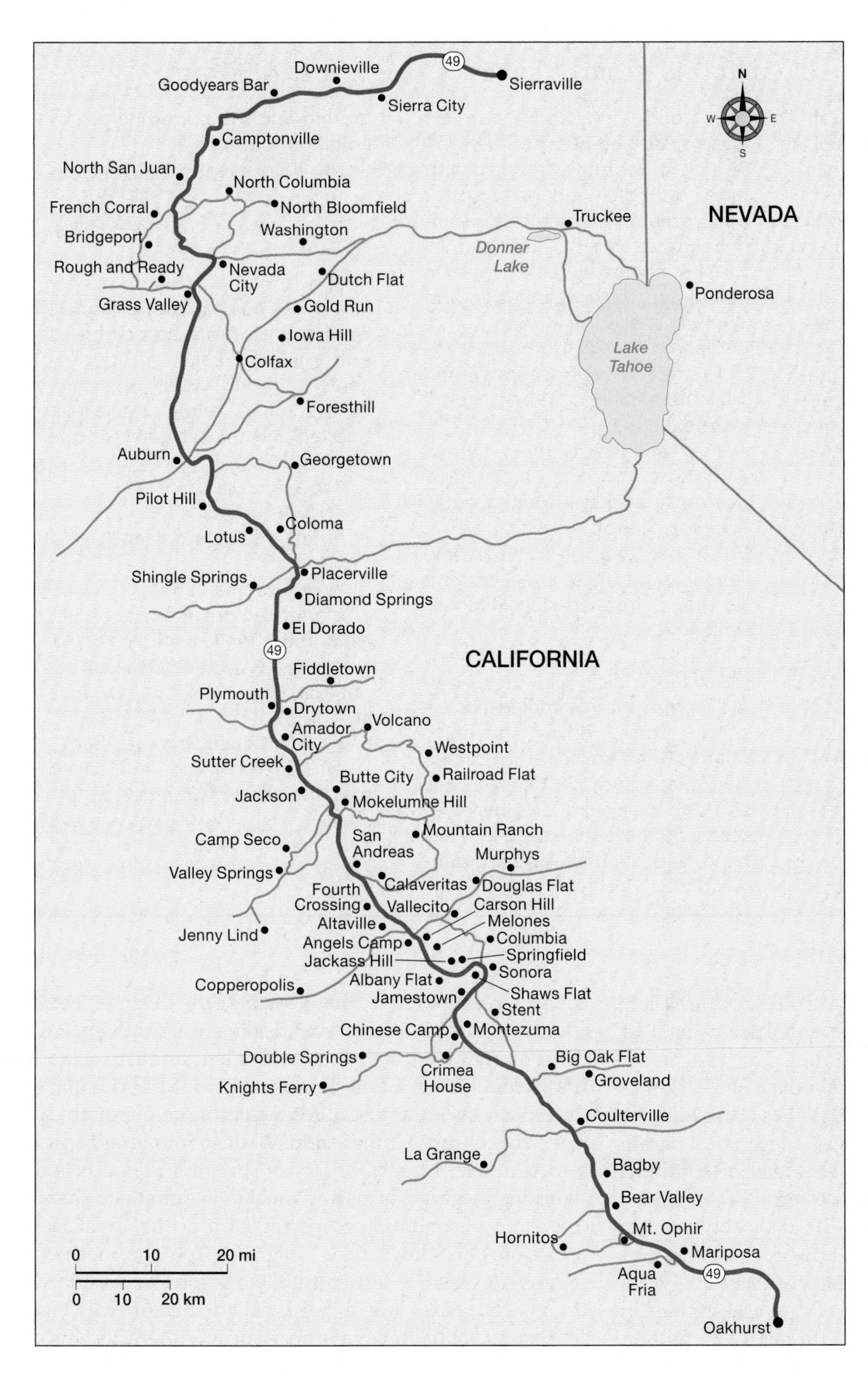

◄ **Figure 15.8** California Gold Rush settlements

became the major metropolitan center in the western United States. Numerous other new settlements were established in gold camps (Figure 15.8). These towns were, in turn, supplied by new and growing agricultural towns and cities in the fertile valleys below. Many settlers who had originally intended to move to Oregon decided instead to try their luck in California. The Gold Rush also contributed to the federal government's decision to route the transcontinental railroad through Sacramento to the port of San Francisco.

Political Economy

Why California?

Although the Gold Rush brought in settlers, and the transcontinental railroad brought steady growth to the Golden State, much of the non-urban agricultural land in the state remained isolated and relatively empty throughout the 19th century. The arid and semiarid climate that hindered agricultural development was one

BOX 15.2 A First Glimpse of California (from Steinbeck's *The Grapes of Wrath*)

American author John Steinbeck captured the trauma of the Depression-era migration of desperate farmers from the drought-ridden Great Plains states to California more dramatically than any other American writer. In his classic book, *The Grapes of Wrath*, Steinbeck chronicles the story of the Joad family as they traveled from their home in Oklahoma to California's Central Valley. In one of the most dramatic moments in the book, the Joads reacted like this when they first looked out across the immense valley they would soon call home:

> "I want ta look at her." The grain fields golden in the morning, and the willow lines, the eucalyptus trees in rows.
>
> Pa sighed, "I never knowed they was anything like her." The peach trees and the walnut groves, and the dark green patches of oranges. And red roofs among the trees, and barns—rich barns. Al got out and stretched his legs.
>
> He called, "Ma—come look. We're there."
>
> Ruthie and Winfield scrambled down from the car, and then they stood, silent and awestruck, embarrassed before the great valley. The distance was thinned with haze, and the land grew softer and softer in the distance. A windmill flashed in the sun, and its turning blades were like a little heliograph, far away. Ruthie and Winfield looked at it, and Ruthie whispered, "It's California."
>
> Winfield moved his lips silently over the syllables. "There's fruit," he said aloud.
>
> Casy and Uncle John, Connie and Rose of Sharon climbed down. And they stood silently. Rose of Sharon had started to brush her hair back when she caught sight of the valley, and her hand dropped slowly to her side.
>
> Tom said, "Where's Ma? I want Ma to see it. Look, Ma! Come here, Ma." Ma was climbing down slowly, stiffly, down the back board. Tom looked at her. "My God, Ma, you sick?" Her face was stiff and putty-like, and her eyes seemed to have sunk deep into her head, and the rims were red with weariness. Her feet touched the ground and she braced herself by holding the truck side.
>
> Her voice was a croak, "Ya say we're acrost?"
>
> Tom pointed to the great valley. "Look!"
>
> She turned her head and her mouth opened a little. Her fingers went to her throat and gathered a little pinch of skin and twisted gently. "Thank God!" she said. "The fambly's here." Her knees buckled and she sat down on the running board.
>
> "You sick, Ma?"
>
> "No, jus' tar'd."
>
> "Didn' you get no sleep?"
>
> "No."
>
> "Was Granma bad?"
>
> Ma looked down at her hands, lying together like old lovers in her lap. "I wisht I could want an' not tell you. I wisht it could be all—nice."
>
> Pa said, "Then Granma's bad?"
>
> Ma raised her eyes and looked over the valley. "Granma's dead."

Source: John Steinbeck, *The Grapes of Wrath*. 1939. New York: The Viking Press. pp. 309–311. Used by permission of the Penguin Group (USA), Inc.

obstacle. In addition, the state boasted few natural harbors. Moreover, California was isolated from most of the other major population centers of North America and the rest of the world. It is 6000 miles (9656 kilometers) from East Asia, and over 1000 miles (1610 kilometers) of rugged mountains and deserts separate California from heavily populated areas of southern Mexico and the eastern and midwestern United States.

In 1900 Los Angeles had barely 100,000 people; today Los Angeles is the second largest city in North America and one of the 20 largest cities in the world. By the 1960s, California surpassed New York as the largest state in population in the United States with nearly 30 million residents. One of every 9 Americans today lives in California and soon, one of every 13 Americans will live in the Los Angeles area. If California were an independent country, its economy would rank seventh in size in the world, topping the economies of major powers such as Italy, Brazil, and Spain. The emergence of California as an important center for economic growth, innovation, productivity, and prosperity is one of the major stories of global economic development in contemporary times.

Major factors have contributed to California's tremendous growth, including its agricultural and natural resource base, the growth of the defense sector during and after World War II, the strong influence of the entertainment and high technology industries, and its Pacific Rim location. Immigrants to California have also been essential to California's economic growth. In the 19th century, Chinese and Mexican immigrants provided much of the labor for constructing railroads and working the large farms and ranches beginning to shape the state's economic prosperity. During the Great Depression, Dust Bowl "Okie" migrants from the Southern Plains provided an agricultural work force as described in Box 15.2. After World War II began, they provided an industrial labor force, as did African Americans from the rural southern states. Since 1965, when American immigration laws were liberalized, large numbers of immigrants from Mexico, Central and South America, Asia, and elsewhere have moved to California.

California Agriculture and Natural Resources

California's many natural resources were critical to its early economic growth in the 20th century. First and foremost, California valleys contain large tracts of flat

◄ **Figure 15.9** Harvesting tomatoes on a large corporate farm in the San Joaquin Valley

farmland with high-quality soils and long growing seasons. Once irrigation water was brought to the Los Angeles basin in the early 1900s, and to other parts of the state after 1930, farmers began to specialize in fruits, vegetables, and other frost-intolerant crops. These crops have been instrumental in making California the leading agricultural state in the United States for many years.

California's agriculture is highly diverse. The Golden State produces all of the United States' dates, figs, kiwi fruit, olives, and nectarines. It produces nearly all of its apricots, almonds, walnuts, and avocados and a large share of its grapes, lettuce, plums, strawberries, and tomatoes. In fact, for each of these crops the leading producing county in California outproduces each of the other 49 U.S. states. From south to north, the largest and most important production areas are the Imperial-Coachella Valley in southeast California; Los Angeles and other south coast basins; San Francisco and mid-coastal regions; and the Central Valley. The heavily irrigated Imperial-Coachella Valley specializes in producing tomatoes, carrots, and other vegetables that cannot tolerate frost such as dates, despite the fact that this region averages only three to five inches (7 to 13 centimeters) of rain per year.

From 1900 until 1950, Los Angeles County was the leading agricultural county of the United States. Today, most of the agricultural land in the Los Angeles Basin has been converted to urban use, although the area still produces significant vegetable and citrus crops. Indeed, small areas like those beneath power lines are cultivated, with fruits and flowers sold to local consumers. Much of the farmland near San Francisco has also been urbanized, although the Salinas, Napa-Livermore, and Santa Clara valleys remain important agricultural areas.

Today, the leading agricultural area of California is the Central Valley, which includes the valleys of the San Joaquin River to the south and the Sacramento River to the north. Fresno County, in the San Joaquin Valley, is now the leading agricultural producer in the United States. The San Joaquin Valley produces 90 percent of the United States' grapes and many other crops, including cotton, grains, fruits and vegetables (Figure 15.9). To the north, the rainier Sacramento Valley is less dependent on irrigation and produces large quantities of rice, almonds, tomatoes, sugar beets, wheat, and other crops.

Agriculture in California is characterized by a unique set of problems relative to other North American settings. Whereas the large majority of farms east of the Rockies are operated by their owners, only about half of California's farms are owner operated. The remainder are owned by corporations or absentee landowners and operated by hired employees. Since the 19th century, agricultural interests have relied on labor from Latin America, East Asia, and other parts of the United States, including the workers who moved from the Great Plains states during the Dust Bowl of the 1930s. During that period, the state of California tried unsuccessfully to ban the in-migration of "Okies". Not until World War II, with its need for a labor force in defense plants, did Dust Bowl migrants achieve a modicum of acceptance in California. More recently, California growers have turned to Mexico for agricultural labor. By the 1960s, thousands of Mexicans migrated from farm to farm over the course of the agricultural year, cultivating and picking grapes, lettuce, and other fruits and vegetables. The activities of labor leaders such as Cesar Chavez along with media accounts called attention to the substandard housing, substandard wages, and inadequate

▲ **Figure 15.10** Out of town visitors to San Francisco's Lombard Street, which is claimed to be the "crookedest street in the world"

views across the bay to San Francisco, the Golden Gate Bridge, and the Pacific Ocean.

Cities of the Central Valley

While the large majority of urban Californians live near the coast, cities in the interior are also experiencing rapid growth. Indeed, such cities boast cheaper housing, shorter commutes, and less smog and traffic than do coastal cities, attracting many new residents. Agriculture was responsible for the initial development of urban areas of inland California, especially in the Central Valley. The largest interior city of California is the state capital of Sacramento, whose metropolitan area contains about 1.5 million people. While government has replaced agriculture as the major economic base of Sacramento, food processing remains important, as does the aerospace industry. Nearby Davis is the site of the University of California at Davis and a leading center for agricultural research. Many other towns of the Central Valley have all experienced rapid growth in recent years. While maintaining their agricultural roots, each city is working to develop a separate identity to differentiate itself from the crowd. Thus we see Bakersfield, which was a leading destination for Dust Bowl migrants prior to World War II, using this heritage of Southern Plains culture to establish itself as the leading country music center of the state—and the home of stars such as Buck Owens, Merle Haggard, and Dwight Yoakam. Nearby Delano celebrates its history as the community from where Cesar Chavez began organizing migrant farm laborers in the 1960s. And Merced won the state political "lottery" by gaining the right to host the newest University of California campus. The decades to come will be a fascinating time to watch the evolution of these and other interior cities as they continue to change and redefine themselves in order to capture and maintain the business, wealth, and culture of specific target populations. Also playing a major role in these trends are the Sacramento Valley cities of Redding, Chico, and Marysville/Yuba City.

The Future of California

California has evolved into one of the wealthiest and most productive regions of North America. Its participation in the ever expanding global economy of the

Pacific Rim, the diversity of its people, landscapes, resources, and products; and its role as the world's leader in pop culture make this part of the United States unique and economically powerful. For almost a century, California has been a trendsetter in terms of popular culture. The California way of life has affected popular slang, hit songs, fashion, movie preferences, and even food choices in the world at large. For better or worse, this trend will continue. The powerful impact of media networks, still centered in California and heavily controlled by Californians, will ensure that this cultural impact continues well into the future. Despite a slowdown in economic growth in recent years, a lingering sense of doom about the potential for destruction by a catastrophic earthquake, and the ongoing struggle to find new ways to live in a diverse society and cope with high housing costs, smog, and traffic congestion, California expands and thrives.

There has been a continuous swing of the pendulum among power centers in this region of North America. During the Mexican era between 1822 and 1846, the Spanish villages of southern California held the edge. The Gold Rush swung the pendulum northward to San Francisco and Sacramento. Then, health seekers, the discovery of oil, the success of citrus production and other irrigated agriculture, and land development efforts once again swung the power to the south to the Los Angeles Basin. It remains there at the present time and, at least in terms of overall population, will probably remain there for the foreseeable future.

The majority of the residents of this populous state remain deeply divided into two distinctive groups—"northern Californians" and "southern Californians." No agreement has yet been reached on where the exact boundary between these two parts of the state lies. Some say the Tehachapi Mountains, others put it at Santa Barbara. Still others insist it is located at Monterey. Some have marked this important boundary according to newspaper preferences. Residents who read the *Los Angeles Times* are true southern Californians and those who read the *San Francisco Chronicle* are northerners.

These strong feelings of subregional identity are based on the fact that the bulk of the state's population is in the south, while the bulk of its natural resources are in the north. Feelings about this issue run so deep that there have been regular attempts by certain groups to secede from the state over the years. As early as 1859, southern Californians submitted a separatist plea that actually passed the legislature but was buried in Congress because of the crisis of the impending Civil War. During the 1990s, a bill to separate California into three states was introduced into the California legislature. Large signs just off Interstate 5 near the Oregon border still cry out for the approval of a separate state in far northern California and southern Oregon (Figure 15.11)!

▲ **Figure 15.11** State of Jefferson sign on hay barn in Siskiyou County, California

Despite these regional differences, there can be no doubt that Californians are inexorably linked by water systems, a unique climate, economic interdependence, and cultural identity. And these internal linkages are intensified by connections between Californians and the outside world. Today, transnational linkages also shape the lives of the region's newest immigrants. E-mail and other Internet connections and family visits, for example, regularly unite California-based Mexican residents with their families south of the border. These dual citizens also have created a hybrid mix of identities and perceptions about being and becoming a Californian.

In addition, continuing population growth, especially in the state's largest urban areas, will undoubtedly continue through this century and contribute to changing the identity of the region. Much of this change will be driven by arrival of new immigrants from many nations of the world. But the perception of California is based on more than its diverse racial and ethnic groups and landscapes. Its economy, its urban landscapes, and its cultural and social systems stamp California's own identity on its residents, as they adapt to the culture and practices of the region. In this way, California will continue to play a pivotal role in altering the culture of newcomers to the United States and to the mainstream of American culture for many years to come.

As we have seen, California's location on the West Coast made it the end of the road for many Americans who heeded Horace Greeley's admonition to "go west." The discovery of gold shortly after California's annexation to the United States reinforced the perception that California is the promised land. Later, the image of California's romantic past was reinforced in various novels and movies. In the 20th century, the arrival of the entertainment industry gave California the opportunity to manipulate its own image. This reinforced California's status as a trendsetter, from the development of suburbia after World War II, to the arrival of the "Beat Generation" in San Francisco in the 1950s, to the hippie era of a decade later, to the "buzz" sustained by Hollywood media moguls today. Thus more than any other North American region, California's image shapes and is shaped by the reality of California life.

Image aside, the reality is that Californians will continue to seek ways to solve the challenges of continued crowding, natural hazards (wildfires, drought, flooding, and geological movement) and, at times, racial and ethnic discord. For more than a century and a half, Californians have pioneered the development of an innovative, creative, and high-intensity lifestyle. Undoubtedly, 21st-century Californians will continue to approach the problems associated with the reality of day-to-day life from new and innovative perspectives. Many of their decisions will influence life in the state, throughout North America, and around the world.

Suggested Readings and Web Sites

Books and Articles

Bakker, Edna. 1984. *An Island Called California: An Ecological Introduction to its Natural Communities.* Berkeley: University of California Press. An enjoyable guide to the unique flora and fauna of California.

Barich, Bill. 1994. *Big Dreams: Into the Heart of California.* New York: Vintage Books. A wise, funny and sometimes heartbreaking meditation on California.

DeLyser, Dydia. *Ramona Memories.* 2005. Minneapolis: University of Minnesota Press. A beautiful book that provides evidence that the title character of Helen Hunt Jackson's fictional book, *Ramona,* may have been one of the most significant women in the history of early southern California.

Hardwick, Susan Wiley and Donald G. Holtgrieve. 1996. *Valley for Dreams: Life and Landscape in the Sacramento Valley.* Lanham, MD: Rowman and Littlefield. A historical and environmental geography of the evolution of the Sacramento Valley up to the mid-1990s.

Holiday, J.S. 1981. *The World Rushed In: The California Gold Rush Experience.* New York: Simon and Schuster. A classic story of the global impacts of the Gold Rush on California culture, environments, landscapes, and economic development.

Johnson, Stephen, Gerald Haslam, and Robert Dawson. 1993. *The Great Central Valley: California's Heartland.* Berkeley: University of California Press. A coffee table book filled with spectacular photographs and abbreviated text on California's Central Valley.

Johnston, Verna R. 1998. *Sierra Nevada: The Naturalist's Companion.* Berkeley: University of California. A hiker's guidebook for exploration of a mountain range John Muir called "the range of light." Indispensable!

Kelley, Robert. 1989. *Battling the Inland Sea: American Political Culture, Public Policy, and the Sacramento Valley, 1850–1986.* Berkeley: University of California Press. A close-up look at one of the major environmental and economic challenges facing the state of California and other parts of the West—water.

Kirsch, Robert and William S. Murphy. 1967. *West of the West.* New York: E.P. Dutton and Co., Inc. Popular book on the history of California's people, cultures, and landscapes.

Margolin, Malcolm. 1993. *The Way We Lived: California Indian Reminiscences, Stories, and Songs.* Berkeley: Heyday Books. A collection of essays, stories, and other descriptions of the challenges and beauty of Native American groups and lives in California.

Reisner, Marc. 1987. *Cadillac Desert: The American West and Its Disappearing Water.* New York: Penguin Books. Best-selling book and television series on water issues in the West. Tells the often dramatic story of the impacts of the decisions of politicians, environmentalists, and land developers on California's water control systems in the 20th century.

Starr, Kevin. 2002. *Embattled Dreams: California in War and Peace, 1940–1950.* New York: Oxford University Press.

______. 1985. *Inventing the Dream: California through the Progressive Era.* New York: Oxford University Press.

______. 1981. *Americans and the California Dream, 1850–1915.* Santa Barbara and Salt Lake City: Peregrine Smith, Inc.

A trio of books chronicling the history and geography of water, politics, and economic development in California—and the state's impacts on the rest of the nation.

Selby, William A. 2006. *Rediscovering the Golden State: California Geography.* New York: John Wiley.

Web Sites

California Energy Commission
http://www.energy.ca.gov/index.html

Population Statistics
http://quickfacts.census.gov/qfd/states/06000.html

Los Angeles City Home Page
http://www.ci.la.ca.us/

San Francisco City Home Page
http://www.ci.sf.ca.us/

California History
http://www.californiahistory.net/

Tourism
http://gocalif.ca.gov/

California Parks and Recreation
http://ceres.ca.gov/ceres/calweb/Natl_Parks.html

▲ **Figure 16.5** Willamette Valley farmland

Ketchikan, are oriented to fishing, although tourism is increasingly important as well. Valdez, located between Anchorage and the southeastern panhandle, is a major port for exporting petroleum from Alaska's North Slope to the conterminous United States and to East Asia. Valdez was the site of one of the worst oil spills in U.S. history in 1989 and was previously damaged by an earthquake in 1964.

The Pacific Northwest and northern California contain some of the largest trees in North America. Indeed, the redwoods that grow along the coast in southwestern Oregon and northwestern California are the tallest trees in the world. Spruce, Douglas fir, and many other species of large and commercially significant trees also grow in abundance. Logging and the production of paper and wood products became major contributors to the economy of the Pacific Northwest in the 19th century and are still very important today.

Lumber production remains a major economic activity in coastal Washington and Oregon and in British Columbia. In some places, helicopters are used to extract logs from steep, slippery slopes. Logs, wood products, and paper are exported from ports such as Coos Bay and Astoria in Oregon and Prince Rupert, British Columbia. Because many logs are used in construction of houses and other buildings, demand for timber tends to increase and decrease with the strength of the economy. During recession periods, fewer houses are started and demand for logs decreases. However, the cyclical nature of timber demand for housing within North America has been offset somewhat in recent decades by steady increases in demand for Pacific Northwest timber in Asia.

Rivers in the Pacific Northwest also produce abundant supplies of fish. Their deep year-round waters also contribute to the generation of hydroelectric power that is so critically important to this region and other parts of the western United States and Canada. Box 16.3 provides additional information about the development of early water power along the Columbia River in Oregon and its surprising link to popular American music.

The Pacific Rim Connection

The political economy of the Pacific Northwest has also been influenced by the region's isolation from major population centers elsewhere in North America and in Europe. Seattle, for example, is nearly 1000 miles (1610 kilometers) by highway from San Francisco and nearly 3000 miles (4830 kilometers) from New York City. The rugged Sierra Nevada and Cascade Mountains have impeded overland transportation, and heavy winter snows still often force the closure of rail lines and highways. Especially north of Vancouver, many communities in the Pacific Northwest are completely isolated from the rest of North America by land. The scenic and historic communities of Alaska's southeastern panhandle, including Sitka, Skagway, Ketchikan, and even the state capital of Juneau, can only be reached by air or water (Figure 16.6). In fact, many Alaskans have argued for and have voted to move the state capital to a site closer to Anchorage. But the voters have so far refused to allocate the funds needed to pay for construction of a new capital city and so the capital remains at Juneau.

BOX 16.3 Woody Guthrie and the BPA: A Bargain for American Music

It was one of the strangest alliances ever forged in the northwest corner of the U.S.: the bureaucratic Bonneville Power Administration (BPA) and its mission to produce and sell hydroelectric power and Woody Guthrie, Oklahoma Dust Bowl loner, restless troubadour, populist, activist, and inspired and prolific composer of both protest songs and anthems in praise of America's beauty, bounty, and the common man.

The year was 1941. The nation was emerging from the Great Depression's gloom but teetering on the brink of World War II. The BPA, still under fire as a New Deal boondoggle, was looking for some kind of positive spin. Guthrie was fleeing the New York and Los Angeles music scenes because, as he lamented, "The monopoly on music pays a few pet writers to go screwy trying to write and rewrite the same old notes."

Woody was recruited by the BPA and arrived in Portland in his old beat-up Pontiac with his wife and three small children. The BPA gave him a 30-day contract as an "information consultant" with instructions to write a song a day. He piled into the shiny new government Hudson and was off on a road trip up the Columbia Gorge filled with inspiration. Bonneville Dam had been completed in 1939, and Grand Coulee Dam was due to be finished in 1942.

After a two-month trip driving and hiking through the mountains and small towns of the Pacific Northwest, Guthrie wrote 26 songs (not quite fulfilling his contract). They were recorded "in a clothes closet down there at the BPA house in Portland" and "played at all sorts of sizes of meetings where people bought bonds to bring the power lines out over the fields and hills to their own little places." Among these songs were several destined to become Woody Guthrie classics, including "Roll on Columbia," "Pastures of Plenty," and "Ramblin' Blues." Several years ago, the Hood River Electric Co-op Board voted to remove Woody's name from the substation that had been dedicated to him many years before. Their rationale was that his radical views rendered him too politically incorrect for the co-op's clientele. The board apparently let slip from its collective mind the lyrics to one of Woody Guthrie's most famous songs:

This land is my land.
This land is your land.
From California to the New York Islands.
From the redwood forest to the Gulf Stream waters
This land is made for you and me.

Source: Adapted from John Terry, *The Oregonian*, May 1, 2005, A20.

Historically, isolation, coupled with difficulties in maintaining surface transportation linkages, has discouraged the development of large-scale industry in the Pacific Northwest. Prior to World War II, most industry in the Pacific Northwest involved processing of local raw materials such as salmon canning and paper products, or producing goods for local consumption.

During and after World War II, changes in the world economy reduced the relative isolation of the Pacific Northwest. In 1942, the Japanese invaded the western Aleutian Islands—the only U.S. soil occupied by Japan during the war. Concerned that the Japanese would invade northwestern North America along the Aleutian chain, the Allies sent some 35,000 U.S. and 7000 Canadian troops into the Aleutians to defend against a possible invasion. In fact, the threatened Japanese invasion proved to be a diversionary tactic, in part because the consistently rainy, windy weather would play havoc on

Figure 16.6 Juneau, Alaska

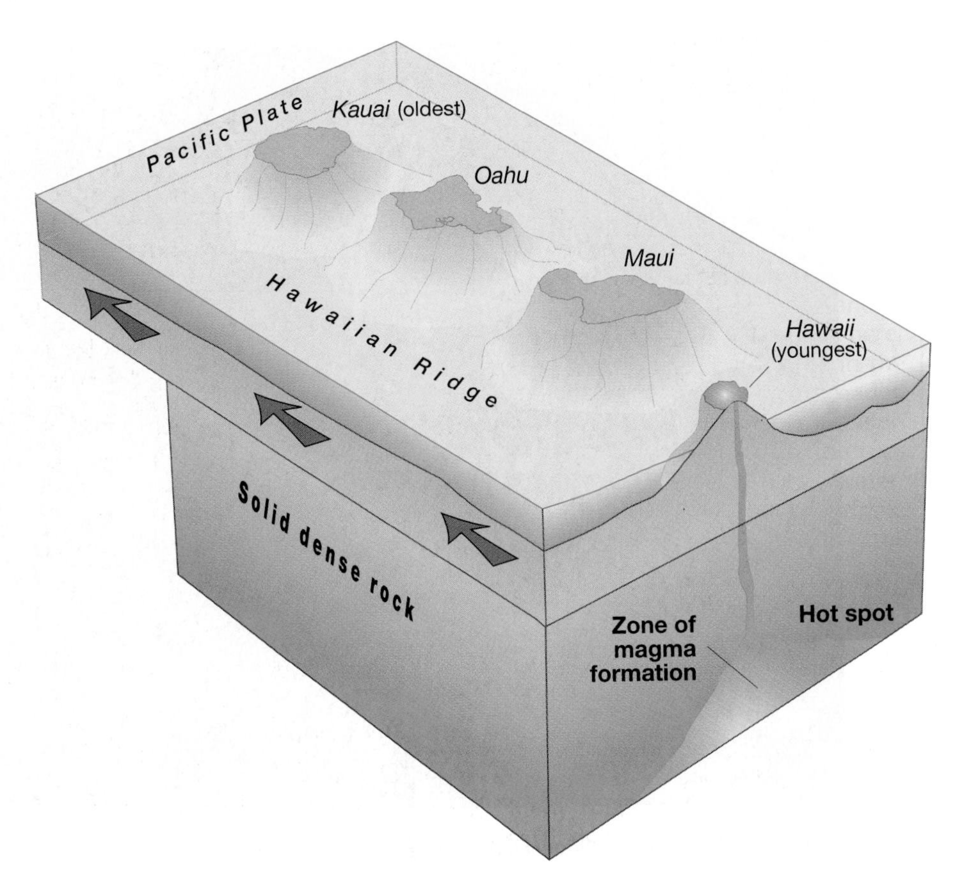

▶ **Figure 17.3** Hot spot plume and formation of islands

are among the most active volcanoes in the world (Figure 17.4). Maui consists of two extinct volcanoes: Haleakala at over 10,000 feet (3050 meters) in elevation to the east, and the smaller but rugged West Maui Mountains to the west. Likewise Oahu includes the Waianae Mountains paralleling the west coast and the Koolau Range along the east coast. Kauai is dominated by Mount Waialeale, another extinct volcano.

The volcanic origin of the Hawaiian Islands means that flat land is scarce and the topography is rugged. The hilly topography has impeded ground transportation and reinforced Hawai'i's dependence on transportation by sea and by air. However, Hawai'i's volcanic soils are rich. In flat or gently sloping areas, these volcanic soils have supported the cultivation of many different crops.

▶ **Figure 17.4** Kilauea volcano erupting

The Hawaiian Islands are also famous for their coastal landforms. Dramatic cliffs exist where mountains or recent lava flows form the shoreline. The 3250 foot (990 meters) cliffs along the northeastern coast of Molokai are described by the *Guinness Book of World Records* as "the highest sea cliffs in the world." Sand beaches, sometimes with coral reefs, are situated along more shallow sloping shorelines. Hawai'i's coral reefs are generally better developed toward the west, where the islands are older and reefs have had more time to grow. Reefs are also more developed in sheltered bays, where they are protected from pounding waves. Unfortunately, portions of the reefs are in decline because of damage done by boat anchors, "trampling" by snorkelers and waders, overfishing, water pollution, and sedimentation from land-based development and agriculture. Many efforts are now underway to try to preserve these fragile ecosystems.

Climate

Hawai'i's tropical location in the middle of the vast North Pacific Ocean gives it a warm and equable climate with few seasonal extremes in temperature. Both temperature and precipitation can vary dramatically with elevation and orientation of topography to the northeast. It can be snowing on Mauna Kea with totally tropical conditions occurring at the same time in Hilo only 20 miles (32 kilometers) away.

Trade Winds

Trade winds are persistent winds that blow from east to west in tropical oceans throughout the world. They are driven by warm, moist air rising at the equator. This warm air travels at high elevations before cooling and sinking in the subtropics at about 25 degrees north and south latitude. A vast zone of persistent low pressure is created where the air rises at the equator, and an equally vast zone of persistent high pressure is created where the air sinks in the subtropics. Because winds are created by air moving from high to low pressure, the persistent pressure cells create steady winds moving from the subtropical high-pressure zones towards the equatorial low-pressure zones. These winds are then deflected by the Coriolis effect from east to west. Prior to the development of steam-powered ships, mariners relied on these persistent winds for long-distance voyages in the tropical Atlantic and Pacific—hence the name "trade winds." The trade winds blow year-round in Hawai'i, but are strongest in May through September when the high-pressure and low-pressure cells move north, placing the islands in the heart of the trade wind belt.

North of the equator, trade winds generally blow from northeast to southwest. When they encounter the volcanoes of Hawai'i, the trade winds are forced to rise, depositing copious amounts of rainfall along the east and north coasts of the islands. The city of Hilo, on the east coast of the Big Island, receives over 150 inches (380 centimeters) of rain a year—the highest total of any American city. Areas in the foothills inland from the city receive 250 inches (635 centimeters) annually. The eastern slope of Mount Waialeale on Kauai is the rainiest spot on Earth, averaging 486 inches (1230 centimeters)—over 40 feet (12 meters)—of rain annually. Counter to this trend, however, is the strong temperature inversion that accompanies the trade winds and keeps clouds from climbing past 4000 to 8000 feet (1200 to 2450 meters). It is this inversion that explains the relatively low cloud cover and annual precipitation of only 20 inches (50 centimeters) on the higher summits—and the placement at 11,500 feet (3500 meters) on Mauna Kea of a major astronomical observatory that requires clear skies.

In contrast, the west and south sides of the islands lie in the rain shadow of the trade winds and are thus much drier than the windward coasts. For example, downtown Honolulu near Waikiki Beach averages only 21 inches (53 centimeters) of rain a year. The Kona Coast on the west side of the Big Island is also dry, with about 20 inches (51 centimeters) of rain per year. Indeed, places only 20 miles southwest of the summit of Mount Waialeale on Kauai are equally dry—in less than 20 miles (32 kilometers), amazingly, one can travel from the wettest spot on earth to the town of Kaumakani, which gets little more rainfall each year than the deserts of New Mexico and Arizona. The smaller islands of Kahoolawe, Lanai, and Niihau lie in the rain shadows of their larger neighbors and also have relatively drier climates. Rainfall at elevations above 2000 feet (610 meters) generally is greatest during the summer when the trade winds are strongest and there is a greater flow of air up the mountains. Below 2000 feet, topography is less important in determining precipitation levels, and so rainfall at these lower elevations is usually greatest in winter when large storm systems generate islandwide precipitation.

The oceanic location of Hawai'i results in relatively little temperature variability among places at the same elevation. Temperatures in most populated areas of Hawai'i are mild to warm throughout the year, with little seasonal variation. In Honolulu, for example, the average high temperature ranges from 82°F (28°C) in January to 88°F (31°C) in August. Nevertheless, Hawaiians recognize two seasons. "Kau," or summer, is the season from April to October when trade winds are most consistent. "Hoo-ilo," or winter, occurs from November to March when temperatures are slightly cooler, the trade winds less consistent, and the likelihood of major storms is greater. Extreme temperatures are unusual: The highest temperature ever recorded in Hawai'i was 100°F (38°C) and most coastal locations have never experienced a temperature below 50°F (10°C). Rainier windward locations are a bit cooler than sunnier and drier leeward locations. Temperatures also decline with altitude, on average about 2.5°F (1.6°C) for every 1000 (305 meters) feet of elevation. As a result, snow falls on the 13,000-foot (3960-meter) summits of Mauna Kea and Mauna Loa each winter.

Biogeography and Biodiversity

The Hawaiian Islands are Earth's most isolated archipelago. Nowhere else on earth is a significant land mass located at a greater distance from other land areas. This isolation has given Hawai'i a unique flora and fauna. Once volcanic activity created the islands, plants and animals began to colonize them. Not surprisingly given their power of flight, birds became the prominent terrestrial vertebrates. The only native Hawaiian mammal is the Hawaiian bat, which can also fly. Birds brought some plant seeds; others were carried by the wind or by ocean currents.

Over several million years, these arrivals thrived, creating a unique biogeography. Prior to the arrival of human beings, more than 95 percent of all plant and animal species on the islands were **endemic**—that is, found nowhere else on earth. Indeed, many native Hawaiian species are endemic to only a single island. Among the best-known endemic species is the Hawaiian or nene goose (*Branta sandwicensis*), which is the state bird of Hawai'i (Figure 17.5). The nene goose is closely related to the widespread Canada goose, but is much smaller and, as an adaptation to the rugged volcanic topography on which it lives, lacks the webbed feet characteristic of most geese and ducks. Honeycreepers, which are descended from North American finches, are another unique type of bird found only in Hawai'i (Box 17.2).

The isolated islands exhibit substantial levels of **adaptive radiation**. Adaptive radiation refers to a mechanism of evolution in isolated areas that occurs in response to otherwise unfilled ecological niches. In isolated environments such as Hawai'i, closely related species have evolved to fill ecological niches that, on continents, would likely be filled by unrelated plants and animals.

The arrival of humans from Polynesia, Europe, and elsewhere disrupted the natural ecosystem of the Hawaiian Islands. Settlers in Hawai'i, as in other isolated islands around the world, modified the environment by cutting down trees, converting forests and marshes to agricultural land, and killing animals and birds for food. Humans also introduced plants and animals from other parts of the world. Some, including pigs, dogs, cats, and cattle, were introduced deliberately. Others, such as rats, stowed away on ships and were introduced accidentally. Both deliberate and accidental arrivals thrived in Hawai'i, killing off or outcompeting native plants and animals. As a result, many native Hawaiian plant and animal species have become extinct, and 78 animal species and 379 plant species are listed as threatened or in danger of extinction. Several honeycreeper species are now extinct, for example, and others are restricted to isolated areas where the native vegetation upon which they depend for food still thrives. Fortunately, conservation efforts to preserve remaining native Hawaiian plants and animals are taking place. Ships, aircraft, and baggage of travelers coming into Hawai'i are carefully checked and inspected to ascertain that they do not deliberately or inadvertently contain plants or animals that might further disrupt the Hawaiian environment.

▶ **Figure 17.5** Nene geese

BOX 17.2 Adaptive Radiation in Hawai'i

The Hawaiian islands' isolation illustrates an important biogeographic principle, that of adaptive radiation. Adaptive radiation refers to biological change in isolated areas that occurs in response to otherwise unfilled ecological niches. In isolated environments such as Hawai'i, closely related species have evolved to fill ecological niches that on continents would be filled by unrelated plants and animals. Thus these animals and plants evolve characteristics not normally associated with their relatives.

The honeycreeper family illustrates the concept of adaptive radiation. Honeycreepers are closely related to finches. Biologists believe that the ancestors of Hawai'i's honeycreepers were North American finches that flew or were blown off course to the islands thousands of years ago. Over the years, their descendants evolved into a family containing 22 distinct species (Figure 17.A). Each species evolved to feed on particular types of native Hawaiian plants. The honeycreepers' ancestors had the stout beaks typical of seed-eating finches throughout the world, but many species developed in unusual ways to take advantage of vacant ecological niches. For example, the I'i'wi (*Vestiaria coccinea*) has a curved beak, which it uses in order to extract nectar from flowers. The apapane (*Himationae sanguinea*), which feeds on insects that live inside these flowers, has a small, thin beak. On the North American mainland, the ecological niches associated with the I'i'wi and the apanane are filled by hummingbirds and wood warblers, respectively; but in the absence of such species, honeycreepers evolved to fill these ecological niches.

Unfortunately, species such as honeycreepers are especially vulnerable to competition from introduced species. Not only were honeycreepers victims to predation by introduced cats, dogs, and other predators, but the native vegetation upon which they depend for their food has been replaced by exotic plants. Several honeycreeper species are extinct, and others now live only in remote areas where native Hawaiian vegetation still thrives. Strict protection measures have been enacted in order to preserve the remaining honeycreeper species.

▲ **Figure 17.A** Honeycreepers: Apapane (left) and I'i'wi (right)

Hazards and Water Resources

Hawai'i is commonly thought of—and portrayed—as a paradise where every day is balmy, rain falls only at night, and light breezes lull one into a reverie of tropical dreams. It therefore can surprise people when they discover that Hawai'i's natural environment has a nasty side. The active volcanoes on the Big Island—and potentially the presently inactive Maui volcanoes—pose hazards from lava flows, tephra (airborne lava fragments), volcanic gases, ground settling, earthquakes, and tsunamis. Over 40 percent of the Big Island has experienced fresh lava flows in the last 1000 years, and substantial portions of the island are still at risk. Maui has experienced ten eruptions over the past 1000 years but none in the last 400 to 500 years. Fortunately, the Hawaiian shield type volcanoes do not generate catastrophic explosions like the volcanoes of the Pacific Northwest. The lava from Hawaiian volcanoes is therefore generally a threat only to property rather than life.

Earthquakes also pose danger to property and potentially to life, especially on the Big Island with its active volcanoes. Earthquakes in Hawai'i are caused by rising magma that wedges the earth apart. Since 1868, 14 earthquakes on the Big Island have exceeded 6.0 on the Richter scale, with three of these exceeding 7.0. **Tsunamis** pose a much greater threat to life. Tsunamis are large waves that are caused by earthquakes, submarine landslides, or volcanic eruptions. Many travel long distances across the Pacific.

About 50 tsunamis have hit Hawai'i since the early 1800s. A 1946 tsunami originating in the Aleutians killed 176 Hawaiians with waves up to 55 feet (17 meters) in height. In 1960, a tsunami formed off the coast of Chile claimed 61 lives in Hawai'i. Today, government officials issue tsunami warnings well in advance of the 5 to 15 hours it takes tsunamis to reach the islands from the Aleutians or Chile. However, tsunamis generated by local earthquakes or submarine landslides are still a major worry because they reach the Hawaiian coast almost instantaneously, before warnings can be delivered. The U.S. Geological Survey states that, "Any earthquake strong enough to cause difficulty in standing or walking should be regarded as a tsunami warning by people in coastal areas, who should immediately head for higher ground."

Weather can also pose hazards to Hawai'i. Cold fronts bringing gusty winds, intense rainfall, and occasionally tornadoes or waterspouts cross Hawai'i, especially the more northerly islands, six to eight times

The Pacific Islands

The Hawaiian Islands are but a few of the thousands of inhabited and uninhabited islands in the Pacific Ocean. Here we focus briefly on those islands that have been U.S. possessions, along with those characterized by substantial trade and population exchange with Hawai'i.

Guam is the largest and southernmost of the Mariana Islands, which are located some 3700 miles (5920 kilometers) west of Hawai'i. The island is 217 square miles (87.7 sq kilometers) in land area (about three times the size of the District of Columbia) and has a population of about 163,000. The island was first settled by persons of Indonesian and Philippine origin about 3500 years ago; their descendants who live on Guam today are known as Chamorros.

Magellan landed on Guam in 1521. Spain colonized Guam in the late 17th century, using the island as a stopover point for ships traversing the Pacific between Mexico and the Philippines. Spain ceded Guam to the United States in 1898. Guam is a self-governing, organized unincorporated territory of the United States. It is home to U.S. military bases and personnel. Tourism, banking, and agriculture are other important activities. The population of Guam, like that of Hawai'i, is highly diverse. Chamorros make up about 47 percent of the population. About 25 percent are Filipino, 18 percent are East Asian, and 10 percent are of European ancestry. Guam's nickname, "Where America's Day Begins," refers to its location just east of the International Date Line.

American Samoa includes seven islands located 2300 miles (3680 kilometers) southwest of Hawai'i. They are part of a larger group of islands which include the independent country of Samoa. The combined land area of the islands comprising American Samoa is about 90 square miles (39.6 sq kilometers), with a population of about 75,000. American Samoa has been under U.S. administration since 1899. At that time the United States, Britain, and Germany agreed to divide sovereignty over the Samoan islands. The United States obtained control over the easternmost islands. Samoans, like the native Hawaiians, are of Polynesian origin. An estimated 20,000 people of Samoan ancestry live in Hawai'i. Other Samoan communities are located on the mainland, notably in California and Washington State. Dwayne Johnson (The Rock) and Olympic diving champion Greg Louganis are perhaps the best-known Americans of Samoan ancestry.

The **Commonwealth of the Northern Mariana Islands** is an archipelago including 14 inhabited islands that stretch some 300 miles (480 kilometers) north of Guam. The islands have a combined land area of 189 square miles (76,350 sq kilometers) and a population of about 80,000. The population of the Commonwealth, like that of Guam, is primarily of Indonesian and Philippine origin. As in other island groups in the Pacific, the economy of the Commonwealth is oriented to trade, tourism, services, and tropical agriculture.

The Future of Hawai'i

What does the future have in store for Hawai'i? Certainly, the scenic beauty, natural wonders, and cultural heritage of the Aloha State charm migrants and tourists alike. However, Hawai'i faces a special set of economic, social, and environmental problems associated with its small size, isolation, and unique and fragile ecosystem.

Any assessment of Hawai'i's economic future must begin by considering its location as an isolated outpost in the Pacific Ocean without sufficient natural resources to support its population. Its economy is therefore highly dependent on far distant, outside forces. For example, the tourist industry upon which Hawai'i depends is highly tied to global and national economies. A significant decline in tourism would have major impacts on the Hawaiian economy.

Another problem faced in Hawai'i is the scarcity of land and the fragility of its ecology. Hawai'i's small size gives it a relatively high population density, which is increasing. Moreover much of Hawai'i's land is either off limits to development or unsuitable for development. The resulting population pressure has caused considerable increases in land values, especially in the most popular tourist destinations such as the Kona Coast, the Waikiki area of Oahu, and northwestern Maui.

In 2000, the U.S. census gave Americans the opportunity to identify themselves as being of mixed-race ancestry. More than 20 percent of Hawai'i's residents did so—a proportion much higher than any other state (California, Alaska, and Oklahoma were next with between 4 and 6 percent each). Hawaiians have long prided themselves on what some call the "Aloha spirit," which emphasizes tolerance and pride in cultural and ethnic diversity. However, there are marked differences among Hawai'i's ethnic groups in standards of living. A disproportionate number of poor persons in Hawai'i are of Native Hawaiian ancestry. Concerns are also being raised that native Hawaiian culture and language are being lost in the rush to continue integrating Hawai'i's economy with that of mainland North America.

Some Hawaiians, in fact, have actively advocated secession of Hawai'i from the United States. Many supporters of secession have argued that the 1893 overthrow of the Hawaiian monarchy—an action supported by American economic and military interests—occurred illegally and in violation of international law. Although secession is unlikely, many secession advocates call for greater local autonomy and recognition of Native Hawaiians as a unique ethnic group with a status similar to that of Native American tribes. In 2005, Senator Daniel Akaka of Hawai'i introduced a bill that would "reaffirm the special political and legal relationship between the United States and the Native Hawaiian governing entity for purposes of carrying on a government-to-government relationship." Proponents of this measure have argued that Hawai'i should be regarded as the crossroads of the Pacific not only from west to east (that is from East Asia to North America), but also from north to south. Such a view emphasizes the linguistic and cultural affinities between native Hawaiians and other indigenous peoples of the Pacific including the Aleuts of Alaska, the Maori of New Zealand, and the Tahitians and other Pacific Islanders.

The future of Hawai'i, like that of all regions of North America, will be affected by processes of globalization and continuing changes in North America's relationship to the rest of the world. Such changes are far beyond the control of local residents. Given Hawai'i's isolation and its economic dependency on the mainland, the impacts of global economic and social change will no doubt be felt particularly keenly on the islands. The Aloha spirit, the state's unmatched ethnic diversity, and increased awareness of environmental issues within both traditional and contemporary cultures are signs that point the way to a hopeful future.

Suggested Readings and Web Sites

Books and Articles

Allen, Robert C. 2004. *Creating Hawai'i Tourism: A Memoir*. Honolulu: Bess Press. A history of the tourist industry in Hawai'i throughout the 20th century.

Herman, R.D.K. 1999. "The Aloha State: Place Names and the Anti-Conquest of Hawai'i." *Annals of the Association of American Geographers* 89: 76–102. A fascinating analysis of the role of place names as indicators of political and economic "conquests" in Hawai'i.

Juvik, Sonia and James O. Juvik, eds. 1998. *Atlas of Hawai'i*. Honolulu: University Press of Hawai'i.

Michener, James. 1959. *Hawai'i*. New York: Random House, 1959. A classic novel that describes Hawai'i's historical geography from early Polynesian settlement more than a thousand years ago to the statehood period in the late 1950s.

Moore, James G., et al. 1995. "Giant Blocks in the South Kona Landslide, Hawaii," *Geology*, 23:125. An analysis of the impact of regional and global volcanic processes on massive landslides in Kona.

Nordyke, Eleanor C. 1989. *The Peopling of Hawai'i*. Honolulu: University of Hawai'i Press. Ethnographic study of the major groups of people who reside on the Hawaiian Islands.

Padeken, Ahiokalani Maenette Kape, Ah Nee Benham, and Ronald H. Heck. 1998. *Culture and Educational Policy in Hawai'i: The Silencing of Native Voices*. Mahwah, NJ: Lawrence Erlbaum Associates. Describes education policy in Hawai'i, emphasizing the impacts of American education on Hawai'i's culture.

Sanderson, Marie, ed. 1993. *Prevailing Trade Winds: Weather and Climate in Hawai'i*. Honolulu: University of Hawai'i Press. Helpful overview of the atmospheric processes involved in understanding Hawai's's weather and climate past and present.

Stone, Charles P., Clifford W. Smith, and J. Timothy Tunison. 1992. *Alien Plant Invasions in Native Ecosystems of Hawai'i*. Honolulu: University of Hawai'i Press. Excellent coverage of some of the most important details in understanding the impact of plant invasions on natural systems in Hawai'i.

Tucker, Richard P. 2000. *Insatiable Appetite: The United States and the Ecological Destruction of the Tropical World*. Berkeley: University of California Press. Chapter 2 (pp. 63–120) describes the sugar industry and its impacts on Hawai'i and the Philippines.

Atlas of Hawai'i, 3rd ed. 1999. Honolulu: University of Hawai'i Press. A comprehensive atlas of Hawai'i's physical, cultural, and social environments.

Web Sites

Hawaii Weather
http://www.hawaiiweathertoday.com/

U.S. Department of Labor–Hawai'i Economy
http://www.bls.gov/eag/eag.hi.htm

Hawai'i Tourism
http://www.gohawaii.com/

Surfing Information
http://www.surfguidehawaii.com/

Hawai'i National Parks
http://www.nps.gov/havo/

U.S. Geological Survey Hawaiian Volcano Observatory
http://hvo.wr.usgs.gov/

Miscellaneous Web-based materials:

Coral Reef/land management by native peoples:
http://www.coralreefnetwork.com/reefs/ecology/ecology.htm

18
The Far North

PREVIEW

Environmental Setting

- The Far North extends from east to west across the North American continent and includes those areas too far north to permit commercial agriculture and continuous settlement.
- The Far North's physical geography is divided into three large subregions: the Northwest, the Laurentian Shield, and the Arctic, including Greenland.
- The Northwest part of this large northern region is mostly an extension of the high mountain Cordillera and the Pacific coastal mountains, while the land around the Arctic Ocean and the vast expanses of the Laurentian Shield are areas of gentle relief with few hills to interrupt its expansive vistas.
- Global climate change is having a major impact on the peoples and environments of the Far North.

Historical Settlement

- Settlement of the Far North is discontinuous and is oriented to the location of resource-extractive activities.
- Some parts of the Far North are affected by tensions between Euro-Americans and residents with native ancestry.
- Despite the promise of creating the province of Nunavut in Canada (the world's first political territory governed entirely by indigenous people), there remains an uneasy tension on the part of both the Native Americans in the United States and the First Nations people in Canada who are trying to find their place in the modern world while retaining their traditional cultural values.

Political Economy

- Because the climate is too challenging for commercial agriculture in almost all of the Far North, the traditional economy has depended on primary-sector economic activities, including mining, fishing, and lumbering.
- During World War II and the cold war, the Far North's economy was affected by its location relative to Asia, the Arctic Ocean, and the former Soviet Union.
- Increasingly, residents of the Far North are becoming oriented to tourism and other tertiary and quaternary activities.

Culture, People, and Places

- Despite its isolated, peripheral location and harsh climate, Greenland has begun to open up and become known to people in other parts of the world in the first decade of the 21st century, partly because global warming has melted icecaps and glaciers along some of its coast.
- Numerous recreation-oriented communities for tourists, retirees, and residents are located on the southern edge of the Far North, adjacent to the large cities of southern Canada and the northern United States.
- Anchorage, on a cusp between the Far North and the Pacific Northwest region, is the largest city in North America's Far North region.

Peter Erneck was on the phone from Iqualut, dealing in facts and figures about Nunavut, when he stopped and said: "Why don't I just send you an e-mail?" And then he stopped again and said: "You know, sometimes I'm amazed. Forty years ago I was living in an igloo. Now I'm sending e-mails."
(Lanken and Vincent, 1999, p. 39).

Introduction

This chapter focuses on the Far North, a vast expanse of open land, ice, and water located at the northernmost portion of North America. It is hard to imagine how large this region really is. The Yukon Territory and the Northwest Territories alone (not counting Alaska, Nunavut, and Greenland, which we also include in this chapter) contain 40 percent of Canada's land surface. And the coastline of the Arctic Ocean is longer than the Canadian Atlantic and Pacific coastlines combined. In this chapter we will discuss some of the different perceptions of these vast, relatively unknown expanses of the Far North, the impacts of global climate change in this part of North America, and the characteristics of its ever changing environmental, cultural, and political-economic systems.

◀ The Alaska Pipeline is still symbolic of natural-resource issues in the Far North.

▲ **Figure 18.2** Endangered polar bears in Labrador

to even subtle variations in amounts of sunlight, temperature, and precipitation. Thus the Arctic and other parts of the Far North region, including Alaska and Greenland, are particularly vulnerable to the ongoing impacts of **global climate change**. For example, polar bears depend upon being able to hunt seals on the sea ice sheets, which are melting due to increasing sea temperatures. With less hunting habitat available, polar bears populations have steadily been decreasing (see Figure 18.2). Global climate models indicate that global warming induced by the **greenhouse effect** will be most acute in polar regions. Ecosystems in Alaska are also threatened by climate change. The Bering Sea and Gulf of Alaska are among the most productive in the world and also the most highly susceptible to global climate change.

As global temperatures rise, the polar ice covering the Arctic Ocean is melting at a rate of 8 percent per decade. This means that within ten years the Northwest Passage, now only open to reinforced-ice breaking vessels, could be open to conventional craft at least a few weeks per year in the summer. This possibility has engendered contrasting opinions about potential future navigation in this area. The United States calls the passage an international strait open to all. Canada claims control because it considers the passage an internal waterway, like the Mississippi River. An ice-free northern passage could be a far shorter route than use of the Panama Canal for Pacific–Atlantic ocean shipping. At this time both nations have agreed to disagree on the issue.

More details about the impacts of climate change on the environments and economies of the Far North based on data from climate models are presented in Box 18.1.

These harsh environmental conditions have a major impact on agriculture and settlement patterns in the Far North. Although many people assume that the extremely cold temperatures limit agriculture in the region, in fact the lack of precipitation is the primary limiting factor in most places. In Dawson in the Yukon, the average summer temperature is above 50°F (10°C) during the three summer months, nearly one month longer than in northern Quebec (see Figure 18.3). Drought conditions are common. This makes irrigation more important for agriculture than fears of frost, at least during the warmer months of the year. In the Arctic and northern Ontario and Quebec, there are really three major problems faced by prospective farmers and other residents: lack of adequate soil to grow crops, lack of precipitation during the growing season, and lack of sunshine. And yet despite these challenges for human settlement, as we will explore in the following sections of this chapter, the Far North is a region of great diversity—not only in terms of its physical geography but also its resilient peoples, cultures, and economies.

Historical Settlement

Aboriginal Peoples

The indigenous peoples of the North include several distinct cultures. Most scholars agree that the ancestors of today's Native Americans in the United States and

BOX 18.1 Impacts of Global Climate Change in the Far North

Hyper-extreme weather conditions and sensitive vegetation make the Arctic and other parts of the Far North especially vulnerable to the impacts of global climate change. These areas also play a critically important role in global processes such as atmosphere and ocean circulation which includes potentially important links to sources and sinks of trace gases. Ozone depletion, increases in atmospheric greenhouse gases, large-scale pollution, and changing patterns of natural-resource use all demonstrate that human activities are altering Earth's system at an accelerated pace. Perhaps nowhere on the planet are the impacts of these changes felt as keenly nor do they pose as much potential damage as in the Far North.

Alaska is a place where the impacts of global climate change have been studied intensively for the past several decades. Not only does Alaska's enormous size encompass extreme climatic differences, its coastal margins provide visible evidence of one of the most dramatic indicators of global warming, melting ice. Tourists, scientists, and residents alike are well aware of the common site of glacial ice melting as huge chunks of ice break off the edges of coastal glaciers and fall into the Pacific Ocean.

Climate trends in Alaska over the past three decades have shown considerable warming. This warming has already led to major impacts on the environment and the economy. If present trends continue, the state's already fragile natural resource-dependent economy will be hit especially hard. Alaska natural ecosystems are also threatened as warmer temperatures affect the habitats of both land and sea animals and fish. Alaska has experienced the largest warming trend of any state in the United States, with a rise in average temperature of about 5°F (3°C) since the 1960s. This has resulted in a melting of glaciers, thawing of permafrost, and reduction of sea ice. Geographers and other scientists who use climate modeling have predicted that this warming trend will continue in future decades with the largest changes happening during winter months.

Adapted from the Center for Global Change and Arctic and System Research, 2005.

▲ **Figure 18.3** Dawson in the Yukon is warm enough for tourists to enjoy the sun

First Nations peoples in Canada migrated eastward from Asia thousands of years ago. Four major cultural groups in the region include the **Algonquin-speaking** Crees and Ojibways who inhabited what is now northern Quebec and Ontario; the **Athabascan cultures** of present-day northern Manitoba, Saskatchewan, Alberta, and the Northwest Territories; the **Aleut** of the Aleutian Islands; and the **Inuit** of the Arctic shores in Alaska, Canada, and Greenland (see Figure 18.4). Over the years, intermarriage has reduced the number of full-blooded **indigenous peoples** relative to persons of mixed ancestry.

The Algonquins and Athabascans were primarily oriented to the wooded **taiga** environment in which they lived. They followed the caribou, hunted other game, and fished in the numerous lakes and rivers that dominate the Canadian Shield. The Aleut and Inuit, in contrast, were predominantly oriented to the sea. They fished and hunted whales, narwhals, walruses, seals, and other marine mammals and birds. Inuit peoples were formerly called Eskimos ("eaters of raw meat" in the Athabascan language) but this name is not used today since it is associated with the judgmental tone of early Euro-American settlers and colonizers. Inuits are the dominant population group in Arctic Alaska, northern Canada, and Greenland as well as on the Chukotka Peninsula in Russia.

The native peoples of Greenland first migrated across the north Atlantic from the Baffin Island about 5000 years ago during a warm climatic period. These early people were able to survive along the barren northern cost of Greenland by hunting muskoxen, Arctic hares, walrus, seals, and other animals. Like the

Figure 18.5 Remote airfield in Greenland

Canadians and Americans of European ancestry who penetrated the North generally came in search of three commodities: animals, trees, and minerals. While some did try to farm, few succeeded. However, a few small areas within the vast North do contain viable agricultural communities. The Matanuska Valley, northeast of Anchorage, Alaska, contains about two-thirds of Alaska's commercial farms. Historically this region produced food for the Anchorage market, but farming in the Matanuska Valley is declining in importance given the region's harsh climate, high production costs, and increased availability of food imported from elsewhere. In Canada, farmers have succeeded in making reasonable livings despite the harsh, cold climate in the Peace River district of northern Alberta and British Columbia and in the Clay Belt of northern Quebec and northeastern Ontario.

Except for very limited areas in the Far North such as the fertile Matanuska Valley in Alaska and Peace Valley in northern Alberta, agriculture is far less important to the economy than are other resource-extractive activities, including trapping, logging, and mining. Mammals living in the North often have thick, luxurious fur to survive the cold northern winter, and their pelts have been valued by Americans, Europeans, and Asians for centuries. Fur trapping has been significant to the economy of the North for centuries. Some fur trappers were French Canadian ***couriers du bois*** (see Chapter 6) or other European-Americans who moved northward temporarily, whereas other Europeans bought or traded for furs from Native fur trappers while not actually trapping animals themselves. Some trappers still catch wild animals, while other fur-bearing animals, especially foxes and mink, are raised on fur farms for their pelts. During the past several decades, the value of the fur trade has declined considerably, in part because demand for fur coats has declined in association with international pressure and concern that trapping is cruel and painful to animals.

In coastal areas of the Far North, whaling has long been practiced. Red Bay in Labrador and Baffin Island in present-day Nunavut were important whaling stations in the 19th century until whale stocks depleted to the point that whaling was no longer profitable (Figure 18.6). The presence of whalers, along with that of fur trappers in inland areas, was significant in that it represented the first integration of native peoples into the global economy; fur traders and whalers introduced guns, knives, metal tools, and other Western technologies and commercial products to the North for the first time.

Fishing is another important resource extraction industry in the Far North. The North Atlantic coast has supported a fishing industry for more than 500 years, as we discussed in Chapter 5. Atlantic cod and related species fueled the exploration and colonization of North America. On the opposite side of the continent, the Gulf of Alaska and other parts of the northwest coast of North America are also areas of bountiful fishery harvests.

Industrial fishing at the Bristol Bay sockeye fishery, in fact, was at one point the largest fishery in the world by dollar value (although farmed salmon has reduced the role of this fishery in recent years). Fishing rights remain a major issue both in Alaska and in the North Atlantic. Coastal communities have an acute interest in

▲ **Figure 18.6** Relicts of a historic Basque whaling station are still visible in the landscape of Red Bay, a remote harbor on the coast of Labrador

maintaining the fisheries' **resource base** because the health of fishing fleets and commercial fishing, along with fish processing, are major sources of employment and food for local residents. According to commercial fishing reports in Alaska, the presence of a significant fishing industry in the Far North improves the quality of life in local communities in several ways. It provides employment and income, creates municipal revenues, and justifies state funding of capital projects. It also creates a user base for fleet processors, which, in turn, generates service charge revenues for operations and infrastructure.

However, today's fish populations have been drastically reduced on both the Pacific and the Atlantic coasts of the Far North as intense overfishing by domestic and foreign fleets have decimated fish populations, reducing their numbers to historic lows (Box 18.2). Biological diversity, especially of species with the greatest

BOX 18.2 Preserving Forest Ecosystems in the Far North

The authors of a recent "Policy Forum" in *Science* provide a fascinating and controversial "take" on preserving boreal forests in the Far North (Mayer et al. 2005). They argued that boreal forests, which cover 32 percent of the planet, are one of the last relatively intact terrestrial biomes on Earth and are a critical carbon sink in global climate dynamics. Mature and old-growth boreal forests also provide many culturally and economically important products such as wood-based lumber, pulp, and fuel wood and numerous non-wood products. Ongoing debates and decisions about whether or not to continue to allow intensive wood harvesting in the name of conservation of naturally intact forests continues in earnest. Thus, where biodiversity is valued as a desirable outcome, wood harvests have been limited or banned outright in the Far North.

The scientists who conducted this study go on to suggest that increasing domestic forest protection in the boreal forests of the Far North without decreasing demand for wood will necessitate an increase in foreign imports. This, in turn, will create a damaging impact on **forest biodiversity** elsewhere in the world. So, they argue, in some cases the decision to protect local forests and then import needed timber back across the border to meet construction, fuel, and other needs in the Far North may cause **global environmental impacts** to boomerang back into the region's protected forests. The outcome of some conservation programs in this region, therefore, may result in unexpected impacts that could, in the end, prove more damaging to local and regional ecosystems than carefully controlled cutting of the forests in the first place.

Source: Mayer, A.L., P. E. Kauppi, P.K. Angelstam, Y. Zhang, and P. M. Tikka. "Importing Timber, Exporting Ecological Impact," *Science* 15 (April), 2005: 359–360.

economic importance, has been altered by intense fishing. Today, human utilization of coastal resources is increasing and, along with it, demands on coastal resources have continued to climb.

The taiga region of the North, which extends from Alaska eastward to Labrador, contains uncounted millions of spruce, pine, fir, and other trees. Indeed, the Northern taiga is one of the largest forests in the world. For centuries, the woods of the North have been logged, with the wood used directly in construction or converted to pulp or paper. During the mid-20th century, the pulp and paper industry was Canada's largest. Logging has been especially intensive in the southern portion of the region, which has better transportation and is more accessible to markets for paper and wood products (Box 18.2).

The abundant hydroelectric power characteristic of the North has allowed the establishment of numerous pulp and paper mills. Indeed, hydroelectric power produced in the North is used to produce aluminum and electricity for many Canadian metropolitan centers. For example, a major dam on the Peace River in northeastern British Columbia provides much of the power for the Vancouver metropolitan area. Trees in the North grow slowly, however, and logging is nowhere near as prevalent and lucrative as is the case in the American South and other regions where warmer climates allow faster tree growth.

Mining and Mineral Extraction

Mining and mineral extraction are, for the most part, much more lucrative than trapping and logging and indeed these activities dominate local economies in many areas. The mineral resources of the North include iron, nickel, copper, precious metals, and petroleum, all of which occur in large quantities in various parts of the North. The Canadian Shield is one of the most mineral-rich areas on Earth's surface, and it contains substantial quantities of a variety of metals, including iron ore, copper, nickel, and tin.

Canada is one of the world's leading producers of nickel. Many North Americans associated nickel with five-cent coins, but nickel is used to make a wide variety of alloys. In addition to coins, nickel alloys are used in batteries, machine tools, electroplating, sheet metal, tableware, stainless steel, and many other commonly used products. In 2001, Canada produced more than 15 percent of all nickel mined, and it has more than 10 percent of the world's proven nickel reserves. The world's largest nickel deposit was discovered near Nain, Labrador, in 2002 and offers much hope for economic development of this remote part of Canada's northeastern coast. Other large nickel reserves are found in the Canadian Shield, with another of the largest nickel deposits in the world found near Sudbury, Ontario, northeast of Lake Superior. The Sudbury basin is one of the richest deposits of other mineral ores in the world as well. Copper was discovered in the area by workers constructing the Canadian Pacific Railway in the 1880s. Large quantities of nickel and other metals including platinum, silver, and gold were found in conjunction with the copper. By 1905 Canada became the world's leading producer of nickel, and Sudbury became the leading center for nickel production (Figure 18.7).

Over the course of the 20th century, Sudbury, like other mining communities, experienced frequent booms and busts. The nickel industry went into the doldrums as global demand for nickel fell after World War I and remained low during the Great Depression. Demand for nickel increased dramatically after the United States and Canada entered World War II, however, and after the war the United States began to stockpile nickel.

The prosperity and growth of Sudbury after World War II were tempered by increased global competition and environmental problems. **Sulfur dioxide** emissions from nickel smelting killed much of the local vegetation, and during the 1960s NASA astronauts used the resulting barren area around Sudbury to practice lunar landings. Construction in 1970 of taller smokestacks and improvements in the smelting process reduced the effect of sulfur dioxide on vegetation, and concerted efforts to replant trees were undertaken. As a result of these efforts, today's overall environment, especially the regrowth of natural vegetation in this area, has improved dramatically. The Sudbury area has grown to a population of nearly 200,000, making it one of the largest communities of the Far North.

In the 1950s, another major nickel deposit was discovered near the village of Thompson, Manitoba, some 500 miles (800 kilometers) north of Winnipeg. The Inco Corporation, which had been instrumental in the development of nickel mining in Sudbury, contracted with the government of the Province of Manitoba to build a modern integrated mining and nickel refining facility as well as the infrastructure—houses, schools, stores, and other buildings—for a complete community. With a population of over 15,000, Thompson is now the third largest city in Manitoba. Civic boosters call attention to their city's combination of modern conveniences and the availability of hunting, fishing, snowmobiling, skiing, boating, and other recreational activities.

Sudbury and Thompson typify the relationship between urban settlement and economic activity in the Far North. They are cities developed in proximity to resources, or along transportation corridors used to export these resources to North American and global markets. Throughout the Far North, there are many other mining-oriented communities. Gold and copper mines are found in the Clay Belt area near Timmins, Ontario. Uranium is mined at Elliott Lake north of Lake Huron. Copper and zinc are mined near Flin Flon on

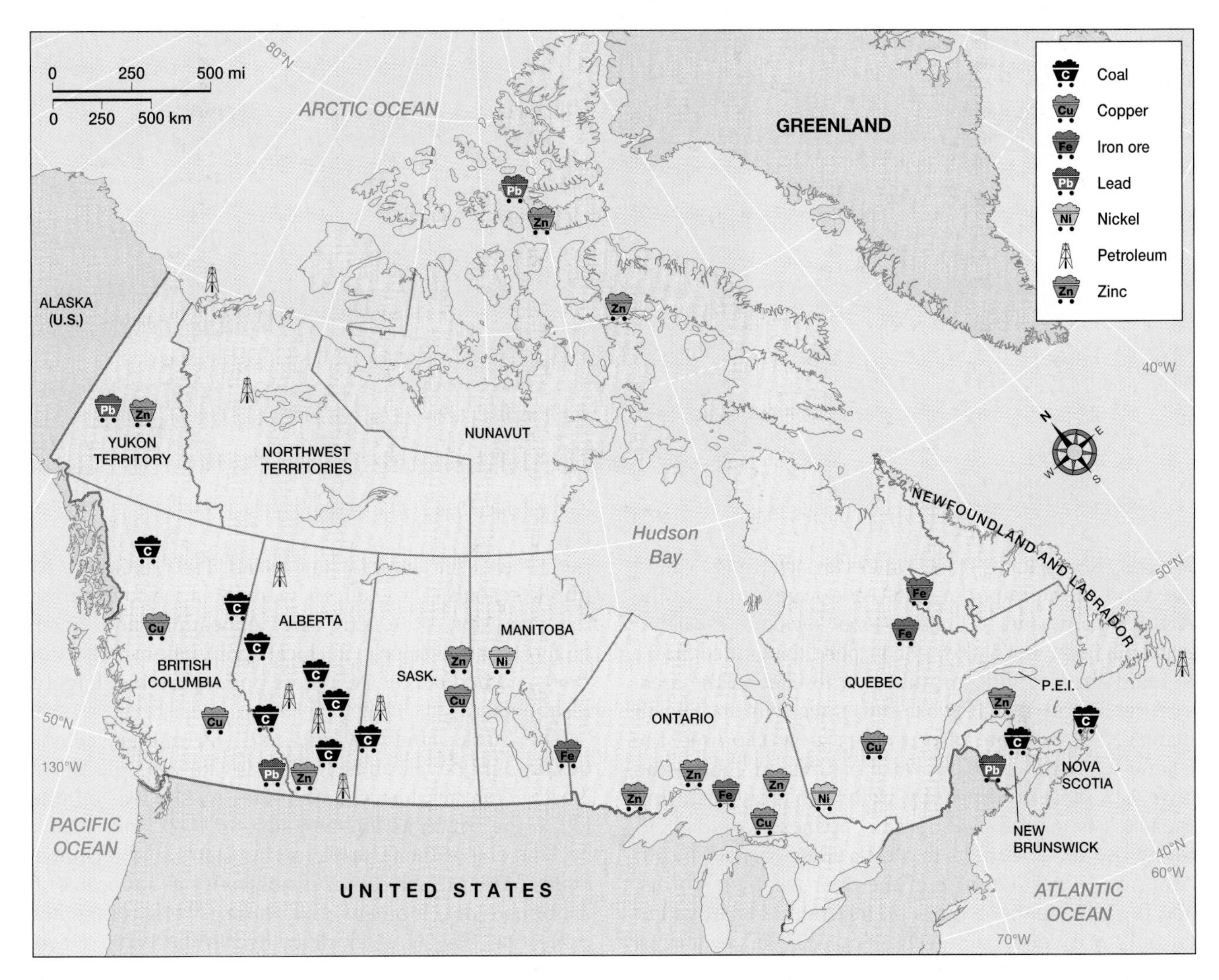

▲ **Figure 18.7** Mineral deposits and mining centers in Canada

the Manitoba–Saskatchewan border, and near Great Slave Lake are several gold, lead, and zinc mining communities. In the United States, the Mesabi Range in northeastern Minnesota, north of Duluth, is North America's leading iron-producing area. Iron and copper are also mined in the Upper Peninsula of Michigan. Local economic conditions in all of these communities are dependent on fluctuating world prices for these respective minerals. Evidence of this global interdependence is apparent today as the growth in China's economy has increased the demand for minerals from this part of the United States and Canada (and subsequently has improved local economic conditions north of the Great Lakes).

Petroleum is another major mineral resource of the Far North. The Prudhoe Bay field in northeastern Alaska is one of the most significant unexploited petroleum deposits in the world. Much of it lies under the Alaska National Wildlife Refuge (Figure 18.8). In 2002, the Bush administration proposed limited oil drilling, but many in opposition have expressed concern that these drilling activities would disrupt the fragile ecology of the area. There are also substantial oil deposits in northern Alberta. In general, oil production in the North is relatively expensive, given the region's isolation, the high costs associated with drilling in permafrost and other aspects of the cold environment, and the high costs associated with transporting the petroleum to markets in more densely populated areas.

Integrating the North into North America and the World

The Allied victory in World War I and the development of aviation helped integrate North America into the global economy, thereby encouraging foreign policy strategists to focus on the Arctic region. As long-distance flight developed, it soon became evident that the shortest route from the United States to either Europe or Asia followed the great circle over the polar

Figure 18.8 Alaska National Wildlife Refuge

regions. Recognizing this, American strategists called attention to the importance of American control of the Arctic region, and as international tensions escalated prior to World War II they encouraged the United States to maintain a significant military presence in the area.

After World War II broke out, interest in the North increased even more after the Japanese captured the islands of Attu and Kiska in the Aleutian chain—the only U.S. soil occupied by a foreign power during the war. Some predicted that the Japanese would eventually use the Aleutians to attack other parts of North America. In part because of the poor weather, storms, and fog, the Japanese attack on the Aleutians proved to be only a diversionary tactic. Some 35,000 American and 7000 Canadian troops were stationed in Alaska and the Aleutians to prevent any further attacks. In order to expedite the movement of troops and supplies to Alaska, the U.S. and Canadian governments cooperated to build the Alaska Highway, which connects Anchorage and Fairbanks with the Yukon Territory, Northwest Territories, and the Alberta cities of Calgary and Edmonton. The Alaska Highway remains the major land route through the western portion of the North today. To the east, the United States Air Force constructed an airfield at Frobisher Bay on Baffin Island to expedite shipment of personnel and war materials to the European theater. Today, Frobisher Bay, since renamed Iqaluit, is the capital of the new Canadian territory of Nunavut.

During the cold war, the perceived threat of nuclear attack from the Soviet Union encouraged the Americans to establish military bases throughout northern Canada, Alaska, and Greenland. The Distant Early Warning system (DEW Line) was a joint U.S.–Canadian project to establish radar stations for tracking potential attacking Soviet aircraft and missiles. During the same period, the Canadian and U.S. governments actively encouraged local Inuit and other native people to give up their traditional nomadic lifestyles, move into permanent settlements, and send their children to public schools. These efforts were intended to improve the provision of health care, education, and other government services to local populations. By the 1960s, nomadism by the native population had largely disappeared.

The Alaska Highway and cold war military activities brought development and many new people to the North. This trend accelerated after Alaska was admitted to the Union as the 49th state in 1959, and Alaska became one of the fastest-growing states in the United States. With growth came tension between advocates of continued development and those of environmental protection. This tension intensified in the 1970s, when oil drilling in Alaska increased in response to the energy crisis of the United States. During that decade, the **Alaska pipeline**, connecting the Prudhoe Bay fields with the port of Valdez in southern Alaska, was constructed despite opposition from many environmentalists, some of whom pointed out that much of the oil produced and transported along the pipeline was actually exported to Asia rather than used within the United States.

In 1979, U.S. President Jimmy Carter signed the **Alaska Wilderness Act** over the objections of many Alaska political and business leaders. The Act created the **Alaska National Wildlife Refuge** (ANWR) and several national parks, putting a large majority of Alaska's land off-limits to developers and mineral exploration interests. The current controversy about drilling in ANWR is another example of the continuing tension between environmentalists and supporters of economic development, not only in Alaska but throughout the Far North. Some insist that drilling for petroleum here is necessary with dwindling oil reserves in other parts of North America and the world. Despite a recent report by the National Research Council that argues that ecological impacts of petroleum drilling actually would be fairly minimal in the region,

BOX 18.3 Saving the Arctic National Wildlife Refuge?

Drilling for gas and oil on the pristine Arctic National Wildlife Refuge (ANWR) is currently being debated in Congress. The coastal portion of these protected 19 million acres in northeast Alaska are viewed as one of the most likely drilling sites. According to the U.S. Geological Survey, the reserves on ANWR land could produce enough recoverable oil as the giant field at Prudhoe Bay just west of ANWR, a place estimated to have 11–13 billion barrels of oil.

But the refuge is home to a variety of plants and animals that would die or be driven off by the drilling. Caribou, polar bears, grizzly bears, wolves, birds, and many other species of undisturbed animals has caused this place to be called "America's Serengeti." A proposal has been made to link ANWR with two Canadian parks that adjoin it to form an international park that would protect these and other animals and plants.

The conflict continues as this book goes to print. *Should Congress open up the area for oil and gas development or should the area's ecosystem be permanently protected from development*? And if the area is opened up for development, how can damage be avoided, minimized, or mitigated? More information that helps answer this question can be found at the U.S. Fish and Wildlife Service Web site: *http:// www. protectthearctic.com/*. Maps of the coastal plain showing existing oil development sites can be found at: *http://www.dog.dnr.state.ak.us/oil/ products/maps/maps*

Source: Adapted from a report given by M. Lynne Corn, Bernard A. Geib, and Pamela Baldwin, Congressional Research Service, 2005, and Justin Blum, "Alaskan Reserve Opens Up for Potential Oil Development," *The Register-Guard*, Eugene, OR, January 12, 2006: A10.

the key issue here is the precedent of opening up preserved land to resource extraction. If ANWR is opened up for development, why not Yellowstone? And reserves in ANWR land are actually fairly small compared to global reserves, but quite large in terms of American reserves. In addition, some argue that as oil and gas prices continue to increase dramatically at gas pumps and elsewhere, extraction from these more marginal reserves has become more economically viable. Box 18.3 provides additional information on the political football that the Alaska National Wildlife Refuge has become in recent years. It is clear that decisions made by politicians and other decision-makers who live a great distance from the Far North region in the capital cities of Ottawa and Washington, D.C., remain critically important in terms of finding ways to maintain the environmental and economic balance of this fragile North American region.

Culture, People, and Places

Lingering Cultural and Ethnic Divides

Despite efforts to resolve problems created by tensions between indigenous peoples and Euro-Americans in the Far North in both the United States and Canada, uneasiness remains. As in other parts of North America, First Nations peoples remain deeply concerned that their lands were taken from them illegally in years past and that they should be compensated for these lands. In 1971, the United States Congress granted Alaska's approximately 50,000 Native Americans, Inuit, and Aleuts title to about 44 million acres (114 million hectares) of land and also paid them a large cash settlement. Today, this native-owned land and its resources are managed by more than 250 native village corporations and 12 regional corporations. Because these native corporations also own subsurface rights, many have become very wealthy from petroleum sales.

Analogous settlements have taken place in Canada, where Native Canadians, Inuit, and Metis, who are persons of mixed native and European ancestry, have pushed for more self-determination. As was the case with Alaska prior to 1959, most of the Canadian North did not enjoy provincial status. Those portions of Canada north of the 60th parallel of latitude and west of Hudson Bay are **territories** of the Canadian federal government. The Yukon Territory and the Northwest Territories elect representatives to the Canadian parliament but otherwise do not enjoy full **provincial status**.

As we have discussed, the new territory of Nunavut was created in 1999. The Nunavut territorial government was empowered by the **Canadian Native Land Claims Act** of 1993 to have a critical voice in the management of land and resources, including offshore resources. The Canadian government guaranteed shares of federal royalties from mineral exploration, title to thousands of square miles of traditional Inuit land, hunting and fishing rights, and "procurement preference" programs giving preference to Inuit-owned contractors. The creation of Nunavut has contributed dramatically to **native self-determination** in the North.

Other than traveling to the Far North to speak with Inuit peoples about the challenges of their day-to-day survival in Canada's Far North, the most dramatic way to understand more about this remote place is to watch the world's first totally indigenously produced movie, *The Fast Runner*. The arts and media world was stunned when this dramatic film won the award for "Best Foreign Film" at the prestigious Cannes Film Festival in

▲ **Figure 18.9** The summer landscape in Illuissat, Greenland

2002. To gain more of an understanding of the problems facing Inuit people today, one of the authors of this text conducted interviews with local residents of an Inuit community in eastern Canada in 2004. According to one 46-year-old woman who lives in Nain, Labrador:

> Our lives are very hard here. After television came into our village about ten years ago, many people, (especially the kids), learned about the outside world and they wanted to be part of it. Now we have a really big alcohol problem here because people are feeling so depressed, you know?

Cities and Small Towns in the Far North

The largest communities in the North are oriented to mining or are port cities, which serve as transportation gateways between the Far North and the rest of North America and the world. Anchorage, on the cusp between the Far North and the Pacific Northwest region, is the largest North American city north of continuous agricultural settlement. Valdez is an oil port. The small city of Churchill, Manitoba, is the major port on Hudson Bay.

Sudbury is the largest of many mining-oriented communities that dot the northern landscape. Others in Ontario include Timmins (home of Shania Twain), Cochrane, and Kirkland Lake. Hibbing, Minnesota, perhaps best known as the birthplace of Bob Dylan, Roger Maris, and Kevin McHale, is at the center of America's leading iron-producing region, the Mesabi Range. Other medium-size communities in the North are primarily regional commercial and/or administrative centers. These include Marquette, Michigan; Prince Albert, Saskatchewan; and Fairbanks, Alaska, along with the administrative capitals of Yellowknife, Northwest Territories; Whitehorse, Yukon; and Iqaluit (previously called Frobisher Bay), Nunavut. In the far eastern part of this region, Greenland's capital city of Nuuk now has more than 15,000 people and it and other small towns along the west coast of Greenland are becoming budding tourist centers (Figure 18.9). Far to the west, the small cities of Barrow, Nome, and Kotzebue, Alaska, along with the planned community of Inuvik, Yukon Territory, are regional centers for Inuit commercial activity and culture.

Greenland

With a total area of 2,175,600 square miles (3,501,290 square kilometers) and a 24,855-mile (40,000-kilometer) coastline, Greenland is the world's largest island. It also contains one of the last remaining **continental glaciers** on Earth along with Antarctica (Figure 18.10). Most of the island is covered by a vast ice sheet that can be seen from trans-Atlantic airplane flights. All but the southern tip of Greenland is located north of the **Arctic Circle**. In fact, Oodaaq, a tiny scrap of rock off its northern coast, is the world's most northerly land, located at 83°40′ north latitude.

Despite its isolated, peripheral location and harsh climate, Greenland has become more accessible to people living in other parts of the world in the first

▲ **Figure 18.10** Icebergs off the coast of Greenland

decade of the 21st century. Global warming has melted icecaps and glaciers, exposing bright green forests and meadows along the coastline, bays, and inlets of the island. This warming is encouraging new tourists and settlement by year-round residents, thereby increasing town populations and use of mountain campsites and hiking trails. Physical geographers and other scientists continue their research on the impacts of global climate change at the Arctic Station in Qeqertarsuaq on Disko Bay, (shown in Figure 18.11). Greenland's mysterious and complex human and environmental pasts as well as the weight of its still unknown future are drawing attention from ecotourists, business investors, and other curious travelers.

The Future of the Far North

During the past several decades, the cultures, environment, and economy of the Far North have changed dramatically. These changes build upon the region's ongoing development as traditional Native American and First Nations societies have given way in some places to economies based on resource extraction. Today, tertiary and quaternary activities are of increasing importance in some areas, although the explorer Stefansson's prediction that the Arctic would become a fully integrated and densely settled region is unlikely to come true in the foreseeable future.

High costs associated with adaptation to the harsh climate along with high transportation costs are important factors slowing the continued development of the Far North. Much of the region has little or no access to long-distance surface transportation. The two major railroads of Canada, the Canadian National and Canadian Pacific, traverse the continent south of the Far North region (Figure 18.12). However, rail transport is limited farther north. The Alaska Highway is one of very few all-weather highways in the region, most of which support alternative modes of transportation.

Aviation has greatly affected day-to-day life in the North. Even small communities have frequent commercial air service, and small planes, which are ubiquitous throughout the North, provide local residents with essential goods and services and connect them to the larger world. Many small planes are equipped with pontoons and skis so that they can land or take off in water and on snow (Figure 18.13). Of course, the cost of transporting people and freight by air is very high relative to surface transportation over comparable distances, making the overall costs associated with even basic necessities very high indeed. Locally, many residents of the North rely on snowmobiles and all-terrain vehicles for transportation.

Despite these concerns, some areas of the North have begun to develop new ways of surviving in this remote and harsh land. Tourism is on the increase in some locales. Hunters, fishermen, and boaters are attracted to the vast expanses of wilderness. Some of these tourists travel from cities to resort areas on the southern fringes of the North. Adjacent to the large cities of southern Canada and the northern United

▲ **Figure 18.11** Arctic research station, Qeqertarsuaq, Greenland

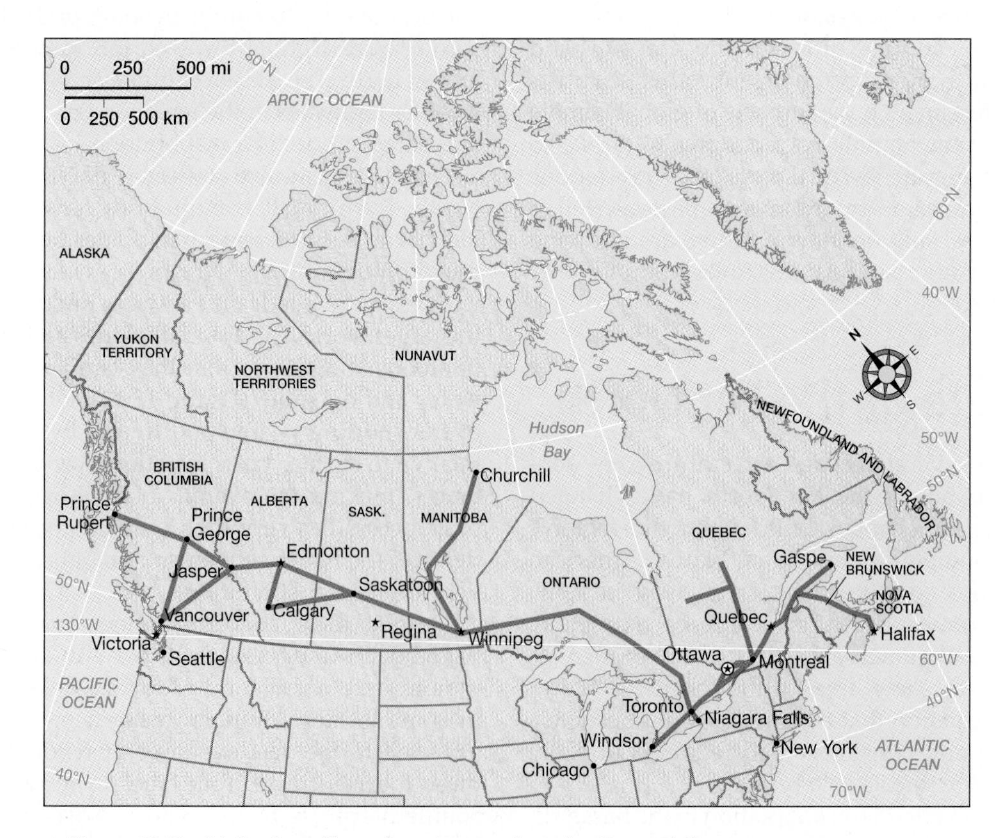

▲ **Figure 18.12** Major Canadian railroad routes (see also Figure 4.6)

◀ **Figure 18.13** Ski plane landing in Arctic snow

States are numerous recreation-oriented communities that are popular with tourists, retirees, and second-home owners from nearby metropolitan areas. Many of the houses are second homes that are located on lakes or rivers, built and maintained by residents of nearby cities, and used for skiing and skating in winter, fishing and boating in summer, and hunting in the autumn. Every weekend, highways between Detroit and northern Michigan, the Twin Cities and northern Minnesota, Montreal and the Laurentian wilderness, and Toronto and the Algonquin Park area northeast of Sudbury are jammed with persons heading to "the lake." In areas such as Brainerd, Minnesota, and Orilla, Ontario, the year-round population is increasing as second-home residents tired of the traffic and congestion of the city decide to retire and permanently move to the North.

Other tourism is focused farther north and involves long-distance travel by vacationers not only from nearby cities but also from farther away. Hunters and fishermen travel long distances by air to hunt moose, bears, and caribou and fish for trout, pike, and other species. Many people in northern Canada, Alaska, and Greenland earn their living by guiding outdoorsmen and tourists from southern Canada and the "Lower 48," and others run hotels, restaurants, and bed-and-breakfast operations catering to these outdoor-oriented tourists (Figure 18.14).

◀ **Figure 18.14** Ecotourists in Nuuk, Greenland

Despite the hope and hype of these new changes in local economies, cultures, environments, and lifestyles throughout the Far North, contemporary life is marked by tension between traditional lifestyles and the modern world as well as by strain between ethnic and racial groups. As Euro-American settlement penetrated the region, traditional lifestyles and customs have given way to modern ways of life. The Inuit and the other aboriginal people of the interior traditionally developed cultures oriented to making careful use of the resources of what is one of the most harsh and inhospitable places on Earth. To be sure, modern technology has made day-to-day survival in the unyielding environment of the North a little easier for them and other residents of the region. Modern medical science has lengthened lifespans considerably, and modern technology has expedited the exploitation of local resources. For example, Inuit people now hunt with guns rather than harpoons, live in prefabricated houses with electricity and running water rather than in igloos, and drive snowmobiles and all-terrain vehicles rather than dogsleds. Cable television, satellite dishes, and the Internet have allowed them to view contemporary life in the developed world. Inuit children play video games, hockey, baseball, and soccer rather than participating in traditional pastimes and recreational activities. These new technologies are also bringing new social problems. Alcoholism and drug addiction, family disruption, and loss of traditional cultures are all too prevalent in some communities as modern technology has reinforced dependence on an uncertain world economy in which resource-dependent communities are especially vulnerable.

Suggested Readings and Web Sites

Books and Articles

Berton, Pierre. 1988. *The Arctic Grail: The Quest for the Northwest Passage and the North Pole, 1818–1900.* New York: Viking. The exciting story of the almost 100-year search for the Northwest Passage.

Bone, Robert M. 1992. *The Geography of the Canadian North: Issues and Challenges.* Don Mills, Ont.: Oxford University Press Canada. A geographical interpretation of life and landscape in the Canadian North.

Dale, Kursti A. 1998. "Response of the Central Canadian Treeline to Recent Climate Changes," *Annals of the Association of American Geographers* 88: 183–208. This discussion of climate change at the treeline in central Canada provides interesting arguments about the impacts of global climate change on vegetation at a variety of scales.

Freeman, Milton M.R., Robert J. Hudson, and Lee Foote. 2005. *Conservation Hunting: People and Wildlife in Canada's North.* Edmonton: Canadian Circumpolar Institute Press. Overview of the environmental impact of "conservation hunting" in remote parts of northern Canada.

French, Hugh M. and Olav Slaymaker, eds. 1993. *Canada's Cold Environments.* Montreal: McGill-Queen's University Press. An examination of the physical landscapes and peoples of Canada's Far North regions.

Gayton, Don. 1998. "Healing Fire: Sometimes Saving a Forest Means Setting it on Fire," *Canadian Geographic* 118: 32–42. Geographical interpretation of the important role of fire in protecting and preserving ecosystems in Canadian forests.

Haycox, Stephen. 2002. *Alaska: An American Colony.* Seattle: University of Washington Press. An engaging, scholarly history, neatly divided into Russian exploration and the American period of settlement.

Hunter, Robert. 2004. *The Greenpeace to Amchitka.* Vancouver: Arsenal Pulp Press. Story of the founding and work of Greenpeace as one of the world's most active environmental organizations.

Keith, Darren, Jerry Arqviq, Louie Kamookak, and Jackie Ameralik. 2005. *Inuit Qaujimaningit Nanurnut/ Inuit Knowledge of Polar Bears.* Edmonton: Canadian Circumpolar Institute Press. Information on the polar bears of Canada's Far North as told through the lens of First Nations peoples (who share this part of Canada with them).

Lanken, Dane and Mary Vincent. 1999. "Nunavut Up and Running," *Canadian Geographic* 119 (1): 37–45. Geographic analysis of the environment, politics, cultures, and challenges and potentials of Canada's newest territory.

Mayer, Audrey, Pekka E. Kauppi, Per K. Angelstam, Yu Zhang, and Paici M. Tikka. 2005. "Importing Timber, Exporting Ecological Impact," *Science* 15 (April): 359–360. A surprising look at the potential global environmental impacts of conservation decision making in the boreal forests of the Far North.

Mowat, Farley. 1960. *Ordeal by Ice.* Boston: Little Brown. An adventure novel that brings the immense challenges of life in the Far North home.

Rundstrom, Robert A. 1990. "A Cultural Interpretation of Inuit Map Accuracy," *The Geographical Review* 80 (2): 155–168. A fascinating article on the innate cartographic abilities of Inuit peoples and the surprising accuracy of many of their earliest maps.

Swaney, Deanna. 2003. *The Arctic*. Melbourne: Lonely Planet Publications. A practical and inspiring travel guide for anyone interested in an expedition to some of the known and lesser known places in the Arctic subregion of the Far North.

Web Sites

"Canada's Newest Territory"
http://www.maps.com

Tourism, Parks and Recreation
http://www.uphere.ca/

Indian and Northern Affairs Canada
http://www.ainc-inac.gc.ca/ch/rcap/sg/sj26_e.html

Polar Bears of Churchill
http://www.churchillmb.net/~cnsc/ab-attrac-bears.html

Extreme Points of North America
http://en.wikipedia.org/wiki/Extreme_points_of_North_America

All Things Arctic
http://www.allthingsarctic.com/countries/canada.aspx

TABLE 19.1 Fastest Growing Cities in the United States between 2000 and 2005

City	2005 Population	2000 Population	Percent Change
Gilbert, AZ	173,989	110,061	58
North Las Vegas, NV	176,635	115,488	53
Rancho Cucamonga, CA	169,353	127,743	33
Chandler, AZ	234,939	126,997	33
Henderson, NV	232,146	175,405	32
Irvine, CA	186,852	143,159	31
Fontana, CA	163,860	129,147	27
Moreno Valley, CA	178,367	142,379	25
Bakersfield, CA	295,536	243,213	22
Chula Vista, CA	210,479	173,555	21

Source: USA Today, 12-22-05.

Similarly, Canada contained only seven provinces: Saskatchewan and Alberta were federally administered territories, and Newfoundland would remain a British colony until 1948. Most Canadians lived in Ontario or Quebec; for example, Alberta had only 1.4 percent of Canada's population in 1900 but more than 10 percent by 2000.

Agriculture was the dominant occupation in North America in 1900. More than half of all North Americans lived on farms. Although the number of manufacturing jobs was increasing, most North American industrial activity was concentrated in the large cities of the eastern United States and southeastern Canada. Trade with Europe was extensive, but the amount of trade with other areas of the world was miniscule by today's standards. Relative to 2000, a much lower percentage of manufactured goods were imported or exported. Most American-made products were marketed and sold within the United States, and high protective tariffs were enacted to discourage foreign competition.

Given the agricultural orientation of the North American economy, it is not surprising that a large majority of North American residents lived in rural areas and small towns. The percentage of persons classified as urban, or living in incorporated communities of 2500 or more people, was less than 20 percent in 1900 but over 80 percent by 2000. Buoyed by their industrial bases, such urban areas as New York, Boston, Philadelphia, Chicago, St. Louis, Montreal, and Toronto had already developed into major cities. The populations of these and other cities, however, were concentrated into much smaller areas and a large majority of urban residents lived within the limits of central cities. The ever-expanding suburban rings surrounding every major North American city today were farmland or wilderness.

Culturally and economically, the gap between urban and rural lifestyles was much greater than is the case today. Middle- and upper-class urban residents enjoyed electricity, running water, indoor plumbing, and many other "modern" conveniences. Especially outside the northeastern United States, however, such conveniences were largely absent in rural and farm areas. Farm families generally relied on well water for washing and drinking, coal and wood for heating, and draft animals for labor. Not until the Rural Electrification Administration began operating in the 1930s was the convenience associated with electrical appliances a reality for most farm dwellers in the United States.

On a formal level, the political institutions governing North America were well established in 1900. Both the United States and Canada were functioning democracies at a time when very few other countries of the world had established democratic forms of government. In practice, however, early North American democracy was far more restrictive than is the case today. For the most part, the opportunity to vote and participate in government and politics was limited to white men. Native Americans would not be recognized as United States citizens until the passage of the Dawes Act of 1928. In the South, where a large majority of African Americans resided, "Jim Crow" laws effectively denied the right of most African Americans to vote, serve on juries, or run for public office. Women in 1900 could vote in federal elections in only a few states, mostly in the Rocky Mountain region. Not until 1919 did the passage of the 19th Amendment guarantee women the right to vote throughout the country.

The vast and far-flung North American political economy was tied together by a system of transportation and communication that was advanced relative to the standards of the rest of the world at the time but quite primitive by contemporary standards. North America boasted the world's largest and most extensive railroad network. Completion of the transcontinental railroad in 1869 made it possible for someone to travel from New York to San Francisco—a journey that would have taken months in 1800—in less than a week. Railroad connections were extensive in the Northeast, and increasing in the South, the West, and western Canada. However, "horseless carriages" had only recently been perfected and were still regarded as novelty items. The heyday of the Model T Ford, which put mechanized personal transportation in reach of middle-class Americans throughout the country, would not occur until after World War I. Those few North Americans who owned automobiles often have difficulty finding passable roads on which to drive them, especially outside cities. Rural communities and outlying farms were generally not served by paved roads. Primitive rural roads were often impassable during and after storms or after snowmelt in the spring. The Wright Brothers' first successful flight at Kitty Hawk, North Carolina, would not take place until 1903.

Telegraph lines made cross-continental communication feasible, and North America boasted a modern, efficient postal service which allowed mail to be delivered to urban and rural areas as fast as it could be moved along rail and road networks from city to city. Telephone service was well-established, especially in cities, although long-distance telephone connections were unreliable and expensive. The instantaneous communication that most North Americans take for granted today was a product of the second half of the 20th century.

Like other parts of the world, North America has undergone a dramatic **demographic transition**. Preindustrial North America was characterized by very high birth and death rates. Death rates began to drop rapidly in the 19th century in association with the development of modern medical science and technology. By the mid-20th century, birthrates also began to fall steadily. Families of ten or more children, which were commonplace as late as the early 20th century, are highly unusual today. Thus, the demographics of contemporary North America are characterized by much lower birthrates and much lower death rates than was the case throughout most of North America's history. At the same time, continued advances in medical science, nutrition, and health care have resulted in steady increases in longevity. Thus, the continent we have discussed in this textbook would seem to provide the ideal case study for illustrating the various stages of the classic demographic transition model.

But how will this ongoing demographic transition influence the human and physical geography of North America in the 21st century? First and foremost, these changes are resulting in a steady aging of North America's population. The median age of North Americans is now more than 33 years—that is, half of North Americans are more than 33 years old. As life expectancies increase and birthrates remain low relative to historical trends, we can anticipate that the median age will increase even more. As a result, the ratio of elderly persons to younger persons will continue to increase (Figure 19.1).

What are the impacts of this aging of the population on North American society? The increasing ratio of elderly to non-elderly people means that relatively fewer North Americans will work and earn money to take care of their elders. Currently, political leaders in the United States are debating reform of the Social Security system. When Social Security was established in the 1930s, the ratio of retirees to working persons was about 1:30, and the average length of one's retirement was less than five years. Today, this ratio is about 1:3 and it is expected to rise to 1:2 in the foreseeable future. Many people today are retired 30 years or more before their death. This increased ratio is attributable to longer life expectancies and earlier retirement opportunities. Reform of social security and restructuring of how elderly people are supported financially in the future is a high priority in current political circles.

The aging of North America's population has a distinct geographic dimension. Prior to the 20th century, the close ties between people and the land meant that most elderly people had children or other relatives nearby, and that relatively few people moved to elderly-dominated communities. With the decoupling of people from their land base, many elderly persons no longer live near their children and other relatives. Many move to retirement-oriented places; others remain behind after their children leave home.

In examining migration patterns of the retirement population, it is important to distinguish between the "young" elderly and the "older" elderly. The young elderly include those persons, typically in their 60s and 70s, who remain healthy and active. These people make migration decisions on their own, and many seek out retirement-oriented communities or other places associated with attractive climates, recreational opportunities, a reasonable cost of living, and other amenities. The year 2005, in fact, marked the 50th anniversary of the establishment of Sun City, Arizona, which was the first major housing development catering specifically to the young elderly (Figure 19.2). The founders of Sun City coined the term "golden years" in marketing the concept of retirement living. Previously, retirement was the relatively short period between the end of one's working lifetime and the end of one's life, and it was a period usually marked by illness, infirmity, and dependence on others. In the 1950s, about 60 percent of American men aged 65 were in the workforce; today, this number has dropped to about 48 percent.

The older elderly, on the other hand, are those whose health no longer permits them to lead active lifestyles or to make migration decisions on their own. Many live in nursing homes, in assisted living centers, or with children or other relatives, and their decisions concerning where to move are usually made by, or at least in consultation with, family members and medical professionals. In effect, a major demographic trend over the past half century has been the identification of and marketing to the young elderly, a group whose numbers

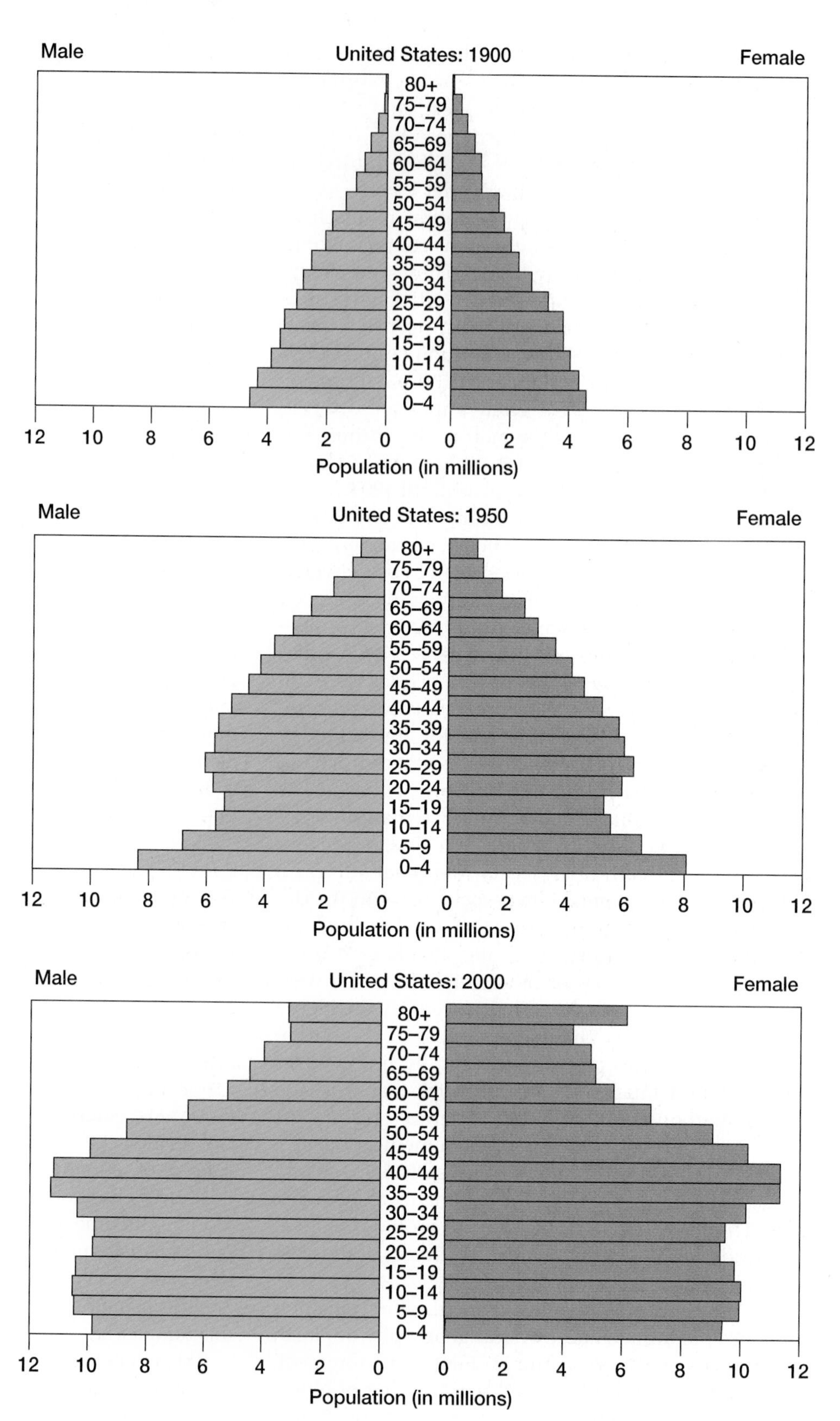

Figure 19.1 U.S. population pyramids, 1900, 1950, 2000

will swell dramatically in the early 21st century as baby boomers leave the workforce.

The result of these demographic and migration trends is that the North American landscape contains two distinctive types of places where population is dominated by the elderly. Places such as central Arizona, central Texas, the Missouri and Arkansas Ozarks, the mountains of North Carolina and Georgia, and parts of Florida contain numerous retirement-oriented communities that attract older people from throughout the country. Sun City, Arizona (and other Sun Cities in other states); Llano, Texas; Mountain

◀ **Figure 19.2** Sun City, Arizona

Home, Arkansas; and Venice, Florida, are but a few communities known throughout the United States to be retirement-oriented. In other places, the median age of the population is high because many younger people move away, leaving their elderly parents behind. With few economic opportunities in many rural areas, young people leave home to attend college or join the armed forces, and few who leave return to their native communities. Dozens of counties in rural Saskatchewan, Manitoba, Montana, North and South Dakota, Nebraska, Kansas, Mississippi, West Virginia, and other rural, slow-growing states and provinces have populations dominated by the elderly. Such communities, as we saw in the Great Plains chapter, can be described as **aging in place** (Figure 19.3).

These two types of communities have very different problems. Retirement-oriented communities and counties tend to be upscale and have growing populations of "young" elderly people who often have substantial disposable incomes. Tension sometimes arises between more affluent, elderly newcomers and long-term residents. For example, in some communities the elderly newcomers have refused to support school bond issues or tax increases to support education, on the grounds that they have already raised their children and are reluctant to pay to educate other people's

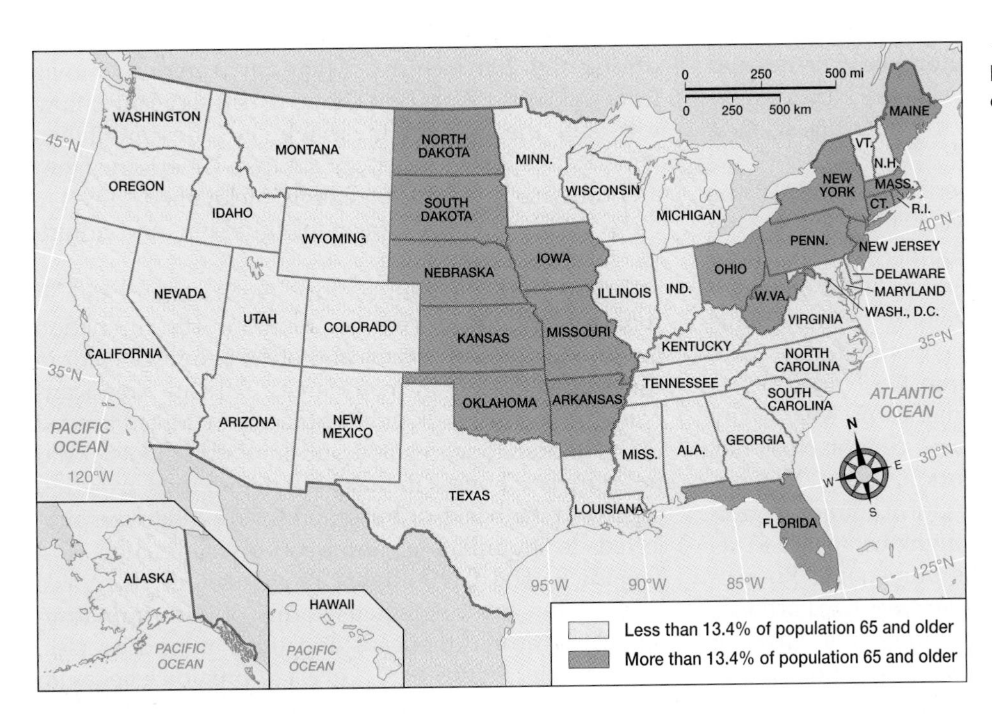

◀ **Figure 19.3** U.S. states with high percentages of persons over 65

children. This is partly age related and partly the result of the Western world's individualistic culture vis-à-vis a moral one. In some such communities, rapid growth and poor long-term planning have contributed to traffic, pollution, overcrowding, and other problems that imperil the very serenity and aesthetics that had made these communities attractive to the retirees in the first place.

Communities that are aging in place have very different problems. Many are isolated and sparsely populated, making it more difficult for the elderly to obtain medical treatment and other services. Lack of access to adequate transportation and to social opportunities has been an increasing issue in some aging-in-place communities. Many elderly persons are reluctant to give up their deep roots and profound place ties to the communities in which they have lived most of their lives. But the problem of providing needed services to sparse, dispersed, and isolated populations in such communities is a very real problem and one that will become exacerbated as the number of elderly people in these communities continues to increase relative to the overall population.

Culture, Race, and Ethnicity

As we have seen throughout this book, North America is a continent of immigrants and most of its residents are descended from persons who migrated, voluntarily or involuntarily, to the continent over a period of 500 years. Both American and Canadian culture have been enriched immeasurably by immigrants past and present who arrived from all over the world. Albert Einstein, Joe DiMaggio, Colin Powell, Alberto Gonzales, Jennifer Lopez, Diana Taurasi, and Tiger Woods are but a few well-known 20th century immigrants, or sons and daughters of immigrants. All are household names and have had profound impacts on the lives of North Americans through their contributions to public life, science, and popular culture.

Most immigrants to North America before 1890 came from northwestern Europe. Between 1890 and 1924, a majority came from eastern and southern Europe. Since the United States and Canada liberalized their immigration laws in the 1960s, a large majority of immigrants have come from other parts of the world. Today, more than 80 percent of immigrants into the United States and Canada come from Latin America or Asia. Many of today's immigrants move to large cities such as Los Angeles, Miami, Atlanta, New York, Chicago, Vancouver, Montreal, and Toronto. Others form ethnic communities in what on the surface might be regarded as unexpected places, including the burgeoning Russian and Ukrainian community in smaller cities like Portland, Oregon, and Vernon, British Columbia; a growing Vietnamese community in Oklahoma City; and expanding Mexican and Central American settlements in the Carolinas and Georgia. Landscape features such as the Cuban market in Portland, Oregon (Figure 19.4) document the ongoing process of migration and adjustment to new communities.

▲ **Figure 19.4** Cuban grocery store in Portland, Oregon

Changing patterns of immigration have been followed by changes in the patterns of acceptance of ethnic identity. During the 19th century, many immigrants faced prejudice from established residents. "No Irish Need Apply" signs greeted Irish immigrants escaping the Irish Potato Famine of the 1840s as they moved to Boston and other East Coast cities. In the late 19th and early 20th centuries, many native-born Americans and Canadians regarded immigrants from southern and eastern Europe as Communists or anarchists and considered their values—in many cases including their Catholic or Jewish faiths—as threats to what they regarded as "American" culture. Such prejudice against immigrant white North Americans began to disappear during the 20th century. Many an American movie from the World War II era shows a drill sergeant calling roll, and the platoon invariably includes an Italian American, a Polish American, an Irish American, a Jew, a Southerner, a Texan (inevitably nicknamed "Tex"), a New Yorker from the streets of Brooklyn, and a midwestern farm boy or two.

Throughout the 20th century, North American culture was characterized by rampant discrimination against African Americans and other people of color in the United States. In many areas, African Americans could not vote or seek public office, were forced to send their children to segregated, substandard schools, could not purchase houses in many neighborhoods, were relegated to the backs of buses, and were otherwise subjected to humiliating forms of discrimination and prejudice. The Civil Rights movement of the 1960s banned the most egregious forms of **discrimination** against African Americans and other minorities (see, for example, Figure 19.5). Levels of prejudice began to

◀ **Figure 19.5** Civil rights demonstration in Birmingham, Alabama, 1964

decline, although surveys of North Americans continue to reveal that many, particularly older people, remain unwilling to accept African Americans and other non-whites as family members, neighbors, or close personal friends.

Declining levels of prejudice have accompanied rising levels of **ethnic awareness** and pride. During the early and mid 20th century, many journalists and social commentators regarded North America as a **melting pot** into which ethnic distinctions would disappear after two or three generations. This melting pot concept was promoted by the mass media. During the early days of radio, movies, and television, numerous performers changed their given names to more "American" names associated with northwestern European origins. For example, Morris Horwitz became Moe Howard of the Three Stooges; Issur Danielovitch became actor Kirk Douglas, and Concetta Franconero became singer Connie Francis. Similarly, actors and announcers were trained to speak "standard American English"—that is, English as spoken by residents of the mid-Atlantic and Great Lakes core region. Performers from Canada, New England, Appalachia, and the Deep South were expected and taught to eliminate their distinctive accents.

By the 1960s and 1970s, however, after the passage of the first legislation in support of multiculturalism in Canada and later in the United States, what had been a source of shame and embarrassment for many foreign-born Canadians, Americans, and their descendants became a source of pride. Singers and actors no longer hid their distinctive accents and backgrounds. Rather than attempt to hide or obscure their ethnic origins, many Americans began to celebrate them as in Figure 19.6. St. Patrick's Day, Columbus Day, and Cinco de Mayo emerged as opportunities for persons of Irish, Italian, and Mexican ancestry, respectively, to express pride in their heritage. James Brown's song, "I'm Black and I'm Proud," inspired many African Americans to take pride in their ethnic background. As persons asserted racial and ethnic pride, however, in some cases conflicts between different ethnic groups arose. Tensions and ugly manifestations of prejudice between both native-born and immigrant ethnic groups continue to fester throughout North America.

The changing nature of ethnic identity and ethnic conflict is influenced by several additional trends, including increasing levels of intermarriage and the increasing spatial separation of ethnic groups. The melting pot model was associated with a distinctive geography emphasizing spatial separation of ethnic groups. During the late 19th and early 20th centuries, many first- and second-generations lived in tightly knit urban ethnic communities. People worked, shopped, played, and worshipped in neighborhoods where populations were dominated by persons of the same ethnic heritage. For an Italian American in Toronto or New York's **Little Italy**, for example, most social contacts—family members, neighbors, parishioners, co-workers, members of the Sons of Italy—were Italian Americans who lived within a few blocks, or, at most, a few miles. Most people married spouses who came from the same place and from a similar background. The second and third generations would be exposed to wider horizons, in part as a result of formal education and military service, and would more likely settle in suburbs where relatively few neighbors were Italian Americans. These persons would have a wider range of social contacts through work, school and college, military service,

▶ **Figure 19.6** Cinco de Mayo celebration in Los Angeles

neighborhoods, churches, and social clubs and would therefore more likely marry persons of different ethnic heritage.

Today, an increasing number of North Americans now identify themselves with more than one ethnic group. In 1997, the U.S. Census began allowing people to identify themselves as **mixed race**. The United States Census in 2000 showed that a substantial number of people did indeed identify themselves in this way. In Hawai'i, more than 21 percent identified themselves as multi-racial (see Chapter 17). Next were Alaska (5.4 percent), California (4.7 percent), and Oklahoma (4.5 percent). The map shown in Figure 19.7 identifies states with very few people who claim a multiple race heritage.

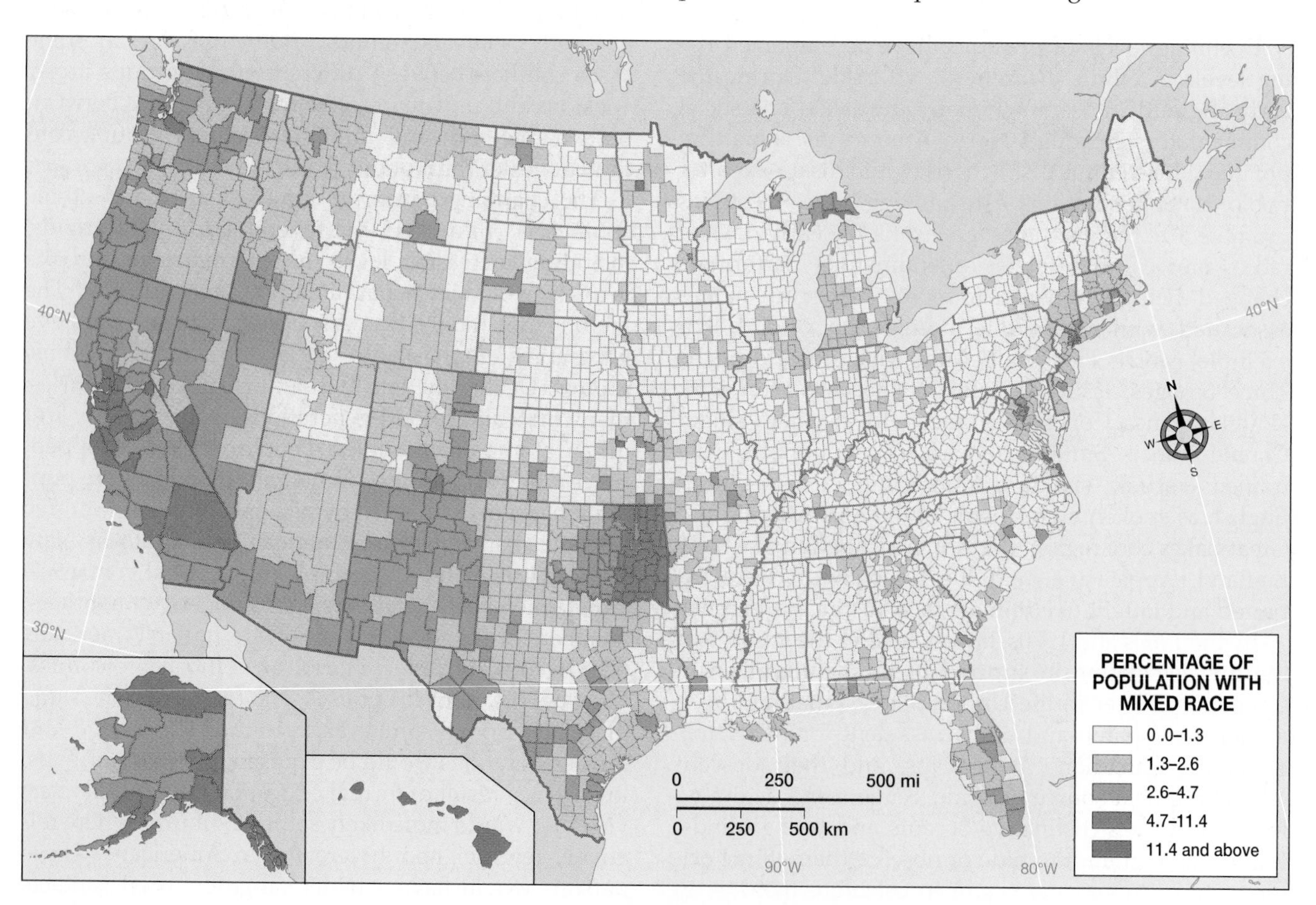

▲ **Figure 19.7** Persons claiming a mixed race heritage in the U.S. 2000

This incorporation process has coincided with a new trend in which ethnic ties are maintained despite spatial separation of group members. Large communities of persons from Mexico, China, the former USSR, India, Korea, Vietnam, the Philippines, and many other places throughout the world have recently moved to the United States and Canada. In contrast to earlier ethnic communities, however, these modern-day ethnic communities are not characterized by spatial proximity. For example, there are large communities of Vietnamese Americans in southern California, Houston, Oklahoma City, New Orleans, and northern Virginia. However, members of these communities generally do not reside in proximity to one another; rather, they are scattered in urban and suburban neighborhoods throughout these respective metropolitan areas. Automobiles, telephones, and the Internet enable community members to keep in close social contact with one another, and members of these communities gather for weddings, funerals, religious ceremonies, festivals, and holidays; however, relatively few live in communities populated exclusively by members of their own ethnic groups. As we discussed earlier in the book, geographers use the term **heterolocalism** to refer to this new form of urban settlement by immigrant groups.

People are increasingly linked across generations and across places by ties to persons with common professional interests and achievements. This trend is evident by looking at well-known politicians, athletes, and entertainers, such as Charlie Sheen, Michael Douglas, Gwnyth Paltrow, Melanie Griffith, Barry Bonds, George W. Bush, Al Gore, and Peyton Manning, all of whom followed famous parents into their professions. This holds true even among professional geographers: One of the authors of this book is a close relative of another geographer who was recently the president of the Association of American Geographers, the premier professional organization for geographers in the United States. The other two authors are married to other Ph.D. level geographers. Many professional geographers maintain close personal as well as professional relationships to fellow geographers around the world. Indeed, like many others in their age, education, and income bracket, many of them are much more apt to confide in or interact with these fellow professionals than they are with their own next-door neighbors. Similarly, many people maintain close contact with persons who share their religious beliefs, political convictions, hobbies, and personal interests even though these people may be separated spatially by thousands of miles. Universities themselves often provide such opportunities as well, as discussed in Box 19.1.

In the foreseeable future, how will these complementary trends in ethnic awareness, identification, and settlement affect North American life? Certainly, we can expect a continued upsurge in ethnic awareness and

BOX 19.1 University Alumni Associations as "Affiliated 'Ethnic' Groups"?

In a globalizing and sometimes increasingly impersonal world, people look to bond with one another on the basis of common interests. Historically, this bonding was facilitated by ties with family, neighbors, religious communities, and ethnic groups, but as we have seen throughout the book, these ties are weakening. People sort themselves out on the basis of attachments to their professions, special interests and political convictions.

Increasingly, university alumni associations have stepped in to play roles formerly associated with ethnic communities. Many university Web sites maintain directories of local alumni associations. These associations invite alumni of the university living in the local area to social events, trips, and fundraising activities. Many of these events are associated with university athletic events such as football and basketball games. If a university sports team plays in or near the city in which the club is located, parties are organized to get university alumni together to support the team; if the university's president or a high-ranking official of the university is visiting the area, events are held to welcome the official and reinforce emotional and, of course, financial ties to the university.

The University of Oregon Alumni Association, for example, has active alumni clubs in many different areas. There are active clubs in several Oregon communities including Eugene (the seat of the university), Portland (the state's largest city), and High Desert (eastern Oregon). Outside Oregon, there are active clubs in Chicago, Denver, Honolulu, New York, Northern California (San Francisco), Orange County (Anaheim), Puget Sound (Seattle), San Diego, Southern California (Los Angeles), Treasure Valley (Boise), and Washington, D.C. Of course, this distribution reflects the migration patterns of many Oregon graduates, many of whom had moved to these communities after graduation.

The distribution of alumni clubs associated with universities elsewhere reflects different migration histories. For example, the University of Vermont has active alumni chapters in many of the places that also have Oregon alumni clubs. In addition, however, there are UVM alumni clubs in Albany; Atlanta; Austin; Boston; Burlington, Vermont; Cleveland; Dallas–Fort Worth; Houston; Manchester, New Hampshire; Miami; Milwaukee; Orlando; Philadelphia; Pittsburgh; Portland, Maine; Providence; Syracuse; Tampa–St. Petersburg; and Toronto, reflecting the greater tendency of Catamount alumni to settle in eastern and midwestern cities relative to former Ducks. Eugene, High Desert, and Treasure Valley are the only places that have active Oregon clubs but do not have active Vermont clubs.

knowledge—knowledge aided greatly by electronic data bases and the Internet. Ethnicity and race will remain important components of one's identity, but it may become less essential in defining the characteristics of many persons. These changes will be associated, it is hoped, with increasing levels of understanding, tolerance, and mutual respect for differences in ethnic heritage as well as other aspects of culture and lifestyle.

Political Economy

Future changes in North America's geography, of course, will be tied closely to changes in local, national, and global economies. Over the past four centuries, the economy of North America has undergone considerable change. During the 17th, 18th, and 19th centuries, North America's economy was dominated by the primary sector. As late as 1900, well over half of North America's people lived on farms, and agricultural products were among the United States' and Canada's major exports. Throughout the 20th century the farm population declined steadily; today barely one percent of North Americans earn their living through agriculture and other primary-sector activities.

The 19th and early 20th centuries saw a spectacular rise in manufacturing in North America. By 1920, the United States and Canada had become major producers of industrial goods that were exported throughout the world. During the mid-20th century, industrial output generated far more revenue into the United States and Canada than did the export of agricultural products and other primary-sector activities. In the late 20th century, the value of manufacturing declined relative to the tertiary and quaternary sectors of the economy. An increasing percentage of the world's output of steel, automobiles, appliances, electronic devices, and other industrial products were manufactured outside North America, in many cases because the cost of labor in less developed countries is substantially less than that in the United States and Canada. By 2000, less than 20 percent of the work force of North America was employed in the primary or secondary sectors of the economy. At the same time, North America remains a dominant player in the world economy.

The domination of the tertiary and quaternary sectors has been accompanied by large-scale economic globalization. North American communities in colonial times were largely self-sufficient; most of the food, clothing, and other goods consumed in these communities were produced locally. Today we live in a highly **globalized economy**. Visitors to any department store or shopping mall can choose from among thousands of products manufactured throughout the world. With continued dramatic decreases in the costs of transportation and communication, we can anticipate that globalization and the integration of North America's economy with those throughout other parts of the world will continue to increase.

What will the continued globalization of the North American economy, and the continued shift from an industrial to a **post-industrial economy**, mean in terms of North America's geography? As has been made evident throughout this book, the economic health of particular places within North America is influenced in part by their economic bases. Those communities whose economic bases are oriented to the primary and secondary sectors often find themselves in economic difficulty as the number of farms continues to decline and as manufacturing jobs are lost to overseas competitors. On the other hand, communities oriented to the tertiary and quaternary sectors, including those oriented to tourism, government, education, research and development, and medical services, are tending to grow and prosper. Large metropolitan areas, with their close global linkages and highly diversified economic bases, are also tending to grow at rates faster than the continental average. But can this growth be sustained without local agriculture or manufacturing?

More generally, we are seeing a decoupling of the links between place and economic productivity. The primary sector of any economy is by definition highly localized; natural resource production must be concentrated where the resources are located. The agricultural productivity of the Great Plains is related to the fact that soils and climatic conditions are highly favorable for large-scale crop production. Likewise, natural resources helped to determine the development of mining economies in Labrador, Montana, and Arizona, or timber production in Maine, British Columbia, and Georgia.

As we have seen throughout the book, natural resource location is an important, but by no means a dominant, factor in the location of industries. Historically, industrial location has been closely tied to the proximity of raw materials and markets. For example, the primacy of western Pennsylvania and eastern Ohio in steel production is associated with this area's location relative to the coalfields of Pennsylvania, West Virginia, and Kentucky; the iron mines of Michigan and Minnesota; and markets for steel and steel products in the large cities of the Atlantic seaboard. Today, however, raw material location and availability is a far less critical component of industrial location. Labor costs are far more critical in determining the location of industries, and the impact of raw material costs, and therefore of raw material location, is relatively minor. This is particularly true with respect to high-technology industry. For example, a newly released CD by a highly popular band might cost $20, but the raw materials used to make the CD itself (along with its plastic case and the cardboard cover) may cost only a few cents. Or through Internet (legal or illegal) transactions there may be no "hard costs" associated with a transaction. Most of the cost of the new release is therefore associated with the creativity of the musicians along with the technical expertise of the recording engineers who produce it for public consumption.

Many tertiary and quaternary sector activities are even less oriented to place. Given modern telecommunications, people in many occupations can live and work anywhere and maintain links and connections with clients and professional associates instantaneously. Ironically, some tertiary and quaternary activities are very highly place oriented. Spending on government and education is concentrated in state and national capital cities and in college and university towns, respectively. The specific location of such spending may be the result of historical accident. Decisions to locate universities and state capitals at their current locations were often made a hundred or more years ago. Yet as we have seen throughout the book, places such as Kelowna, British Columbia; Iowa City, Iowa; Bend, Oregon; Brookings, South Dakota; Morgantown, West Virginia; Lexington, Kentucky; and Gainesville, Florida, are among the fastest-growing communities in their respective states and provinces. However, the locations of some other tertiary- and quaternary-sector activities are determined by the availability of nearby natural features or historical sites.

This situation is illustrated by the example of tourism. Some tourist destinations owe their popularity to their unique locations relative to sites of natural or cultural significance. Examples would include Jackson Hole, Wyoming, because of its proximity to Grand Teton and Yellowstone National Parks, or Honolulu with its beach location and its access to the Arizona Memorial and other historical sites. On the other hand, the most popular tourist sites in North America—Las Vegas and Orlando—do not owe their popularity to natural phenomena or historical significance. Rather, the locations of Las Vegas and Orlando as tourist destinations, as we saw in earlier chapters, are the result of specific decisions made by far-sighted entrepreneurs with little concern for local resources except developable land.

Over the course of time, communities have competed with one another for economic opportunities. As we observed in several chapters throughout the book, plans on the part of major corporations to establish new plants or manufacturing facilities have spurred intense competition among the states, provinces, and local communities that regard these locations as desirable because they are seen as bringing in economic growth, jobs, and people. Of course, in many cases economic growth can be regarded as a mixed blessing. The location of new economic opportunities can result in new jobs and more income, but can also cause environmental degradation, overcrowding, traffic, pollution, and other problems associated with rapid population increases. Nor can we automatically anticipate that new economic opportunities will continue to mean a concentration of local benefits. Constructing a new factory, for example, may create new jobs, but how many of these new jobs will be offered to local residents? Likewise, the building industry itself may produce a local economic boom but it will, most likely, be temporary. The new company may then move employees with specialized training and experience from elsewhere, and some of the local jobs may also be taken by people who commute to work over longer distances. Hence many people in local communities now see the development of new industry as a mixed blessing and would prefer economically sustainable development.

Settlement Patterns and Urban Structure

The foregoing considerations also impact the structure of settlement and urbanization in North America. Today, the fastest-growing places in North America are those located on the fringes of large metropolitan areas. Such places have unique demographic and economic mixes. Do they foreshadow the North America of the mid and late 21st century?

Throughout this book, we have identified several major population movements and examined their impacts on places in various regions of North America. First, settlers of European ancestry moved into lands occupied by Native Americans and First Nations peoples, with the original inhabitants killed, driven away, or resettled onto reservations. By today's standards, this was a slow process. In commissioning Meriwether Lewis and William Clark to explore the Missouri River following the Louisiana Purchase of 1803, President Thomas Jefferson estimated that it would take four or five centuries for Anglo-Americans to occupy the North American continent effectively. Less than a century later, the frontier was considered closed.

The closing of the frontier coincided with the outset of a shift from an agricultural, resource-based economy in North America to an industrial economy. With this shift came large-scale movement from rural to urban areas. Millions of people who had been raised on farms and in rural communities flocked to the cities, where they were joined by millions of immigrants from Europe to provide labor in North America's booming factories.

The third major trend in settlement has been a shift from urban areas to suburbs and, more recently, a move away from more expensive cities in the Coastal South and West (Box 19.2). The rural to urban trend was already underway by the 1920s, but it accelerated dramatically after World War II and the onset of our current post-industrial economy. By 2000, more than half of all North Americans lived in suburbs that continue to grow more rapidly than either urban or more isolated rural areas. In most metropolitan areas, three-quarters or

BOX 19.2 Fleeing Expensive Coastal States in the United States for the Inland South and West

According to a recent analysis of population estimates from the U.S. Census, the search for affordable housing and employment is driving Americans from expensive coastal states to more moderately priced parts of the country. During the first five years of the 21st century, people continue to leave states such as New York and California to start new lives in parts of the Southwest, Southeast, and Rocky Mountains. In the 12 months ending on July 1, 2005, in fact, Florida gained more people (404,434) than any other state for the first time in at least 15 years. Despite four hurricanes that damaged the state during the past year, Florida added an average of 1100 people a day, bringing its population to almost 18 million. If this pace continues, Florida will overtake New York as the second most populous state by 2010. Here are some other highlights from this recent census report:

- New York lost people for the first time since 1980. These losses were mostly in upstate cities, not in New York City and its suburbs.
- California was not the top state for growth for the first time since 1995. Most of the state's gain of 290,109 came from births. The data show that 239,417 more people left California for other parts of the United States than moved in.
- Largely because of strong job growth, Virginia gained more people (86,133) than the nine northeastern states combined.

According to William Frey, a demographer at the Brookings Institution, a Washington, D.C., think tank:

> It's a symptom of the new divide in housing costs between the expensive, congested, urbanized states such as California and New York and newly sprawling suburban states on both coasts. This whole half of the decade housing has been an issue. The question is: Will it continue?

Population shifts ultimately have political consequences because seats in the House of Representatives are reallocated after the full census every ten years. Based on the latest estimates, five states would lose a seat and four other states could gain a seat (because Louisiana is likely to lose one once hundreds of thousands of residents who were displaced by Hurricane Katrina fail to show up in census counts). In sum, according to Robert Lang, Director of the Metropolitan Institute at Virginia Tech:

> The decade began with the economy off track, but the population boom kept rolling along and is on track to nearly match the record 1990s. The country found places to keep booming, shying away from the high cost coasts and seeking the South and the mountain West.

Source: As quoted in Haya El Nasser and Paul Overburg, "People Fleeing Pricey Coastal States for South, West," *USA Today*, posted 12/22/05. *http://www.usatoday.com/news/nation/2005-12-22-census-data_x.htm*

more of the metropolitan population lives outside the central city in **suburbs** or **exurbs**. Increasingly, sprawl is extending urban development farther and farther away from the center of the metropolitan area. Suburban development is taking place on the fringe of metropolitan areas as once-rural landscapes are being bulldozed and converted to suburban housing developments and shopping centers. Such places are known as **edge cities** and they often incorporate to become new cities. The process of suburbanization and exurbanization of North America is associated with continued improvements in communication and transportation. The movement of jobs from central cities to outlying suburban locations is allowing people to relocate further and further from the center of the city and to create new cities with their own employment bases. People continue to be willing to pay to live on the metropolitan fringe, where many feel that they have the best of both worlds—on the one hand, easy access to the culture, amenities, and economic opportunities of the central city; on the other hand, easy access to peace and quiet and recreational opportunities in newer communities.

The popularity of exurban living does not imply that exurbanization is an unmixed blessing. Numerous problems have been associated with suburban and exurban life. In many such communities, the place ties traditionally associated with small-town life are absent or are "reconstructed" by urban planners. As the heterolocalism model suggests, most people are more likely to maintain close ties with fellow ethnic group members and with co-workers or others in their professions as opposed to their neighbors. Exurbanites are much less likely to have deep roots in the communities into which they move, and in some places there have been tensions between newcomers and long-time residents who may regard the newcomers as unappreciative of the culture, values, and histories of their new communities. Environmental problems can also be associated with ongoing suburbanization and exurbanization. The continuing conversion of rural land to suburban housing developments has resulted in habitat fragmentation and, in some places, increased conflicts between people and bears, coyotes, alligators, and mountain lions whose habitats are expropriated for development.

Exurban communities also tend to force people to be car dependent, as public transportation is poor or absent and most people have to rely on cars to get to work, shopping, or recreational activities. Not only is this car dependence a cause of increased pollution, but it may be contributing to imperiled health. A recent study showed that the average American adult living in the suburbs weighs about six pounds more than the average central city resident, and this difference is attributed to the fact that suburbanites are much more likely to drive rather than walk to various places in and near their communities. A United States government report predicted that the epidemic of obesity might, in fact, result in a reversal of North America's long-standing tendency toward increased life expectancy in the years ahead.

These observations and predictions are not inevitable. A growing number of communities, especially those along the American and Canadian Pacific Coast, the Gulf Coast, and in the northeastern United States are employing the principles of **smart growth**. In these communities, new development is located near transit stations and high-density land uses are mixed to help reduce vehicle trips from home to work to schools, work, and shopping. Likewise, land-use densities and intensities are higher and often look like updated versions of early 20th century urban communities.

North America's Role in the World

The tragic attack on the World Trade Center and the Pentagon on September 11, 2001, brought into focus the changing role of the United States in the world of the 21st century. North America is politically and economically the dominant region of the world. One outcome may be the migration to parts of Canada of increasing numbers of Americans as **landed immigrants**.

Americans have been drawn to life "north of the border" for centuries. Beginning with the loyalists to the British Crown leaving the 13 colonies to move to eastern Canada during the American Revolution in the 1700s, then African Americans fleeing their persecuted lives as slaves in the American South on the secret Underground Railroad that ended in southern Ontario, to Vietnam War resisters in the 1960s and 1970s, up to today's migrants from the United States who are moving north to retire and/or escape the Homeland Security–era politics of their birthplace, Americans have relocated in Canada by the tens of thousands.

According to the U.S. Consulate in Vancouver, today there are at least 250,000 legal American immigrants living in the province of British Columbia alone. The largest group came during the Vietnam War years and settled in small towns such as Nelson and Castlegar and in the city of Vancouver. Since that time, immigrants from the United States have continued to move to British Columbia to retire either as full-time residents or to live for part of the year in places like Vancouver Island in their RVs or second homes. Many Vietnam-era and other American migrants now serve as university professors and teachers, city council members, and in other leadership roles in their new Canadian homes.

The northward migration of Americans to Canada provides a fascinating case study of an understudied group of immigrants in North America. What citizenship do most of these new Canadians hold—American, Canadian, or dual citizenship? Do most American-born migrants in British Columbia identify as being Americans or Canadians? How have they changed the landscapes of their new Canadian homeland? These and other questions remain unanswered and deserve further study by human geographers interested in immigration, transnational identities and relationships, and citizenship issues in North America.

As we have seen throughout this book, the history of North America is one of continued increase in power (political, economic, and military) and influence in the global economy. What is now the United States was part of the British, French, and Spanish colonial empires for many years, and Canada remained a British colony until it became independent in 1867. Over the course of the 19th century, North America grew rapidly until by 1900 it had emerged as a leading economic and industrial power. Europe remained at the center of the global political stage, however, until intervention by the United States ensured that the Allies would be successful in winning World War I. After that war, the United States declined to join the League of Nations and did not play a major role in the international community. After World War II, however, the United States became one of the two leading superpowers, and the period between the 1940s and the late 1980s was characterized by worldwide geopolitical conflict between the United States and its allies in Canada, Western Europe, and elsewhere on the one hand, and the Soviet Union and its allies on the other.

All this changed for both Canada and the United States with the end of the cold war in the late 1980s and early 1990s. No longer were global **geopolitics** defined in terms of competition between the democratic, capitalist West (including North America) and the communist governments of the East. As the cold war faded into history, the United States became the last remaining superpower. American values, American popular culture, and American institutions spread throughout the world at a faster and faster pace. The end of the cold war accelerated the pace of **globalization**, and the globalizing world economy was increasingly influenced by the impacts of North American-based corporations (Figure 19.8).

Outside North America, this increased influence has often been seen as a mixed blessing. American technology was welcomed, but American values were in some cases resented. The political scientist Samuel Huntington has argued that future geopolitics may well be defined in terms of a clash of civilizations. Defining a civilization as a group of nations with

▶ **Figure 19.8** Fort Ord, California—once an Army training base for thousands of U.S. troops in three wars, is now the site of California State University, Monterey Bay, and a large wildlife preserve

similar underlying core values, Huntington postulated that future conflicts will involve tension between civilizations characterized by differing core values. Certainly, recent concerns about terrorist activities promulgated by radical, fundamentalist organizations outside North America are consistent with this view. The masterminds behind the terrorists who initiated the 9/11 attacks, and others holding similar views, harbor deep resentment of what they see as intrusion of United States values into their own core beliefs. Recent foreign policy in the United States has been consistent with the clash of civilizations concept; central to the current administration's foreign policy is the view that promoting democratic governance in areas such as Iraq, Afghanistan, and Palestine—which have previously had little experience with democracy—promotes the security of the United States in that the spread of democracy is seen as reducing terrorist threats.

Individual liberty and freedom are cornerstones of North American civilization, but these values clash with alternative conceptualizations of the nature of human existence as seen in other civilizations. Many in North America believe that individual freedom and liberty are fundamental human rights. At the same time, concerns about the unregulated expression of liberty leads to irresponsibility, moral stagnation, and decay as people place their own self-interest above that of **the commons** or Earth's physical systems. In the 1630s, the Puritan philosopher John Winthrop identified North America as a shining city on a hill, or a beacon to those seeking freedom from oppression, material poverty, and spiritual aridity. This sentiment is reflected in the words associated with the Statue of Liberty in New York Harbor: "Give me your tired, your poor, your huddled masses yearning to breathe free. . . ." As we move into the 21st century, are Canada and the United States still the shining cities on the hill envisioned by immigrants from throughout the world? How will differences in social and political values and immigration policies in Canada and Greenland as compared to the United States change the evolving human geography of these very different places? And will the unregulated freedoms and rampant excesses of capitalism expressed in some parts of North America lead inevitably to stagnation, licentiousness, greed, and irresponsibility? These positive and negative attributes will no doubt be reflected in future geographic landscapes in Canada, Greenland, and the United States in different ways and at different times.

Environmental Issues

The regional chapters in this book have introduced relationships between human activities on the land with the natural attributes and resources in places. In spite of all of our technological successes, we are still "bound to the land" directly or indirectly. Part of our concern about unregulated individual responsibility involves concern over the future of Earth's and our continent's physical systems. During the 20th century, unprecedented human population growth and industrial activity resulted in increasing levels of environmental degradation. Water and air pollution have since been addressed and lowered by many "first world" nations, but the majority of the world's population still lives in hazardous environments. Animal and plant species, both aquatic and terrestrial, are becoming **extinct** at ever faster rates as well as transporting to other places as **invading species**. Examples of invading species are the Kudzu vine, fire ants, killer bees, and zebra mussels. Habitat fragmentation, poaching, and encroachment of human civilization into previously uninhabited territory has imperiled the populations of once common wild animals throughout the world. There is mounting evidence to support the concept of global warming and long-run climatic changes associated with global warming could result in the melting of ice caps in the polar regions, higher sea levels, the loss of coastal ecosystems, and impacts on zones of effective agricultural production.

While North America's relatively low population and high incomes have spared the continent from many of the most damaging and immediate impacts of environmental change, by no means is it immune from the impacts of current environmental changes. As discussed in the Far North chapter of this book, global warming is recognized as a contributor to short- and medium-run climatic change in many places, notably in Alaska, Canada, and Greenland, where average temperatures have increased by several degrees over the past three or four decades. In Alaska, for example, in the past few years increased temperatures have resulted in unprecedented forest fires, outbreaks of insect pests that were no longer killed off by colder temperatures, and the forced relocation of Inuit villages because the permafrost on which they were constructed has melted. Native Greenlanders told us in the summer of 2004 that their formerly frozen harbors are now ice free in winter. Human activity elsewhere impacts the North American environment; for example, each spring much of southern and central Texas, including Houston, Austin, and San Antonio, experiences several weeks of intense air pollution caused by smoke drifting northward from southern Mexico and Central America after local farmers clear and burn vegetation on new agricultural land prior to planting crops. Even Hawai'i, the most isolated chain of islands in the world, is receiving dust from erosion in central Asia. Most disturbing, perhaps, is knowing that the activities of North Americans and North American-based corporations are contributing disproportionately to global environmental change. With about 5 percent of the world's population, the United States consumes nearly a third of Earth's natural resources.

Change is inevitable, and it is not possible to reverse the long history of human impacts on Earth's physical system—impacts that go back thousands of years and long predate European occupance of North America. Will continued environmental degradation imperil the quality of life enjoyed by North American residents today? Absolutely. Many would argue that human ingenuity and technology can eventually overcome the deleterious impacts of global environmental change in the United States and elsewhere; others would contest this view and contend that environmental degradation will continue to imperil the quality of life until and unless North Americans adopt a fundamentally different view of the relationships between humans and the environment. Proponents of the latter view often argue that philosophies emphasizing individual liberty and freedom also promote short-term benefit at the expense of ecological responsibility and environmental thinking. As this book has illustrated, one of the first places to look for the outcomes of these issues is on and in our North American landscapes.

Suggested Readings and Web Sites

Books and Articles

Allen, James P. and Eugene J. Turner. 1987. *We the People: An Atlas of America's Ethnic Diversity*. New York: MacMillan. This comprehensive atlas contains the most complete and engaging set of maps, showing distributions of various ethnic groups in the United States by county in 1980, and is based on data from the Census of Population.

Brewer, Cynthia and Trudy Suchan. 2001. *Mapping the Census 2000: The Geography of U.S. Diversity*. Redlands, CA: ESRI Press. More up-to-date maps of ethnicity in the United States.

Brown, Lester R. 2006. *Plan B 2.0.: Rescuing a Planet Under Stress and a Civilization in Trouble*. New York: W.W. Norton & Company.

Cameron, Elspeth, ed. 2004. *Multiculturalism and Immigration in Canada*. Toronto: Canadian Scholars' Press. A recent discussion of multicultural policies and their impacts on immigration in Canada.

Clay, Grady. 1994. *Real Places: An Unconventional Guide to America's Generic Landscape*. Chicago: University of Chicago Press. A very readable paperback book to know about before your next road trip.

Cooper, Andrew F. and Dane Rowlands, eds. 2005. *Canada among Nations*. Montreal: McGill Queen's University Press. An up-to-date global view of Canada's role in the world economy.

Gould, Peter. 1999. *Becoming a Geographer*. Syracuse, NY: Syracuse University Press. If the text in hand sparks an interest in the subject of geography, Professor Gould will answer most questions you may have about becoming a professional geographer.

Janelle, Donald G., Barney Warf, and Kathy Hansen. 2004. *World Minds: Geographical Perspectives on 100 Problems*. Boston: Kluwer Academic Publishers. This is a collection of short essays that demonstrate a geographical perspective on current issues and events happening in today's world.

McBride, Stephen. 2005. *Globalization and the Canadian State*. Black Point, Nova Scotia: Fernwood Publishing. Examination of the role of Canada as an integral part of the global economy.

McKibben, Bill. 2007. *Deep Economy: the Wealth of Communities and the Durable Future*. New York: Henry Holt and Company.

Messamore, Barbara J., ed. 2004. *Canadian Migration Patterns from Britain and North America*. Ottawa: University of Ottawa Press. A focus on the migration pathways, settlement patterns, and experiences of immigrants from Great Britain and the United States who emigrated to Canada.

Randolph, John. 2004. *Environmental Land Use Planning and Management*. Washington: Island Press.

Roberts, Lance W., Rodney A. Clifton, Barry Ferguson, Karen M. Kempen, and Simon Langlois, eds. 2005. *Recent Social Trends in Canada, 1960–2000*. Montreal: McGill Queen's University Press. An edited volume on Canada's changing social and economic characteristics in the final decades of the 20th century.

Zelinsky, Wilbur. 1994. *Exploring the Beloved Country: Geographic Forays into American Society and Culture*. Iowa City: University of Iowa Press. One of the best known cultural historical geographers shares a lifetime of study and observation of American life ways in this fascinating book.

Web Sites

Population Projections
http://www.census.gov/population/www/popclockus.html

Population Pyramids
http://www.census.gov/ipc/www/idbpyr.html

History of Cinco de Mayo
http://www.mexonline.com/cinco.htm

Global Economy
http://www.globalpolicy.org/

Environmental Issues/Regulation in the United States
http://www.epa.gov/

Glossary

Aboriginal peoples Indigenous populations native to a particular area or who settled there during pre-historic times.
Abrasion The physical process of wearing down rock surfaces.
Absolute location Exact position on Earth's surface or on a map, usually given in degrees of longitude and latitude, although several other coordinate systems can be used (see relative location).
Accessibility Ease with which one location may be reached from another.
Acculturation Process of adopting some aspect of another culture.
Adaptive radiation Genetic changes in organisms as they disperse (radiate out) from a central point.
Adiabatic Pertaining to the heating of air as it ascends.
Age of Exploration A period of time in European and world history during the 15th–17th centuries when global exploration began.
Agglomeration Tendency of businesses to locate together for mutual advantage.
Aging in place The decision of elderly people not to relocate during their retirement years.
Alaska National Wildlife Refuge (ANWR) 19-million-acre preserve in NE Alaska at risk due to potential oil and gas extraction.
Alluvial fan Fan-shaped deposits at the mouth of a canyon formed where streams lose velocity and drop sediment.
Alpine Biotic region located in high elevations near or above the treeline.
Alpine glacier A glacier formed at high elevations or in a mountain valley.
Altitudinal zonation Average temperatures decrease as elevation increases resulting in changes in vegetation patterns.
Anglophone English-speaking Canadians who may live in any of the Canadian provinces or territories (but who are a minority in Quebec).
Appropriative water rights (Appropriation) A first come, first served policy of water use; in other words, whoever controls the water in a particular place first, has a right to use it anytime thereafter. Water rights are often gained through the appropriative process.
Aquifer Rock strata permeable to groundwater flow.
Archipelago A line of islands with the same geological structure.
Arete A sharp ridge created from ice movement.
Arid climate A climatic zone where more evapotranspiration occurs than incoming precipitation.
Assimilation Process of absorbing a minority or underrepresented population and culture into the dominant population.
Athabaskan (or Athabascan) The name of a large group of closely related Native American groups in western North America, and of their language family. The Athabaskan family is the largest family in North America in terms of number of languages and the number of speakers.
Atmospheric pressure Pressure exerted by gas molecules as measured by a barometer.
Avalanche The sudden release of gravity-induced snow downslope.
Barrier A physical or cultural obstruction to human travel or migration.
Barrier island A long, narrow, sandy island formed offshore.
Basic Economic activities that attract money into a community from other places.
Bering Land Bridge Prehistoric route of human travel across the North Pacific linking Asia and North America.
Bilingual A person with fluency in two different languages.
Biogeography The study of the location and distribution of plants, animals, and ecosystems.
Biome Complex of plants and animals common to a designated physical region.
Bioregionalism Belief that economic and political decision making should be based on environmental criteria.
Birthrate Number of births per one thousand people in a particular population.
Boom and bust cycle Alternating periods of economic growth and decline.
Boreal forest A northern needleleaf forest.
Boundary Line separating one political or mapped area from another. Reference is sometimes made to conceptual boundaries between ideas.
Break-in-bulk Transfer points where cargos are changed from one transportation mode to another, for example, harbors where products on shipboard are loaded onto trains or trucks.
Buffalo commons The concept of restoring parts of the North American Great Plains to return them to their native habitat recommended by geographers Frank and Deborah Popper.
Cajuns A person of Acadian heritage living in the southern United States.
Caldera The sunken portion of a volcanic crater.
Canadian Native Land Claims Act Policy passed in 1993 that led to the creation of Nunavut in 1999 and self determination for indigenous people.
Cartography The science and skill of map making.
Cash-grain farming A type of primary production based in raising grains for profit.
Center pivot irrigation Irrigation technique whereby water is sprayed over a field from a central location.
Central Business District Traditional high-density core of the city with a concentration of office buildings and retail shops.
Central city The interior, usually older, high-density part of a city often referred to as the city's "downtown."
Central place theory Explanation of city size, based upon distances of cities from one another. A hierarchy of city sizes and number of functions is based on relationships with neighboring places.
Chain migration Migration of people from the same source area to the same new place of residence along specific lines of travel.
Chernozem An agriculturally rich soil type formed in steppe–grassland zones.
Chinook winds North American term for a warm, dry, downslope air flow causing extreme changes in local temperatures.
Cirque A glacially eroded, amphitheater-shaped basin at the head of an alpine glacier valley.
Civil Rights Movement An era in U.S. history in the 1960s and 1970s of political activism in support of African-American peoples, cultures, and rights.
Climate Average atmospheric conditions over a selected time period.
Colonias A node of Latin American settlement in rural and small town places in the Southwest (often located near the Mexican borderlands).
Colony A political entity politically and economically linked to a more powerful nation-state.
Columbian Exchange Transfer of culture, economy, politics, and organisms from Europe to the Americas after 1492.
Composite cone volcano A steep-sided volcano formed by a sequence of explosive eruptions.

Laws of the Indies A 15th-century Spanish guide that laid out specifications for towns and landscape features in colonial areas dominated by Spain.

Leeward slope Slope opposite the windward side of a hill or mountain that is usually drier and causes a rain shadow. Also an area located generally downwind from wind and weather systems.

Loess Windblown soil that is often fertile in nature.

Longitude Measurement in degrees west or east of the Prime Meridian (which passes through Greenwich, England).

Long lot system Colonial survey system laid out in primarily French or formerly French-dominated areas of North America.

Loyalists (known as **United Empire Loyalists** in Canada) People in the original 13 British colonies who remained loyal to the British during and after the Revolutionary War in the United States.

Manifest destiny Nineteenth-century belief in the expansion of the United States to the Pacific Ocean and to the 54th parallel.

Marine west coast climate A region with average climate conditions caused by the moderating effects of the ocean located on the west coast of continents in the northern hemisphere where the westerly winds dominate.

Maritime air mass A moist air mass that blows off a large body of water and is influenced and shaped by it.

Mediterranean climate A type of climate defined by a dry summer and wet winter regime with moderate or hot temperatures located around the Mediterranean Sea, and in parts of California, South Africa, and Australia.

Mediterranean scrub A biome dominated by shrublands located in mediterranean climate zones, characterized by scrubby low-growing bushes and stunted forests.

Melting pot A term first used by a British author in the early 20th century that later came to refer to the notion that people from different ethnic and racial groups would all eventually melt into one homogeneous culture and race in the United States.

Mesa Spanish word for an elevated area of land with a flat top and sides that are usually steep cliffs.

Mesothermal climate zone Areas influenced by large water bodies with moderate temperature ranges.

Metes and bounds Irregular survey system used in Europe and transported to colonial America.

Metropolitan Statistical Area According to U.S. Census definitions: "an integrated economic and social unit with a recognized large population nucleus."

Microclimate Climatic conditions at or near Earth's surface.

Microthermal climate zone Areas distant from large water bodies and with large ranges in temperature.

Mid-Atlantic Ridge A mostly underwater mountain range that runs from about 207 miles (333 kilometers) south of the North Pole to sub-antarctic Bouvet Island.

Migration Movement of people from one place to another.

Mixed farming Multi-crop and animal husbandry system of agriculture.

Mixed race Person who is descended from more than one racial and/or ancestry group.

Mobility Ease and frequency of movement.

Monsoon A stormy rainy season in tropical and subtropical regions.

Moraine Glacial deposits of unsorted and unstratified material at the terminus or edge of a glacier.

Multiculturalism A policy first developed in Canada in the 1970s to define the importance of recognizing and valuing a nation's many races, cultures, and peoples.

Multinational firms Corporations operating in more than one country.

NAFTA Acronym for the "North American Free Trade Agreement," a North American political decision that links economic activities and products in Mexico, the United States, and Canada.

Nation A group of people who share a common identity and who are usually interested in maintaining their own government.

National Geography Standards A critically important publication that defined 18 standards that should be taught in geography instruction in K–12 education.

Nationalism Feeling of commonality by virtue of culture, language, religion, or politics that tends to bind together a population.

Nation-state Political territory occupied by a group having a common identity and a certain degree of unity.

Native American Indigenous people and cultures in the United States who are descended from the first people to reside in North America prior to the arrival of Europeans and who continue to be important today.

Native self-determination The concept of the nation-state first espoused by U.S. President Woodrow Wilson after World War I.

Natural hazards Naturally occurring events such as earthquakes, fires, and floods that may cause damage to people or their created landscape.

Natural resources Naturally occurring things used by humans such as timber, minerals, water, or soils.

Nodal region An identified area characterized with a node or center of importance.

Nomadic herding A type of economic activity dependent on the raising of sheep, goats, or cattle and characterized by almost constantly mobile populations.

Non-basic Economic activities that do not attract money into a community from elsewhere.

Normal lapse rate The average rate of temperature decrease with increasing altitude (usually 3.5°F per 1000 feet in elevation).

Northwest Passage A sea route connecting the Atlantic and Pacific Oceans through the Arctic Archipelago of Canada.

Orogenesis The birth of mountains that results from uplifting of Earth's crust.

Orographic lifting Uplift of an air mass due to physical barrier resulting in cooling temperatures.

Orographic precipitation Rainfall or snowfall caused by the cooling of air masses that are forced to rise over high mountain barriers.

Pacific Rim A political and economic term used to designate the countries on the edges of the Pacific Ocean as well as the various island nations within the region.

Peripheral area A portion of an economic, political, cultural, or geographic region that is guided by decision making elsewhere, which is located at some distance from the core of power.

Permafrost Permanently frozen soil, rock surface, or subsurface.

Physical boundaries A political line of demarcation located with reference to natural features such as rivers or mountain crests.

Physical geography A science that studies the spatial aspects of natural environmental components and processes.

Physiographic region Mapped area with interrelated patterns of topography, geomorphology, and hydrology.

Place Essence of geographic inquiry leading to an understanding of a location's uniqueness and its importance in relation to other places.

Planning Process of anticipating and guiding future land uses in a selected area.

Plaza towns Settlements in North America and elsewhere in the world that were founded by Spanish and other European groups that feature a central square or plaza surrounded by commercial and government buildings.

Polar air mass Atmospheric system originating in polar zones.

Polar climate Atmospheric conditions in areas higher than 60 degrees north and south and on higher mountain peaks, depending on latitude and elevation.

Political economy The study of the relationship between a place's organization of economic systems and issues as they relate to political developments and policies.

Population density The number of people occupying a particular area or region.

Population pyramid A graph that shows the shares of a population's different age groups.

Post-industrial economy An economy in which the quaternary and tertiary sectors dominate economic development and expansion not dependent on labor for manufacturing.

Pressure gradient Differences between areas of higher barometric pressure and areas of lower pressure.

Primary sector Those forms of production dependent on natural resources such as farming, mining, fishing, or forestry.

Public domain Lands in the United States that are owned and controlled by the federal government.

Pueblos Spanish or Native American dwellings or small settlements in the American Southwest.

Push-pull factors Favorable and unfavorable conditions that affect the migration decision-making process.

Quaternary sector A part of the global economy focused on information technology.

Quebecois People who live in Quebec who prefer secession from Canada.

Quiet Revolution The rise of active nationalism among Francophone Quebecers.

Racism Negative and often dangerous attitudes and behaviors based on perceived (and often stereotyped) personal attributes of a person or group of people.

Rain shadow Leeward areas below mountain crests that receive low average rainfall.

Ranches Agricultural system of livestock production.

Ranchos Spanish word for ranches; a system of Spanish colonial settlement.

Rate of natural increase A standard statistic used to describe the total population of an area based on the difference between birthrates and death rates.

Region Area on Earth's surface defined by selected criteria.

Regional geography A method of study used by geographers and others to describe and analyze the spatial and environmental patterns and related processes of particular places on Earth.

Relative location A position on a map or on Earth's surface compared to other positions (see **absolute location**).

Relief Elevation differences in local landscapes.

Resolution Level of detail projected or seen on a map, photograph, remotely sensed image, or computer screen.

Resource Anything used by people, including naturally occurring materials (soil, minerals, trees, and water), labor, or energy sources.

Resource base Natural resources that may become the foundation for economic development in an area.

Riparian vegetation Streamside trees and shrubs that are dependant on nearby water.

Rural Areas of settlement with low population densities and located considerable distances from cities.

Rust Belt Older, former industrial areas of North America near the Great Lakes.

Salinization The process of increasing salt content in air, water, or plants.

Santa Ana winds Hot, dry, winds flowing from east to west in southern California from higher inland locations toward the sea.

Savanna A biome made up of grassland with scattered trees or shrubs.

Scale Relationship between distance on a map and distance on Earth's surface.

Secondary production That part of an economic system focused in manufacturing and distribution of goods.

Seigneurial system A land system used in early French Canada that defined, controlled, and divided up settled areas and agricultural lands.

Semiarid climate A region generally experiencing an average annual rainfall of 15 to 20 inches.

Sense of place Subjective and sometimes emotional connections to our immediate environment.

Sequent occupance A succession of cultures and peoples in one place over time.

Sharecroppers Farmers who rent land with portions of the crops grown on it.

Shield volcanoes A gently sloped symmetrical mountain landform built from many eruptions of lava.

Site A particular location on the surface of Earth often measured in longitude and latitude or according to a street address or other precise measurement system.

Smart growth Sometimes called sustainable development, a plan designed to mitigate the damaging impacts of growth in urban and suburban areas that has minimal impact on natural and human environments or resources.

Soil A dynamic material made up of minerals and organic matter, usually supporting plant growth.

Soil profile A rendering of layered soil types in a specific place, usually labeled as soil horizons.

Solar radiation Energy from the sun impacting Earth.

Sovereignty Legal, governmental control over a country or region.

Spatial interaction The social, economic, and other connections between and among places.

Steppe A biome dominated by grasslands with associated plant and animal communities.

Subarctic zone The area between 55° N and 66.5° N.

Subduction The sinking of one portion of Earth's crust beneath another caused by tectonic activity, and often resulting in the formation of deep ocean trenches.

Suburbanization Spread of settlement in cities outward from the center, usually as a result of transportation improvements and economic development outside the central city.

Suburbs Areas of densely settled residential and commercial land located outside the central city.

Suffrage The right to participate in the political process including the right to vote in governmental elections.

Sun Belt Southern portions of the United States engaged in increased residential and commercial development.

Taiga A northern latitude needleleaf forest.

Technological treadmill The economic process whereby production is increased with the implementation of new technology, which reduces prices which, in turn, creates more demand for new technology to increase production.

Tectonics Large-scale movement of portions of Earth's crust driven by internal energy.

Tectonic uplift A geological process most often caused by plate tectonics that increases elevation.

Temperature inversion Atmospheric occurrence of a layer of warm air on top of a colder air mass.

Tennessee Valley Authority Federal regional water and power resource management agency that built dams and other flood control systems on the Tennessee River in the American South.

Terra incognita Literally, "land unknown;" unrecorded or blank spaces on early maps depicting unexplored places.

Territory Sense of privacy, ownership, or occupation of particular areas.

Tertiary sector That portion of an economy devoted to trade, services, government, or education.

Till Unstratified and unsorted deposits of soil and rocks that are carried into an area by moving ice.

Time–space continuum Relationship between the dimensions of time and space as expressed in travel times between points.

Tombolos Landforms of local sand deposits connecting offshore islands with a coastline.